T0122564

Handbook of Cognitive and Autonomous Systems for Fire Resilient Infrastructures

MZ Naser • Glenn Corbett
Editors

Handbook of Cognitive and Autonomous Systems for Fire Resilient Infrastructures

 Springer

Editors
MZ Naser
Clemson University
Clemson, SC, USA

Glenn Corbett
John Jay College of Criminal Justice
New York, USA

ISBN 978-3-030-98687-2   ISBN 978-3-030-98685-8   (eBook)
https://doi.org/10.1007/978-3-030-98685-8

© Springer Nature Switzerland AG 2022
This work is subject to copyright. All rights are reserved by the Publisher, whether the whole or part of the material is concerned, specifically the rights of translation, reprinting, reuse of illustrations, recitation, broadcasting, reproduction on microfilms or in any other physical way, and transmission or information storage and retrieval, electronic adaptation, computer software, or by similar or dissimilar methodology now known or hereafter developed.
The use of general descriptive names, registered names, trademarks, service marks, etc. in this publication does not imply, even in the absence of a specific statement, that such names are exempt from the relevant protective laws and regulations and therefore free for general use.
The publisher, the authors and the editors are safe to assume that the advice and information in this book are believed to be true and accurate at the date of publication. Neither the publisher nor the authors or the editors give a warranty, expressed or implied, with respect to the material contained herein or for any errors or omissions that may have been made. The publisher remains neutral with regard to jurisdictional claims in published maps and institutional affiliations.

This Springer imprint is published by the registered company Springer Nature Switzerland AG
The registered company address is: Gewerbestrasse 11, 6330 Cham, Switzerland

Preface

We continue to view buildings as gravity-defying structures designed to withstand the adversity of humans, nature, and time. One of the most extreme events a structure might undergo is fire. Unlike other traditional load actions (i.e., wind, seismic events, etc., which are primarily bound to a seasonal or geographical location/region), fire, on the other hand, can virtually break out anywhere and anytime. While fire has been noted as a critical issue over the past few years, research on this area continues to favor dated experimental and numerical approaches. To further complicate this matter, the fire design of structures still heavily relies on prescriptive solutions with little room for innovation or flexibility.

Unfortunately, the past few years have witnessed a drastic surge in both frequency and intensity of fire incidents, both of which are transforming our history—and, more specifically, our fire engineering history. The aftermath of these incidents is a continuous reminder of the serious flaws in our virtually unchanged, decades-old construction philosophy. It is equally concerning and troubling that structural engineers seem to converge on the notion that it is quite impractical, and perhaps unfeasible, to "truly" design fire-resistant structures. This motivates adopting a new look into this challenge, perhaps one that draws inspiration from the advent rise of automation and robotics. This is the primary motivation behind this handbook.

Through this lens, buildings (or structures) are not to be thought of as a series of passive and rigid arrangements of load-bearing members; but rather, we must appreciate the philosophically striking similarities between such buildings and robots. For example, buildings and robots are often subjected to external and/or internal forces. As such, they are designed with embedded structural systems capable of handling such effects to maintain structural integrity. Furthermore, both buildings and robots frequently operate in extreme environmental conditions under a continually complex combination of temperature, pressure, and stress. Nowadays, buildings are commonly being supplemented with sensory instruments, primarily to monitor their "health" as to give insights into deteriorating processes (i.e., corrosion), energy consumption, or into occupants' preferences (e.g., thermal comfort, etc.). Similarly, robots are also embedded with sensors to enable self-diagnostic interaction with surrounding environments and to deliver valuable information to human operators. One can see that there is more to buildings and robots than meets the eye.

The above discussion promotes the following question, what if buildings are designed to incorporate robotic features? What if buildings have cognitive and autonomous abilities? How can we realize such abilities?

With the onset of the fourth industrial revolution, advances in artificial intelligence (AI), internet-of-things (IoT), and robotics are expected to be heavily integrated into the construction industry. Unlike other works, this handbook is not interested in adopting robots as construction workers, nor as 3D printers, but rather seeks to integrate signature features from robots into skyscrapers and infrastructure with especial attention to disaster-induced collapse mitigation.

From this book's perspective, future buildings can be thought of as giant shape-shifting robots with cognitive and autonomous capabilities. The cognitive capability refers to a building's intelligence to understand its surroundings and interact with its occupants. For instance, a cognitive skyscraper will be able to identify the breakout of a disaster (i.e., multistory fire) while at the same time analyzing cascading events in real time using state-of-the-art AI. Such a skyscraper will also be able to pinpoint vulnerable regions or load-bearing components and foresee how damages arising from an extreme event can lead to unwarranted failure/collapse. This is equivalent to a robot assessing its own condition in the event of a mechanical arm impairment or loss. In this scenario, the robot will seek to first understand the impact of such loss, and then it will attempt to devise plausible solutions to overcome such damage.

However, *"knowing without action is useless."* Preliminary efforts have shown how a cognitive building is not only able to convey its predictions to occupants and first responders to facilitate evacuation but also be able to physically turn this information into actions (Naser 2019, 2020). In order to build up the "action" component, future skyscrapers can be supplemented with tentacle-like, and self-deployable (TL-SD) robotic load-bearing structural components. Such components are to be designed independently of the main structural system and hence can be integrated into building façade or compartment boundaries (see Fig. 1). The deployment of a TL-SD structural component can be activated once the cognitive component of the building detects excessive deformations or instability. The deployment of such a tentacle starts by unfolding and inflating a TL-SD structural component into a relatively stiff configuration through pressurized dynamic sliding and shape-memory hinges etc.—similar to those often used in octopus-inspired robots.

Thus, a building equipped with TL-SD structural components will have the capability to autonomously reconfigure its internal structure to divert the adverse effects of extreme events away from its occupants, as well as critical load-bearing members, and possibly towards outside the building; thereby preventing significant damage and collapse. Such events can include extreme temperatures in excess of 800–1000 °C, significant seismic vibration, blast, or impact. As a result, a cognitive and autonomous shape-shifting skyscraper can achieve higher levels of structural resilience (survivability) under extreme environments. This improved performance mitigates failure (collapse), thus allowing first responders to tackle the adverse effects of disasters.

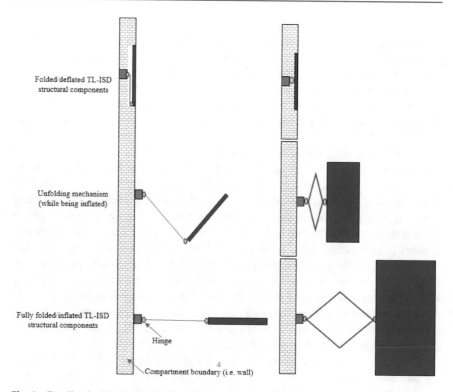

Folded/deflated TL-ISD
structural components

Unfolding mechanism
(while being inflated)

Fully folded/inflated TL-ISD
structural components

Hinge

Compartment boundary (i.e. wall)

Fig. 1 Details of a TL-SD structural component

Despite the above optimistic look, one should note that the construction industry is notorious for its inertia and slow adoption of new technologies. Open discussions and collaborations are needed to enable a smooth transition from theory to design to implementation.

Notably, this forward-looking handbook addresses the critical issue of bringing buildings and emergency responders together, a rather novel approach to fire safety even in the twenty-first century. Incredibly, most building designs today do not consider the interaction between firefighters and the buildings they operate in. This handbook literally fills in this gap by providing cutting-edge research, providing design professionals with an understanding of the needs of firefighters and how that impacts building design.

This handbook homes findings from 13 chapters from leading experts on fire safety, engineering, and firefighting from around the world. These experts share state-of-the-art information with regard to various aspects of modernizing the future of safety, engineering, and firefighting. This handbook starts with Chap. 1 where Brian Meacham presents a look into *Sociotechnical Systems Framing for Performance-Based Design for Fire Safety* to lay the foundation for transforming our ways of tackling fire hazards. Then, enters Chap. 2 by Vytenis Babrauskas, to draw a picture for *A Twenty-First Century Approach to Fire Resistance* that stretches our comfort zone into a new frontier. In Chap. 3, Casey Grant showcases a roadmap to *Integrating Modern Technologies to Realize Fire-Resistant Infrastructures*. It is in this chapter that we start to tie automation concepts often seen in parallel fields to our fire domain. In Chap. 4, *Intelligent Science Empowers Building Fire Protection Technology Development*, Feng Luo goes on to showcase how existing fire protection technologies can tremendously improve by borrowing the concepts of IoT and AI.

LaMalva and Medina present us with a look into how the integration of automation will reshape our building codes and standards in Chap. 5, *Building Codes and the Fire Regulatory Context of Smart and Autonomous Infrastructure*. Then, in Chap. 6, Xinyan Huang and colleagues deliver *Perspectives of Using Artificial Intelligence in Building Fire Safety*, thereby delving into the depth of AI and its role in improving building fire safety. Chapters 7 and 8 are written in collaboration between Brian Lattimer and Jonathan Hodges. These two chapters specialize in *Intelligent Firefighting* and *The Role of Artificial Intelligence in Firefighting*—a key resource for our first responders in the new era of cognitive and autonomous structures. Charles Jennings' Chap. 9, *Implementing AI to Assist Situation Awareness: Organizational and Policy Challenges,* focuses specifically on the use of artificial intelligence to provide real-time situational awareness on changing conditions for emergency responders.

The last four chapters capture intricate details with regard to burning questions we continue to face. For example, in Chap. 10, Wojciech Kowalski and colleagues chart a *probabilistic-based reliability approach to overcoming fire hazards in structures*. Chapter 11 is titled *Autonomous Sensor-Driven Pressurization Systems: Novel Solutions and Future Trends*, where Wojciech Węgrzyński and Piotr Antosiewicz focus on autonomous solutions for smoke control in buildings to maintain smoke-free evacuation routes under fires. Ana Sausa revisits the classical standard fire testing from a new perspective—one that incorporates *Hybrid Fire Testing* to enable us to move from component-level testing to system-level fire testing in Chap. 12. Finally, in Chap. 13, Liming Jiang and colleagues give an approach for *Realistic Fire Resistance Evaluation in the Context of Autonomous Infrastructure*.

This handbook started with an idea that we shared during 2018–2019. Initially, we were hoping to complete this handbook by 2020. Little did we know that a pandemic was on the horizon. While the pandemic challenged our initial deadline, we could not have delivered this handbook without the hard work of our contributors. Our hats go to them. Finally, we would like to thank Springer for lending our contributors and us this platform to showcase a glimpse of what the future holds for our domain. A special thanks go to Paul Drougas, Kritheka Elango and R.Savita for believing in the message of this handbook, continued support, and for taking the lead on editing and formatting this handbook.

References

Naser, M. Z. (2019). "Autonomous and resilient infrastructure with cognitive and self-deployable load-bearing structural components." *Automation in Construction*, Elsevier, 99, 59–67.

Naser, M. Z. (2020). "Enabling cognitive and autonomous infrastructure in extreme events through computer vision." *Innovative Infrastructure Solutions*, Springer, 5(3), 99.

Clemson, SC, USA   M. Z. Naser
New York, NY, USA   Glenn Corbett

Contents

Toward a Sociotechnical Systems Framing for Performance-Based Design for Fire Safety

Brian J. Meacham

1.1 Current Structure/Process for PBD of Fire Safety

This section reflects the author's perception of the current structure and framing of performance-based design (PBD) for fire safety. It is suggested that PBD for fire safety emerged in the 1990s, [1–7] concurrent with the introduction of functional- and performance-based building regulations in several countries [1, 8–11]. The foundations of PBD for fire safety can be found in the following:

- Research into compartment fire dynamics of the 1950s–1970s [e.g., 12–18], which ultimately gave rise to important concepts such as defined stages of fire development, means to estimate compartment fire temperatures, fire growth rate, rate of heat release, and the like, which embody the fire safety science that underpins fire safety engineering [see, for example, 19].
- Concepts of reliability-based design [e.g., 20–24] and quantitative risk analysis and of risk-based and risk-informed design concepts [e.g., 25–31] which emerged in the 1960s and 1970s.
- The application of these concepts to fire design of structure in the 1970s and 1980s [e.g., 32–36].
- Systems constructs and design approaches introduced by pioneers of "fire systems approaches" in the 1970s and 1980s [e.g., 37–45]. Although different terminology is used, the "fire system" components introduced at this time, and still in use, include fire prevention, fire initiation, and development, means to control fire spread, means to control smoke spread, detection, suppression, structural stability, occupant protection, fire service operations.

B. J. Meacham (✉)
Meacham Associates, Shrewsbury, MA, USA
e-mail: brian@meachamassociates.com

© Springer Nature Switzerland AG 2022
M. Z. Naser, G. Corbett (eds.), *Handbook of Cognitive and Autonomous Systems for Fire Resilient Infrastructures*, https://doi.org/10.1007/978-3-030-98685-8_1

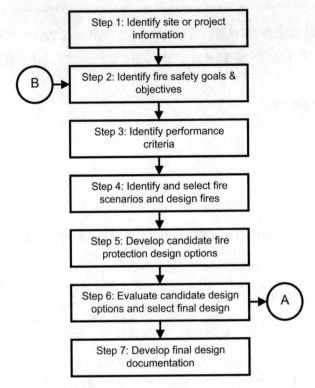

Fig. 1.1 Fundamental steps in the PBD for fire safety process (Source: author, adapted from [5])

- Introduction of computer modeling applications for fire in the 1970s and 1980s [e.g., 46–51].

By the late 1980s and through the 1990s, outcomes of fire safety science and engineering research, structural fire engineering research, the concept of fire systems components, and reliability- and risk-based design began to be integrated into structural fire performance design [32–36], fire safety engineering (FSE) frameworks [52–55], and ultimately into frameworks/process descriptions for PBD for fire safety [2, 4, 5, 8, 56, 57]. Although some differences exist between the various frameworks, they all follow the general structure and process flow as illustrated in Figs. 1.1 and 1.4 (adapted from [5]).

The starting point in the process is identifying the fire safety goals and objectives. These may be embodied in a regulation or may need to be developed for a specific facility, process, or system. When embodied in building regulation, the objectives may be referred to as functional statements, operative requirements, performance objectives, or simply objectives. These are often qualitative statements, which described the intended function of a system and/or the outcome that is desired. Examples include statements such as 'the building and fire safety systems shall be designed to prevent the exposure of those occupants not intimate with first materials burning to untenable conditions during the time required for them to reach a place of safety outside of the building,' or 'the building and fire safety systems shall be designed to limit a fire which occurs in the building to the compartment of fire origin.'

It can be the case that in developing an actual design, the engineer needs to formulate "design objectives," which may still be qualitative, but which reflect a specific function or outcome that can readily be translated into engineering concepts. For example, in the case of the performance objective stated above, "the building and fire safety systems shall be designed to limit a fire which occurs in the building to the compartment of fire origin," there is no particular sense of how one might limit a fire to the compartment of fire origin. To narrow the focus, a fire safety engineer might therefore define a design objective, which will help meet the performance expectation, but can be readily quantified, such as "minimize the potential for flashover or traveling fire in any compartment in which a fire may occur." By expressing design objectives in such a way, metrics used to describe such phenomena, such as upper-level temperatures, or radiant heat flux at floor level, may be used as design criteria, as discussed below. Both performance objectives and design objectives should be formulated to adequately describe the expected outcomes (e.g., achievement of safety, avoidance of failure), and to the extent practicable, the associated timeframe, such that a decision regarding regulatory compliance and/or performance verification can be clearly made.

Once identified and agreed, performance (or design) criteria must then be identified and agreed. These are quantitative measures by which performance will be stated and designs will be evaluated. These too may be embodied in a regulation, a design standard or guideline, or may need to be developed for a specific component, facility or operational process. Performance (or design) criteria may be stated deterministically or probabilistically, will be related to the method of evaluation, and may be closely associated with specific analytical approaches or methods used for demonstration of compliance and/or verification of performance. It will often be the case that several performance and/or design criteria will be needed to adequately assess and verify the target performance expectations. Exemplar quantities that may be expressed as criteria include temperature, heat flux, species concentrations, visibility, probability of success or failure, and exceedance probability. When expressed deterministically, many criteria require some form of qualifier, such as location (e.g., temperature in the upper gas layer or temperature at 1.5 m above floor level; heat flux at floor level or at 1.5 m above floor level), time/duration (e.g., CO concentration of 1500 ppm for 10 min; heat flux level for 2.5 min), or both. When expressed probabilistically, there may be safety factors that reflect uncertainty or variability, or a time component (e.g., 95% likelihood within 10 min), or other. Criteria will be needed for each type or category of targets being considered by the analysis (e.g., people, contents, structural system components, nonstructural building system components, equipment or systems associated with the use of the building). As will be discussed later in the chapter, identifying and selecting appropriate criteria can be one of the most challenging parts of the performance-based design process.

The next step involves identifying fire scenarios of concern and developing quantified design fires that will be a fundamental basis of analysis. Many PBD frameworks for fire safety separate fire scenario and design fire development because they are approached in much different ways, using different data, tools, and methods. However, they are inextricably linked, a concept that some engineers sometime forget. A fire scenario is a description of a specific fire situation from ignition, through established burning, to maximum extent of growth and resulting decay [5]. There is an infinite number of event scenarios possible for any given building. However, if is generally possible to group the scenarios into smaller subsets of scenarios that can be used as part of the design process. These are called design fire scenarios. The range of design fire scenarios selected for consideration should reflect how likely it is that the event will occur, and if it does occur, how it is expected to impact the building or infrastructure.

In the current PBD for fire safety paradigm, there are many factors that one must consider when developing event scenarios and paring them down into design scenarios, including [5]:

- *Pre-fire situation*: building, compartment, conditions.
- *Ignition potential*: credible combinations of temperature, energy, time, and contact with fuel.
- *Initial fuels*: state, surface area to mass ratio, ignition temperature, rate of heat release.
- *Secondary fuels*: proximity to initial fuels, characteristics, amount, distribution.
- *Extension potential:* beyond compartment, structure, area (if exterior fire).
- *Target locations*: people, contents, structure, operational equipment.
- *Critical factors*: ventilation, environmental, operational, building use.
- *Occupant characteristics*: alert, asleep, ability, age, mental acuity, dependencies.
- *Available relevant data*.

Hazard and risk assessment tools are often the most appropriate mechanisms to help identify fire scenarios and clustering them into representative design fire scenarios.

Once a set of design fire scenarios has been developed, a set of corresponding design fires need to be developed. A design fire is an engineering description of a specific fire scenario or fire scenario cluster. Design fires are used to evaluate the candidate design options, much the same as design loads will be used in structural analysis and design to ensure that the strength of a structure is adequate [5]. Design fires may be characterized by such parameters as growth rate, heat release rate, species production and production rate, burning duration, and other such attributes that can be measured, calculated, and estimated. In its simplest form, a design fire may be reflected as a t^2 fire curve or temperature–time curve. A more complete design fire curve would include the realms or stages of fire development: pre-flaming, established burning, growth, peak and/or steady state, and decay. This is graphically represented in Fig. 1.2.

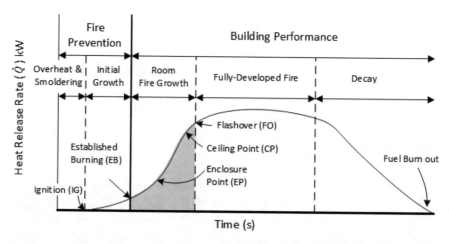

Fig. 1.2 Exemplar representation of realms (stages) of design fire (Source: author, adapted from [59])

In most cases, a design fire curve is actually a composite curve, representing an initial fuel package, and subsequently ignited fuel packages, until fuel has been exhausted, oxygen has been limited, or the fire has been suppressed.

In the current PBD for fire safety framework, design fires should be developed for every design fire scenario considered for a compartment, building, structure, or process. This particularly includes fires that may in some way pose challenges to the building or to fire suppression efforts, including concealed (e.g., in ceiling voids or wall cavities) or shielded fires (e.g., partially covered by a temporary or permanent building feature or content), readily combustible fuels with very fast growth rates and high heat release rates (e.g., flammable liquids, some plastics), areas in which large fuel loads are present (e.g., storage), and the like. Consideration should be given the types of fires that challenge the various fire design options, such as detection devices, sprinklers (where shielded and fast-growing fire can be a concern), large fuel loads (which can burn for long periods and present structural fire performance challenges), and deliberate fires (which may impact paths of fire egress and escape).

Only after the goals, objectives, criteria, scenarios and design fires have been identified and agreed should consideration of alternative fire safety design (mitigation) options be considered. This is particularly true for a full performance-based approach that begins with no specific fire protection measures embodied in the conceptual design. However, in some cases, code- or regulation-specified fire safety design may be suggested or required, with the performance-based process being applied to assess alternative approaches. In this situation, as well as in the case of the analysis and design of fire safety for an existing facility, the analysis should consider all features that are in place.

As noted above, a facility's holistic fire safety system is often comprised of several of the following system components: fire prevention means and measures;

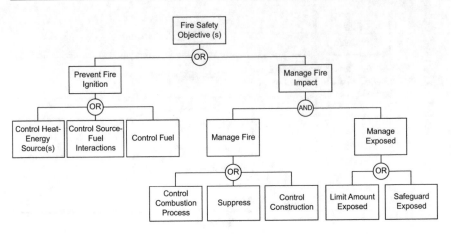

Fig. 1.3 Top Branches of the Fire Safety Concept Tree (Adapted from [60]. Reproduced with permission of NFPA from NFPA 550, *Guide to the Fire Safety Concepts Tree*, 2022 edition. Copyright © 2021, National Fire Protection Association. For a full copy of NFPA 550, www.nfpa.org

means to limit or control fire initiation and development; means to control the spread if fire and fire effluents; means to detect fire, notify occupants and the fire service, and means to suppress fire; means to assure structural stability; means to protect occupants during movement to a place of safety; and means to facilitate fire service operations. A particularly helpful tool to assist engineers in identifying options for consideration is the *Fire Safety Concepts Tree* embodied within NFPA 550 [60], which originally dates to 1985 [40] (Fig. 1.3).

Once an initial set of candidate design (mitigation) options has been identified, the next step is to evaluate their efficacy in achieving the agreed fire safety goals and objectives, as assessed against the agreed performance (design) criteria, when subjected to the set of design fires. This evaluation process is iterative, since there are likely several potential design options which will be technically feasible. The aim is to develop a set of technically credible options, and then apply benefit–cost analysis or other means to compare viable options and select a final design. This process is illustrated in Fig. 1.4 [5]).

There are three major components/considerations of the evaluation process—the analytical concept or framework for evaluation (e.g., the required safe egress time (RSET) must be significantly less than the available safe egress time (ASET) with an appropriate margin of safety)—the evaluation paradigm and level of complexity (e.g., expert judgment, comparative, deterministic, (risk) index, probabilistic, or some combination)—and the data, tools, and methods (e.g., risk assessment models, computational models) required for evaluation at the selected level of rigor. Within each component, due consideration and treatment of uncertainty, variability, and unknowns is required.

Consider the exemplar performance objective presented above, "the building and fire safety systems shall be designed to prevent the exposure of those occupants not intimate with first materials burning to untenable conditions during the time required

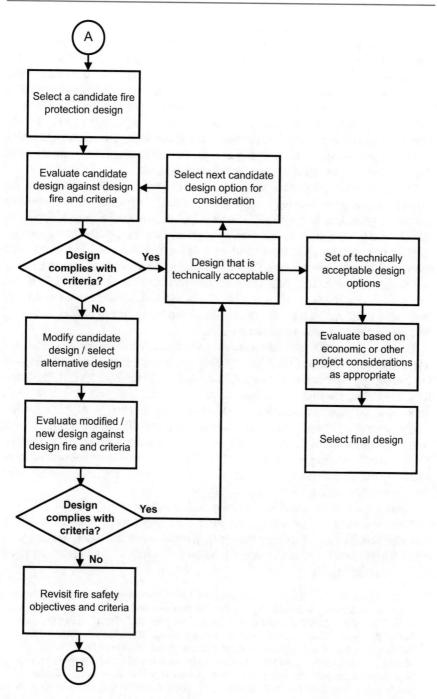

Fig. 1.4 Evaluation of potential fire safety design options (Source: author, adapted from [5])

for them to reach a place of safety outside of the building." In the current PBD for fire safety framework, an appropriate analytical concept/evaluation framework is that RSET (time to reach a place of safety) is significantly less than ASET (time to untenable conditions) with an appropriate margin of safety. How one conducts an ASET-RSET analysis can range from expert judgment to probabilistic or quantified risk analysis (PRA or QRA). The level of complexity varies widely between these approaches, as may confidence in the assessed outcome, especially as related to the level of recognition and treatment of uncertainty, variability and unknowns, which will be integrally related to availability and goodness of data and the appropriateness and appropriate application of specific tools and methods.

The above concepts and more are described in greater detail in any one of the several fire safety engineering or performance-based design for fire safety-related standards, guidelines (e.g., [61–68]), textbooks, and handbooks (e.g., [45, 58, 59, 68–73]) that are available. In addition, through the first two decades of the twenty-first Century, research has been conducted in each of the areas and more, with recommendations for enhancing PBD for fire safety frameworks, scenario analysis and design fire selection, risk-informed and risk-based approaches to FSE and structural design for fire, and more [74–97]. In addition, significant advancements have been made in the area of computational modeling of fire effects, structural response to fire, and occupant evacuation.

However, even with such advancements, the fundamental approach to PBD for fire safety has not changed significantly. This would not necessarily a bad thing—if the current approach is robust, one might argue few changes are needed. However, there are indications that this is not the case and that some concerns exist, even after some 30 years of experience. At present, the acceptance of PBD for fire safety remains challenging in many countries [e.g., 98–101], there has been what some have suggested as "backward" movement with the development of "prescribed performance" based approaches [e.g., 99, 102], and inadequate consideration of technological performance for facilitating consistent building fire performance throughout the building's life [e.g., 83, 88].

There are several potential reasons why such concerns persist [164].

First, fire safety engineering, and PBD for fire safety in particular, are in the adolescent stage of development [7]. Consider the following summary of observations made by Allin Cornell in 1981 on the maturation of a newly emerging engineering area [163].

> In its earliest stages, a new area is characterized by rather uncoordinated relationships between practice and research, between needs and solutions, and between the profession as a whole and those dedicated to the development of the area itself. The problems that have been addressed by researchers tend to reflect personal tastes, ease of formulation or solution, and simple chance. Those applications in practice that do exist tend to be 'local', small parts of larger problems isolated and resolved without reference to a broader framework, because the framework has not been constructed. The major attention is often given to certain minor problems that have been popularized only by the internal dynamics of the small research community itself.
>
> In contrast, at a mature, stable level of development of an area of engineering one customarily finds a smooth interaction between research and practice. A vocabulary has

evolved, a general framework exists, and the capabilities and limitations of the area are widely appreciated. Virtually all practitioners have received exposure to the subject and are accustomed to recognizing the kinds of situations in which the method is applicable and even to articulating their problems in the language of the area. Most research is being conducted in response to obvious needs of practice.

Clearly in between these two stages of growth there is room for the uneven levels of development that characterize adolescence. Stimulated by larger numbers of contributors, growth is very fast and in some problem areas is indeed beginning to follow rather than anticipate the needs of practice. New practice-generated problems for research are being identified not only by growing numbers of experienced researchers but also by the engineers in practice who have begun to appreciate which of their old problems the new area can help. These initial reactions from practice are often poorly articulated and often, unfortunately, too optimistic. Nonetheless one can see the establishment of certain consensus positions that determine both a framework and a viable set of solutions for at least some rather broadly defined problem areas. But the development and internal coordination of the area are still largely incomplete at this stage: some topics are virtually untouched, limits of effectiveness of the parts or the whole are not well understood, some applications are rather naively formulated, and some practical applications have begun to address a larger framework but not yet with the confidence or the wisdom of experience.

Given the current status of fire safety engineering, and especially performance-based design for fire safety, it can be argued that fire safety engineering is a healthy adolescent. Unfortunately, this has not changed since 1999, when it was observed that "research has begun better addressing the needs of practice, the essential elements of a framework and vocabulary have been developed, and many practitioners appreciate where and how the current methodologies can address their problems. However, the field remains largely uncoordinated, it lacks a comprehensive framework where the limits of effectiveness are well understood, and some applications are rather naively formulated." [7].

Second, much of performance-based design for fire safety currently focuses of safety to life of people, and control of human behavior is extremely challenging to assure [164]. Building occupants are highly variable and dynamically contribute to the fire hazards and risk, both in the probability of fire occurrence and in the magnitude of fire consequences, and it is not clear that assurance of life safety is feasible. Furthermore, consideration of "acceptable" levels of risk to life is a difficult construct for many actors in the building regulatory system, but "risk-free" environments are not possible. This confluence of a stochastic environment, in which the absence of risk is difficult for some stakeholders to agree with, remains a concern that must be addressed.

Third, following on from the above, even though fire is stochastic in nature, and therefore risk-informed and risk-based approaches to PBD for fire safety should be recognized as appropriate approaches, it can be difficult to gain acceptance of risk-informed and risk-based approaches to PBD for fire safety because it can be difficult to gain agreement on, and acceptance of, the fundamental premise that there is impossible to remove all risk (i.e., there is no absolute safety or 'zero risk' condition) [164]. Furthermore, even when it can be agreed that some level of risk exists, it can be difficult to gain agreement on how to express or quantify the tolerable levels of fire risk (and/or fire safety performance), especially when expressed in terms of risk to life. This is particularly difficult for the fire service, whose mission is to save people and property from the unwanted effects of fire. Ideally, they would like to reduce life safety risk to zero, but this is impractical. While they understand this, it can be difficult for such an admission to be made publicly, especially as a result may be that the public then questions whose life is unworthy of saving (which is not a particularly fair question).

Finally, performance-based design for fire safety has not adequately recognized and embodied sociotechnical system (STS) constructs [164]. Buildings, infrastructure, and many hazardous operations are complex sociotechnical systems (STS), with many complex interactions, not all of which are generally considered - or used - in fire safety analysis and mitigation, especially once the building, infrastructure or operations are in use. In addition, because fire is a rare event, and there are few "immediate" indicators of poor fire performance of buildings (i.e., they are tested only when there is a fire), fire performance is rarely tested, and feedback on performance into the sociotechnical system is limited and often incomplete. Complicating the situation further, in some parts of the world there is a lack of adequately educated and suitably qualified professionals across the spectrum, from analysis and design to approval and enforcement of designs, inadequate communication and information flows, and lack of feedback into the system on actual performance, all of which can contribute to a lack of confidence in analysis and design, the proper implementation of designs, and of actual performance in use (i.e., inadequate considerations of institutions and actors in STS context).

These are all sociotechnical system (STS) challenges—none can be solved solely through technology (e.g., computational modeling tools, building fire safety systems), institutions (e.g., regulations, certification), or actors (stakeholders)—but only with an integrated STS approach that recognizes and accounts for challenges and opportunities associated with each component. The following outlines concepts of STS and safety systems thinking, and how these approaches might be useful in advancing the discipline of fire safety engineering and the practice of performance-based design for fire safety. Excerpting and building on [164], the following describes how viewing the problem through a STS and safety system lens can result in a different approach to describing the fire safety objective(s), how they might be attained, and how performance might be assessed, which ultimately can lead to safer and better performing buildings.

1.2   Sociotechnical Systems (STS) Concept

The emergence of sociotechnical systems (STS) theory and concepts came from the study of the roles of social and technological components within organizations and the realization that they are integrally linked [103]. STS have been characterized as having three levels of focus: (1) primary work systems within an organization, (2) whole of organization systems, and (3) macrosocial systems, which include systems in communities and industrial sectors, and institutions operating at the overall level of society [104]. It has been argued that building regulatory systems are macrosocial systems, in which systems in communities, industrial sectors, and institutions operate at the societal level [105].

Many variations of STS theory have been advanced since 1993, largely driven by differing perspectives of the disciplines involved, e.g., business/organizational management, economics, psychology, sociology, design/technology/engineering, political science, and risk and safety management [105]. The characterization of infrastructure and buildings (i.e., the built environment) as STS came into focus with research into such areas as accident analysis, risk management, and system safety [106–108], building-hazard-regulation interactions [105, 109, 110], innovation in construction [111–114], and critical infrastructure interdependencies and vulnerabilities [115–119].

While most systems require interaction between technology, institutions, and actors, not all have the same level of interdependency. For example, Ottens et al. [117] characterize three types or categories of engineering systems: "(1) engineering systems that perform their function without either actors or social institutions performing a sub-function within the system [e.g., the landing gear of an airplane], (2) engineering systems in which actors perform sub-functions but social institutions play no role [e.g., an airplane], and (3) engineering systems that need both actors and some social/institutional infrastructure to be in place in order to perform their function [e.g., an airport]" (pp. 134–135), and conclude that "type 3" systems are STS. This is equivalent to Trist's macrosocial STS domain [105] (Fig. 1.5).

The "type 3" engineering or macrosocial systems, which rely on numerous actors and social/institutional infrastructure, have the added complexity of numerous actor involvement. Kroes et al. [118] note that: "At the sociotechnical level many stakeholders are involved that all have their own goals and visions, and normally none of these actors can impose their decisions on the other actors. For this reason, STSs cannot be designed, made, and controlled from some central point of view, as for instance a car. Instead, the STS is continuously being redesigned by many actors from within the system" (p. 813).

Fig. 1.5 Simplified representation of STS interactions [Source: author]

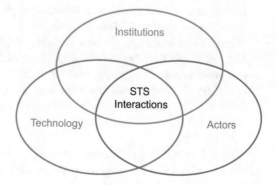

Arguably buildings, and the building regulatory process (including allowable design processes), are STS, as the system is continuously being redesigned by many actors from within the system [105]. As such, design of any building "system," including a buildings' fire safety system (the aggregate of all fire safety features), is a STS design challenge. If considered in this way, the design focus becomes controlling fire hazards and the safe operation of the STS—which is the building, system or infrastructure—throughout its intended life. Such a focus would result in a paradigm shift in thinking about PBD for fire safety.

1.3   Evaluating PB Design for Fire Through a STS Lens

Summarizing briefly the framing of buildings and building regulatory systems as STS [105, 110, 120, 164]:

- Buildings (complex buildings in particular) are macrosocial, or type 3 "systems of systems" that need both actors and some social/institutional infrastructure to be in place, and working integrally with technology, in order to deliver their intended performance. Buildings have multiple types and levels of technology which need to work together for the building to operate, such as heating, ventilation, and air-conditioning systems, façade/envelop design and construction, and sensor and control networks, or passive and active systems, sensors and controls for fire safety, and occupant evacuation. These technology systems are designed to work for the users—and rely on users to maintain them. The performance of these systems is regulated by and controlled via complex regulatory and private sector institutions, from building regulation to insurance. Ultimately, all components—actors, institutions, and technology—must work together holistically for the building (as a complex system-of-systems) to meet its societal and private sector objectives.
- Building regulatory systems are STS since they must consider the roles that institutions, stakeholders, and technology play in design, construction, and operation of buildings. This includes (a) the need to establish clear societal objectives for buildings, (b) how the technologies must work at each phase, and in total, in meeting the societal objectives, and (c) how to balance the myriad stakeholder input from those involved in the building design, construction, and operation phases of a building's lifespan. Failure to adequately consider any of these aspects, and their multiple components, can result in building regulatory system failure.

A model of building regulatory systems as STS [105], based on Petak's system model for building risk management and resiliency to earthquake hazards [109], is illustrated in Fig. 1.6.

With this as a base, a sociotechnical building regulatory system assessment model (STBRSAM) was created, with the aim being a tool with which building regulatory systems could be evaluated [110, 120]. A truncated representation of the STBRSAM is presented in Fig. 1.6, focusing on the system control component. The

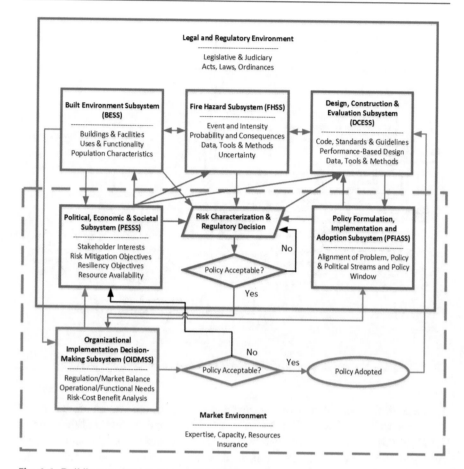

Fig. 1.6 Building regulatory system as STS [105]

STBRSAM is based on Rasmussen's Abstraction Hierarchy as applied to safety control [106, 107, 121, 122] and Leveson's System-Theoretic Accident Model and Processes (STAMP) construct for safety control [108, 123].

The safety management system, Leveson [108] argues, has two basic hierarchical control structures—one for system development and one for system operation— with interactions between them. An aircraft manufacturer might only have system development under its immediate control, but safety involves both development and operational use of the aircraft, and neither can be accomplished successfully in isolation: safety must be designed into the system, and safety during operation depends partly on the original design and partly on effective control over operations (the system control).

Underpinning Leveson's STAMP model is that system theory is a useful way to analyze accidents, particularly system accidents, and that in this conception of safety, "accidents occur when external disturbances, component failures, or dysfunctional interactions among system components are not adequately handled

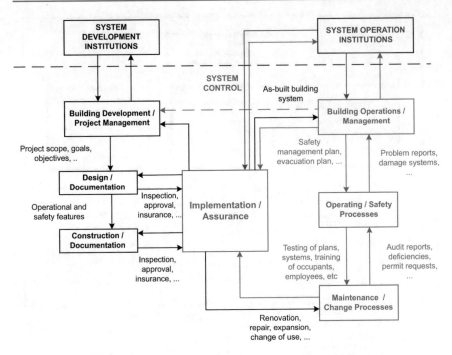

Fig. 1.7 System control component of STBRSAM (Source: author, adapted from [120])

by the control system, that is, they result from inadequate control or enforcement of safety-related constraints on the development, design, and operation of the system" [108]. In this paradigm, safety can be viewed as a control problem, and safety is managed by a control structure embedded in an adaptive socio-technical system, the goal of which is to enforce constraints on system development (including both the development process itself and the resulting system design) and on system operation that result in safe behavior. In this conception, understanding why an accident occurred requires determining why the control structure was ineffective, and preventing future accidents requires designing a control structure that will enforce the necessary constraints.

Bjelland et al. [84] introduce this type of systems thinking for fire safety design in the context of buildings as STS. Applying similar thinking, the STBRSAM framework [110, 120] was proposed as a tool to assess building regulatory systems for potential failure points. In Fig. 1.7, the "system development institutions" and "system operation institutions" reflect the legislation, regulations, standards and such which govern building design and operation; the technological components of the STS are the building and its fire safety systems; and the building fire safety system is controlled via the interactions of actors and systems within the "system control" framing. A main point behind the STBRSAM [110, 120] is that well-performing buildings depend on far more than just building regulations, which direct design requirements, and the fire and occupational health and safety regulations, which address buildings in use ("system development" and "system operation" institutions). There is a strong dependency on extra-regulatory ("system control") interactions between: the design, including required functionality of critical systems and the documents which define these requirements; requirements around safety in use, including evacuation and related operational requirements, inspection, test and maintenance (ITM) testing, and related assurances of critical systems performance; and the actors in the systems (e.g., designers, engineers, technicians, building owner/manager, tenants). Performance expectations for buildings derive from experience of buildings in use, over time, as subjected to different stressors, such as fire or other hazard events. The building regulatory system thus reflects the need to focus on STS interactions between institutions, actors and technologies to define (describe, reflect) acceptable/tolerable building performance, and in the case specifically reflected in Fig. 1.6, acceptable/tolerable building fire safety performance. This aligns well with the thinking of Bjelland et al. [84].

Likewise, the PBD for fire safety process requires the interactions of actors, institutions and technologies to develop and deliver well-performing buildings. How FSEs undertake designs, following which institutional norms (standards, regulations), using what technologies (e.g., computational modeling), and with what expectations (i.e., use, operation and maintenance of building to continually achieve performance expectations), should be reflective of a STS design process. FSE need to recognize that buildings could be regarded as complex systems of systems, which must work integrally and holistically to meet societal expectations for building fire performance, but not just during design and construction, but throughout the buildings' lifetime as well. Gaps in any area of the STS can lead to failures in achieving performance expectations.

If one accepts the premise that PDB for fire should reflect a STS design process, one can look to core concepts of STS design to understand perhaps why further advancement or acceptance of PBD for fire safety has not been realized, and how one might facilitate change. Starting with representative STS design principles from Cherns [125] (in italics), and applying them to PBD for fire safety, some key issues emerge [164].

- *The process of design must be compatible with its objectives.* Performance-based design for fire safety is an objective-driven activity, the aim of which is to design buildings that perform well during fire events, taking into account people and technology interactions. The PBD for fire safety process should therefore be structured around the ability to demonstrate achievement of objectives considering STS interactions. In principle, this is the current process, but in practice, it seems to fall well short in many applications. One aspect that warrants reconsideration is the extent to which people directly are considered as a target of PBD for fire safety, as compared with the target being the building, including the design of systems to limit fire (which in turn impact life safety), as the focus. That is, if building fire performance is the societal objective, this means that fire safety design objectives and criteria should be expressed in terms of building performance in case of fire. It also means that analysis of the fire performance of building should be focused on achieving the building performance expectations (with impacts on life safety being an attribute of building performance).

- *No more should be specified than is absolutely essential, but the essential must be specified.* This is an aim of performance-based design in general—focus on functional requirements, performance objectives, and performance criteria, and not on specifications—but it also reflects one of the major challenges with performance-based design for fire safety—how to assess tolerable (acceptable) performance under continually changing conditions. It is not apparent that we have yet clearly identified what is *absolutely essential* and how to reflect this in a way that can be analyzed, designed, and suitably assessed. Also, establishing objectives is not for the FSE alone—in must be undertaken in a STS interactions environment. As noted above, this might manifest as a refocus on building system performance. Furthermore, focusing on the "absolutely essential" may mean a shift in thinking about fire scenarios from a focus on "subsystem" verification to overall systems performance..

- *For groups to be flexible and able to respond to change, they need a variety of skills.* This is critical area that remains far from the level that is necessary for a well-functioning PBD for fire safety system approach. This applies to all actors in all sectors of the building fire performance arena: FSEs, technicians, building officials, fire officials, insurers, and more who are directly working with fire performance aspects, but also those working within the broader sociotechnical building regulatory system. Even if a particular job function does not utilize education and training associated with performance analysis or design every day, a robust understanding of the concepts, and of how the performance system works, is required for actors to work effectively in a performance system. Appropriate education and training has been identified as a need for some 30 years [e.g., 5, 7, 9, 80], and yet remains unfulfilled [e.g., 99, 101, 126].

- *Information must go, in the first instance, to the place where it is needed for action.* This has been a challenge throughout the history of PBD for fire safety—information flow. Challenges exist within the design team, with FSE often being engaged at the end, and without full information. This translates into a lack of consideration of key issues, and subsequent communication of limitations to the end user. For the past 30 years there have been concerns that the ultimate building owner, managers, and tenants do not get the PBD documentation and critical information aimed at assuring the target performance in maintained. This "golden thread" of information, or lack thereof, was identified as a major concern coming out of the Grenfell Tower fire inquiries and investigations. Proper and adequate information flow is essential. Ignoring this, or not addressing it comprehensively, can leave significant gaps in understanding needed for the PBD to work as intended. Rasmusson and Svendung [107] and Leveson [108, 123, 124] provide several examples where lack of good information flow and feedback were significant contributors to major losses.

- *Boundaries should facilitate the sharing of knowledge and experience. They should occur where there is a natural discontinuity—time, technology change, etc.—in the work process. Boundaries occur where work activities pass from one group to another and a new set of activities or skills is required. All groups should learn from each other despite the existence of the boundary.* This principle is closely tied to the above two. Boundaries exist each time information relevant to a PBD for fire safety is transmitted through the system, e.g., client to architect, architect to engineer, engineer to technicians and manufacturers, manufacturers to contractors, contractors to owners/managers, owners/managers to clients. These boundaries are attributes of the system. However, they should be opportunities for teaching and sharing knowledge and experience—so as to help broadening and deepening of skills needed to work effectively in a PBD system—and to facilitate proper and appropriate flow of information.

- *Systems of social support must be designed to reinforce the desired social behavior.* In the PBD for fire safety realm, this relates significantly to assumptions made about people and buildings—whether owners and managers expected to maintain buildings to design parameters—or occupants, for whom certain behaviors and responses are assumed by FSEs. It also significantly relates to social behavior in terms of professionals. There is a critical need for high levels of competency, ethics, and professionalism to responsibly and comprehensively undertake performance-based design for fire safety. Social constructs in which market behavior forces a "race to the bottom" works against PBD for fire safety—and confidence in resulting designs. Underpinning this is the need for appropriate education across all actors in the systems—not just FSEs—but building and fire officials, manufacturers, and others involved in the process.
- *The recognition that design is an iterative process. Design never stops. New demands and conditions in the work environment mean that continual rethinking of structures and objectives is required.* This is a significant opportunity and challenge for PBD for fire safety. Starting with the last principle—continual rethinking of structures and objectives is required—it can be argued that this has occurred to some extent regarding the structure of PBD for fire safety, but it can also be argued that this has occurred in the absence of consideration of changing societal expectations, institutional constraints, and technological evolution. Furthermore, the fact that design is an iterative process should be well accepted, and in fact is built into the PBD for fire safety process. However, increasingly there are examples of practitioners shortcutting the iterative process, especially when it comes to considering perspectives and needs of the various stakeholders.

Gehandler [90, 127] and Bjelland [128] have identified many of these same concerns, and have suggested that PBD for fire safety be viewed as an iterative problem-solving process between the designers and the stakeholders, the overall aim of which starts with an objective of an inherently safer (cannot fail) and fail-safe (forgiving to errors) design of the building fire safety system. The framing should be that of a STS, where solutions consider technology as well as users, operators, and institutional elements, with the aim to keep the system within safe limits, given the variability in human behavior. Deliberation should Include any objectives the stakeholders value, be they technical, social, ethical, political or societal, with decisions made through the application of appropriate decision theory to structure and decompose the decision into manageable parts.

Many insights can also be gained from the system safety perspective, especially those of Leveson, who outlines a systems approach to engineering safety [124], and lays out definitions and principles of what she refers to as "Safety-III, the third phase of evolution of systems safety," (Table 1, pp. 27–29) many of which can be applied to building fire safety systems design, especially when considering buildings and infrastructure systems as complex STS.

- Safety is defined as freedom from unacceptable losses as identified by the system stakeholders. The goal is to eliminate, mitigate, or control hazards, which are the states that can lead to these losses.
- Safety management concentrates on preventing hazards and losses, but does learn from accidents, incidents, and audits of how system is performing.
- Accidents are caused by inadequate control over hazards. Linear causality is not assumed. There is no such thing as a root cause. The entire socio-technical system must be designed to prevent hazards; the goal of investigation is to identify why the safety control structure did not prevent the loss.
- The system must be designed to allow humans to be flexible and resilient and to handle unexpected events.
- Design the system so that performance variability is safe and conflicts between productivity, achieving system goals, and safety are eliminated or minimized. Design so that when performance (of operators, hardware, software, managers, etc.) varies outside safe boundaries, safety is still maintained.

Checkland's soft systems methodology (SSM) [129–131], which outlines a systems thinking approach, is also pertinent for PBD for fire safety. As noted in [164], SMM was developed as a result of perceived shortcomings in early systems thinking, which focused on systems as individual representations of a particular worldview, and arguably had shortcomings in being able to address complexity associated with different worldviews and changing conditions over time. The introduction of the notion of "worldview," which was considered essential in dealing with human social complexity, resulted in a change in thinking of systems models not as descriptions of something in the real world, but simply as devices (based on worldview) to organize a debate about "change to bring about improvement" ([130], p196). A critical component of SSM is the acceptance that complexity requires adaptation, and adaptation comes often from emergent properties of the system, which may be unknown at the outset of system planning or design. Acceptance of the concepts of complexity, adaptation, and emergent properties are important to the STS formulation of buildings and infrastructure as STS, and of how PBD for fire safety needs to operate in such systems, especially in decision-making about system objectives.

In PBD for fire safety, it is already a part of the approach that the measure of successful performance should be a result of stakeholder deliberation and agreement. However, that still does not occur to the extent it should, and key stakeholders (actors and institutions) can be left out of the decision-making process.

It is also that case that fire safe operation of buildings focuses on reducing hazards, and should learn from audits of how "the system" is performing, but the latter seems to be missing in many instances. As is argued here, to develop robust PBD for fire safety, the entirety of building as a STS must be considered, and design for fire integrated into the overall building performance objectives [164]. While it is possible to design building fire safety to allow for human flexibility in managing fire events, that is often not feasible, given the lack of focus on education and training, in particular responses to emerging situations (as compared with predetermined scenarios). Finally, while building fire safety design could be focused more on a "fail-safe" approach, that is often not the case. Rather, emphasis is often placed on engineering out robustness and redundancy (so-called value engineering), which if in place could be helpful if primary system components or subsystems fail. All of these factors should be reconsidered in the PBD for fire safety construct.

Looking ahead, Leveson [124] also identifies several areas where addition focus is needed as System Safety evolves (pp. 105–106). Again, the case can be made that these are applicable to the evolution of PBD for fire safety as well, especially in consideration of PBD of buildings and other systems as STS.

- "Create new holistic hazard/risk modeling and analysis techniques that include all facets of the sociotechnical system and how they can operate together to prevent losses. These tools should assist in making difficult conflicts and tradeoffs. Create tools to provide qualitative safety design information to engineers and decision makers. De-emphasize probabilistic methods, which are not accurate for complex systems.
- Create techniques and approaches that emphasize building safety into a system from the very beginning and provide better systems analysis and design tools that use these approaches. Decrease the current emphasis on adding protection to a completed design and trying to assure safety after-the-fact, i.e., after the design is complete: Neither is possible in complex systems.
- Create and use true sociotechnical approaches. Such approaches will require more integration of human factors and hardware/software engineering. It will also require better communication and ways for human factors engineers and hardware/software engineers to work together. Using STAMP as a basis, new modeling and analysis tools have been developed that provide a common model and communication process for all the engineers on a complex design project.
- Develop top-down, holistic approaches that allow us to handle the complexity of today's high-tech, complex systems. Closely related is to make sure that software and new technology can be handled by all our engineering and safety approaches.
- Improve our ability to create effective safety management systems and organizational and industry safety cultures.
- Create improved techniques for dealing with safety during operations and for using the massive amounts of data that can now be collected.
- Create new approaches to certification of safety for highly automated systems.
- Provide better education about safety for everyone."

In the system safety construct, one could argue that the analysis and design is less about the scenarios that could occur, and more about the ones that could lead to system failure, and then to design "fail-safe" systems. This would deviate from current PBD for fire safety practice in considering all possible fire scenarios, and reducing them to likely scenarios through probabilistic analysis, and instead focusing on scenarios or conditions that could lead to failure of the fire safety system, regardless of probability, and designing for prevention of failure, taking into account other stakeholder objectives (such as cost, operability, and so forth). In the broader STS context, the "system" for which fail-safe status is sought is the building as a STS, not specific fire safety subsystems (e.g., suppression systems), which need to deliver on their performance objectives as part of the whole.

A change to include more of a systems safety approach would not mean the quantitative or probabilistic risk analysis for fire would go away; rather, its use would be more for informing decisions, which have other considerations as well (in the sociotechnical system), as compared with producing a risk estimate that would be the sole determinant in a decision. This is the foundation for the U.S. Nuclear Regulatory Commission's (USNRC) "risk-informed performance-based" regulatory and design approach [132], which was brought into some design standards for fire safety [e.g., 133], and has been suggested as being more widely applicable for fire safety regulation and design for some time [e.g., 74, 79, 83, 88].

Pulling things together, one can again look to Checkland [131], who outlines four concepts that help to define, minimally, the concept system (pp. 446–447):

Firstly, and obviously, any entity called 'a system' may also contain within itself, functional subsystems, and may itself, as a whole, be a functional part of a wider system. So, a system will, in principle, be part of a 'layered structure' making a hierarchy of systems. Which level is that of 'system,' which that of 'subsystem,' or 'wider system' is a matter of judgment made by the person making use of the concept.

Secondly, to achieve adaptation to change, there will have to be "processes of communication." These will have to involve both the system and its environment. These processes will enable performance to be monitored so that a decision to adapt or not can be taken— either via automatic processes or by human beings.

Thirdly, if action to adapt is to be taken, the system will have to have available a number of possible control processes (responses to the shocks from the environment and internal failure), which can be appropriately activated to bring about change.

Fourthly, there will be definable "emergent properties" that characterize the particular system or systems of interest, this being the pre-eminent systems idea.

The first three parts are completely consistent with Leveson, including subsystems, controller, and communication feedback loop, and arguably the emergent properties are built into Leveson's control system, albeit not defined as such. For PBD for fire safety, consideration of the emergent aspects of the design process is important as system objectives are developed during the iterative design process.

A final attribute of STS that is incredibly important to consider in the context of the slow evolution and acceptance of PBD for fire safety is the rate at which technology advancements are understood and applied, and knowledge is gained and applied, within different elements of STS systems [164]. Rasmussen and Svedung [107] observe that if the pace of change of technology is much faster than the pace of change within legislation, regulation and management structure, that the potential for system failure exists. In many respects, performance-based building regulation, and PBD for fire safety, is a response to the challenge of the time lag associated with updating building regulations and standards during periods of rapid changes. In the 1980s and 1990s, technology and innovation was outpacing the ability for approval and adoption for use into prescriptive building regulation, and performance-based design was seen as a mechanism to demonstrate that the new technology and innovation could be used without negatively impacting building fire safety [5].

However, the rapid introduction of PBD for fire safety faced similar challenges that Rasmussen and Svedung [107] identified with the rapid pace of change in the hazardous process industry, which they studied, but with different institutions and actors, importantly those responsible for review and approval of performance-based designs. Emerging tools and methods used by FSEs, such as computational modeling, was not familiar to enforcement officials. Many lacked the education or training to understand well how they worked (as that was not needed before), the process that FSEs were following (i.e., PBD and FSE frameworks) were largely "high-level" processes and lacked details (such as specific scenarios, design fires, or performance criteria to consider), and their ability to confidently verify compliance (or performance) was limited. This created a disparity between engineers excited about new technologies and opportunities and enforcement officials concerned with the ability to assure designs would perform as stated. Perhaps more than any other sociotechnical issue, this mismatched set of expectations, amongst integrally linked actors and institutions in the system, as related to the technology being used, resulted in a lack of confidence by authorities of the new PBD approach.

Furthermore, as Perez [134] argues, when a technological revolution occurs, it does not just add new industries to the structure, but through the techno-economic paradigm that results, can provide the means for modernizing all the existing industries and activities, but over different time periods. The first decades are the period of implementation, during which a critical mass of "agents of change" become established and put pressure on existing institutions. During this period, there is tension between investment in existing technologies and structures versus those associated with revolutionary change. After some decades, the transformation potential of the technology revolution spreads across the economy, creating the potential for it to realize widespread adoption, and with it, modification to the institutions which have some bearing on it. The decadal lag can be seen with PBD for fire safety.

Such challenges with the rapid introduction of technology resulted in the formulation of the so-called hype cycle [135], illustrated in Fig. 1.8 [136], which was developed to represent the maturity, adoption, and social application of specific technologies.

Although developed with a focus on high-tech developments, and criticisms of the veracity of the hype cycle for product development exist (e.g., it is not actually a cycle) [137], it provides a useful graphical/thought structure with which to look at the evolution of PBD for fire safety [164].

- Technology triggers—the development of computational models to explore fire effects in compartments, the response of structural elements to fire, and movement of people, as accelerated by the widespread availability of the personal computer (1980s–1990s). The availability of ready access to fire safety

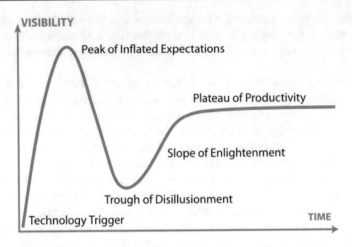

Fig. 1.8 Gartner research's hype cycle diagram [136]

science and engineering knowledge through the first edition (1983) of the *SFPE Handbook of Fire Protection Engineering* [138] played a major role as well. The introduction of functional- and performance-based building codes was an enabler. So too were analogies drawn from structural engineering, such as the formulation of the concept that available safe egress time (ASET)—required safe egress time (RSET) must be greater than zero, with some margin of safety. Such developments provided inspiration to "engineer" fire safety in a way that had not been done to a large extent previously.

- Peak of inflated expectations—arguably the "peak" occurred in the 1990s–2005 timeframe, as the introduction of performance-based codes in several countries took place, and several PBD for fire safety concepts and frameworks were introduced and advanced. Some of the excitement about PBD for fire safety was facilitated by development of PBD concepts in related areas, such as seismic engineering [139], and the rapid development of lower-cost, higher-power computers, the explosive availability of information through the internet, and the sense that perhaps the discipline of fire safety engineering was maturing [7].

- Trough of disillusionment—it is suggested that the fall began sometime after 2005, perhaps bottoming-out around the 2010–2012 timeframe. During this period, PBD for fire safety was highly scrutinized by building and fire enforcement officials, in particular, in part because the "process-oriented" frameworks lack sufficient detail for critical components (e.g., scenarios, design fires, performance criteria, modeling parameters) to result in consistency in application of PBD for fire safety approaches and the verification of post-design fire performance of buildings. This led to uncertainty in approvals, which led to additional time and cost of projects, which diminished its attractiveness to the market (e.g., developers). Furthermore, while building technology was advancing, with increased use of sensors, "smart" building controls, and integration of continuous environmental learning into building environmental control [e.g., 140–142], the design of fire safety systems did not significantly change. While advancements in thinking and approaches to PBD for fire safety were made [e.g., 74–97], as well as early thinking about better fire technology integration in buildings, including integration with sensors [e.g., 143, 144], there were few gains in terms of either raising the level of confidence by authorities and the market, or in developing truly innovative solutions. In this way, the lack of due consideration of institutional, stakeholder, and technology concerns effectively stagnated the advancement of PBD approaches and their acceptability, resulting in "prescribed performance" verification methods and "traditional" design solutions.

- Slope of enlightenment—the current state of practice in 2021 is arguably advancement up the slope of enlightenment. This is surmised considering recent contributions regarding systems thinking [127, 128], risk-informed approaches [e.g., 78, 79, 83, 92–94], and the need to make better use of building technologies as part of both designing and maintaining expected building fire performance [145–147]. While one hopes that PBD for fire safety reaches the plateau of productivity within the decade, that remains to be seen. However, there are signs of momentum. In addition to new thinking around the PBD for fire safety framework, performance-based building regulation is being "reimagined" in several countries, and the fire safety engineering community is being challenged to step up with updated and more robust methods. If the regulatory system expectations, market expectations, and PBD approaches converge, that would be a significant achievement along the path to a plateau of productivity for PBD for fire safety and innovation in building fire safety systems technologies.

In summary, PBD for fire safety has neither evolved as anticipated nor gained the broad acceptance as was expected since its introduction in the 1990s. This is not a failure of the core concepts, per se, but more so a function of the failure to consider STS attributes that are critical to the application and acceptance of new technologies within a sociotechnical building regulatory system, within which PBD for fire safety aims to design safety systems to mitigate fire hazards in buildings, infrastructures, and operations to an acceptable or tolerable level. However, by applying STS thinking, helpful changes can be made.

1.4 Advancing PBD for Fire Safety by Incorporating STS Concepts

Evaluating PBD for fire safety through a STS lens has illustrated why perhaps PBD for fire safety has not yet reached its full potential, regardless of having been accepted in concept for some 30 years [164]. Key requirements for fire safety of buildings and other systems as sociotechnical systems have not been addressed, and institutional and market lag associated with uptake of new approaches has played a role as well. However, in the coming decades, it can be expected that more technological advances will occur within the realms of fire safety performance buildings, systems, and infrastructures, and that a STS framing for PBD for fire safety can both advance the use of PBD concepts and the use of innovative technology to achieve fire safety of STSs. The following steps are proposed to help facilitate this advancement [164].

Step 1—Recognition and acceptance of PBD for fire safety as a STS challenge. PBD for fire safety cannot be undertaken in a vacuum. First, it must be accepted that buildings and infrastructure, and building regulatory systems, can be viewed as STS, and that there are institutional, technological, economic, and human factors which must be considered in designing for safety from fire in these systems. A potentially helpful framing is to consider the whole of the building fire safety approach (i.e., active and passive system design, ITM, safety management, etc.) as being the "system control" that must work holistically and integrally to result in building fire safety. The development of the fire safety "system controller" for the building or infrastructure is the aim of PBD for fire safety. To move in this direction, not only will more systems safety thinking be required, but better consideration of STS attributes will be required, including:

• developing process guidance that better addresses institutional concerns, particularly those from the building and fire authorities, with respect to how data and technology are used in the analysis and design process,
• reconsidering and potentially restructuring the fire safety objectives for design,
• reconsidering the role of risk and uncertainty in the process, and how best to address these constructs,
• more explicit consideration of how building fire safety technology performance will be used, and assured throughout the lifetime of the building or infrastructure, to assure that building fire safety performance objectives will be met.

Getting agreement of stakeholders within the process remains a key component, and more engagement in the broader environment is critical. Embracing the iterative nature of STS design is important, and there can be no shortcutting of deliberation with stakeholders on appropriate means of analysis and on the range of potential design options. Figuring out, and focusing on that which is essential, will be vital.

Step 2—Revisit design goals and objectives—start by asking if we are solving the right problem—and focus on the building as a STS. Buildings are sociotechnical systems, in which people and buildings interact. The focus needs to be on those interactions, with a focus on what PBD for fire safety can control for. While human behavior must be considered, it cannot necessarily be controlled. Implementing building technology, which helps identify and control fire hazards, that in turn limits occupant risk, is within the realm of the FSE. Earlier in the paper a representative objective was stated as "the building and fire safety systems shall be designed to prevent the exposure of those occupants not intimate with first materials burning to untenable conditions during the time required for them to reach a place of safety outside of the building." Stating an objective in this manner should be reconsidered.

For example, is it possible to "prevent" exposure, or should the objective be to "limit" exposure, or perhaps to "limit the spread of fire effluents from entering protected exit components?" Can the "time required" to reach a place of safety be accurately estimated, given the variability in population and unknowns that may exist at the time of an actual fire, or should some time limits be developed for different building uses and occupant characteristics? This is not to say that the ASET and RSET comparison is not a viable method of evaluation, but that if used, the uncertainties need to be clearly addressed, and in the STS thinking for design, a "fail-safe" design option should be the aim. (From a system safety perspective, if you adequately control the hazard, you control the loss potential.)

The area in which the FSE has most control is in managing the creation and extent of the hazard, followed by implementing technologies to help guide escaping occupants. FSE has no real control over occupant behavior. If certain behaviors are assumed, requirements for educating and training occupants to facilitate expected responses should be part of designs. Finally, for what targets in a building is society most concerned, and how are they being addressed? If vulnerable populations are a major concern of society, objectives should be recast or reprioritized to provide appropriate focus (e.g., redefining an intolerable hazard), for which building fire safety features can be designed to address. Overall, this requires reconsideration of: what are the goals of the system for which fire safety needs must be addressed, what are the fire safety objectives within the system and how should they be defined, and how can fire safety be built into the system to achieve the goals and objectives.

Perhaps a way to restructure the evacuation objective above might be: "provide a holistic building fire safety system design that provides early warning, affordances that facilitate wayfinding in fire conditions, and means to control the rate and spread of fire and fire effects to a degree appropriate of the potential fire hazard and characteristics of the building occupants." In such a structure, it is acknowledged that the target of protection is building occupants, but the design (which the FSE can control) is focused on building fire safety systems, which can be used to influence human behavior and provide for safety, but does not intimate that occupants will not be exposed to untenable conditions. It also suggests that occupants, as part of the STS, have a role to play in their own safety in the case of fire.

Step 3—Consider the appropriate balance of risk and safety as bases of performance objectives, criteria and performance evaluation. This is a critical area, and perhaps the one in which least socio-institutional acceptance has been achieved over the past 30 years. Fire is a stochastic event. Human behavior and reaction to fire and other hazard events is highly uncertain but may be characterized probabilistically, within clear boundaries. Reliability of fire safety systems is not 100%. The framing of PBD for fire safety as strictly a deterministic problem, based on a very small number of nonrepresentative scenarios and limited size design fires, is inappropriate and contributes to challenges in obtaining agreement on design verification. However, a push for quantitative risk assessment methods, when there is a lack of data, and lack of understanding of the concepts, is inappropriate as well. A risk-informed framing can provide opportunities to overcome these challenges.

At present, because statistical data are often lacking for quantitative fire risk and reliability analyses, it can be difficult for some authorities to accept risk-based fire engineering design. However, if a broader analytic-deliberative approach is taken to characterizing fire risk [e.g., 148, 149], it is possible to obtain agreement on acceptable means to estimate and incorporate risk data into fire safety decisions. For system safety, Leveson [124] suggests that there is a need to "create new holistic hazard/risk modeling and analysis techniques that include all facets of the sociotechnical system and how they can operate together to prevent losses. These tools should assist in making difficult conflicts and tradeoffs. Create tools to provide qualitative safety design information to engineers and decision makers." For PBD for fire safety, this means careful thinking about which aspects of the building of infrastructure fire safety system design problem can benefit from risk analysis for the benefit of informing decisions.

However, more than just new tools are needed, it is also necessary to address the general lack of knowledge and experience in probability, statistics, risk, and risk-informed design *for fire*, both by FSEs and authorities. If one looks to other disciplines, such as reliability-based structural engineering, in general, or performance-based design for earthquake or wind loads in particular, the risk underpins many of the design approaches in use and are accepted by building authorities. This did not happen overnight, but is an outcome of years of research, development of codes of practice and standards, and of education. The discipline of FSE, and approaches to PBD for fire safety, have been slow to incorporate risk characterization and risk-informed design concepts, and slow to educate and prepare institutions and the market for risk-informed design. A significant step came with the publication of ISO 23932 in 2018 [65], in which it is recognized and stated that all FSE analyses, including PBD for fire safety analyses, are risk analyses— the issue is simply who identifies the tolerable risk and how (e.g., as embodied in prescriptive requirements or some form of design criteria). In this regard, one approach to advance PBD for fire safety is to think about how risk might be used as a basis for establishing desired/tolerable levels of safety performance. This is not a new insight, but in the STS framing, it becomes a critical need.

Step 4—Revisit scenario construction and application within a STS context. This might require the biggest shift in thinking about PBD for fire safety as compared with current practice. As outlined in the introduction, the current approach is largely to consider all possible fire scenarios that could occur, and through scenario reduction techniques, group them into representative design fire scenarios. It is often suggested that risk analysis techniques can be helpful in this process. In deterministic analyses, the approach may be shortcut to simply picking a few (or in some cases, a single) design fire. In either approach, the resulting design fires are ultimately used to "test" the proposed fire safety system design. A long-standing concern with the "common" deterministic approach is that there is often not a robust approach to identifying "the" design fire (sometimes just a single fire), and thus it may not be representative. For probabilistic approaches, the concern is that key data are often missing, such as fire ignition frequencies, for estimation of probability of scenario occurrence. In either approach, it is typically unknown what the actual fuels and ventilation parameters will be at the actual instance of a future fire event, what the state of proposed fire safety systems will be at this time, and whether other system parameters will be as assumed (e.g., occupant characteristics). These approaches, which focus on picking scenarios/fires that may occur, and then testing potential mitigation for their efficacy in controlling the event, depend significantly on the adequacy of the scenarios and resulting design fire representations. Arguably, the risk analysis approach should result in more representative scenario and design fire selection; however, the unknowns can be significant, and uncertainty may not be treated adequately.

In system safety thinking, the consideration is more on how the system can be engineered to prevent unwanted scenarios (accidents, losses) from occurring, as compared with identifying the scenarios. In the case of PBD for fire safety, the system is the building or infrastructure of interest. Since buildings and infrastructure are STSs, a systems approach would consider how the whole of the STS can help in preventing unwanted scenarios from occurring, and likewise, should consider the whole of the system design for fire safety. In this case, the focus shifts from considering all scenarios that could occur, to focusing on those which could cause significant disruption of the STS, and designing to prevent it (or at least lower the probability of occurrence).

A possible approach is considering fire scenarios as "learning tools" more so than "verification tools." That is, the purpose of fire scenarios is to gain insight about how the system performs under different loads, rather than assuming the scenario (and load) will be the "right one" for the building or infrastructure. In such a construct, some learning may be relevant for design of the building, some for the design of the safety organization, while other learning is relevant for the emergency responders.

Step 5—Embrace innovative and emerging fire safety technologies in performance-based analysis and performance-based design for fire safety—in an appropriate and robust manner. Over the past 30 years, numerous advancements have been made in the area of "smart building" technologies (sensors, controls, algorithms) for energy performance, with research needs continuing to be advanced [e.g., 139–141]. Arguably, fire safety systems in buildings have been slow to embrace similar opportunities. As noted previously, the integration of sensors, fire safety technology, and computation modeling has been explored to forecast fire development in buildings [142, 143] but not advanced. New efforts are underway in this area [144–146], but it will likely be several years before mainstream technologies are available.

Similarly, concepts such as occupant self-evacuation elevators were explored and promoted following the 2001 attack on the World Trade Center towers in New York City, but it took several years for many building regulations to recognize such systems and they are still not widely used, with researchers noting that there is still a reluctance to incorporate such systems into building fire safety designs [e.g., 150]. There are several reasons for this, including incomplete understanding of human evacuation decision behavior and institutional concerns, clear STS components. Likewise, dynamic evacuation signage, in which exit signs are connected to intelligence about location of a fire in a building and directionality is dynamically indicated has been researched and is available [e.g., 151, 152], but remains largely still a research issue, as the technology is not yet readily accepted in some building regulatory systems. Recently, it has been hypothesized that an intelligent adoption of autonomous structural components (ASCs) can institute a platform to enable resilient performance in critical infrastructures, such as buildings, that allows self-deployment mechanism when triggered by external actions [e.g., 153, 154].

Gaining acceptance of new technologies in building fire safety is a STS challenge. Like PBD for fire safety, a convergence of technology acceptance, institutional acceptance, and stakeholder acceptance is needed. In this regard, performance-based design for fire safety as a STS system can, and arguably should, be a facilitator in bringing the realms into convergence as part of building fire safety design, in promoting innovation in design remains an objective.

Step 6—Place equal emphasis on assuring continued delivery of performance in use as on design. In general, performance-based design of buildings should be user-centered. Unfortunately, the current paradigm is largely regulatory compliance. Performance-based design for fire safety should explicitly consider and address how buildings will be used throughout their normal life—not just in emergencies—and assure that the fire safety systems will be ready when needed. Alvarez [83, 88] refers to this as designing for the chronic use and not just the acute. To do this effectively, FSEs need to consider those attributes of normal building use that can render ineffective the fire safety systems and features that may only be used on occasion. A classic example is the use by building occupants of door-hold-open devices to effectively negate smoke (and sometimes fire) doors to ease normal building flow. Blocking exit doors, which are rarely used in some building occupancy types, is another example. There are many more.

It is not appropriate for FSEs undertaking PBD for fire safety to say, "fire safety management is the owner's responsibility and not mine." In a STS, it is joint responsibility. It cannot and should not be assumed that the owner/manager/tenant will know what is critical for fire safety performance if not adequately informed. Likewise, it cannot be assumed that systems will adequately be maintained without appropriate information or that proper evacuation training will occur without specific guidance. Significant loss of life and damage to property can result from poor or inadequate information flow. One can look to the 2017 Grenfell Tower fire disaster to see the impact of poor information flow to end users [e.g., 120, 155–157]. Design guidance and/or engineers that consider operations and maintenance manuals, or adequate description of critical systems, features and assumptions that may impact building fire performance, are doing a disservice to their clients and to the aims of performance-based design for fire safety. As we see post-occupancy changes to buildings, some of which may be extra-regulatory, such as energy performance retrofits, it is important to consider fire safety impacts. Such potential changes should be considered in the PBD for fire safety process, as it is a critical component of buildings as STS, as reflected in Fig. 1.7.

Step 7—Recognize and address the risks within the system associated with inadequately prepared, unqualified, and/or noncompetent actors. In the current paradigm, it is presumed that only suitably educated, competent, and qualified fire safety engineers will undertake PBD for fire safety; however, it has been identified that this is not always the case, and that as a result, the reputation of, and confidence in, FSEs and PBD for fire safety has suffered [e.g., 99, 100, 101, 126]. While this is a significant concern, the magnitude of the concern increases when one expands the scope to PBD for fire safety design of buildings and infrastructure as STS. Within the broader STS realm, numerous other actors also play significant roles in achieving well-performing buildings and infrastructure, and if there are failures by other actors as well, this too can influence fire safety performance. Examples of system failures, involving many actors with some responsibility for design and delivery of appropriately fire safe buildings, include the 2017 Grenfell Tower fire in London [e.g., 120, 155–157] and the 2014 Lacrosse Building fire in Melbourne, Australia [158–160]. In both situations, incomplete/incorrect interpretation of regulations, incomplete specification of product, incomplete understanding of fire performance in the intended use, and incomplete knowledge and training associated with fire safe building use, contributed to significant fire events. The failures were evidenced across numerous actors involved in the assessment, design, construction and approval/verification process, as well as by, in the case of the Lacrosse building, occupants of the building. By failing to consider these buildings as STSs, and the associated interrelationships and requirements of all actors to appropriately fulfill their roles in achieving a safe building, gaps emerged that ultimately resulted in failures of expected building fire safety performance. While it is not within the realm of fire safety engineering to control all other actors and aspects of the design, construction, and use stages, one can argue that fire safety engineers should consider implications of failure by these actors, and other aspects of buildings or infrastructures as STSs, in the fire safety strategy and PBD for fire safety processes. This would contribute significantly to the system safety concepts overviewed above.

Step 8—Broaden participation of institutions and actors in establishing—and achieving—building fire safety performance goals. At a regulatory level, and often at the design level as well, decisions regarding building or infrastructure performance are made within "silos" by a wide range of subject matter experts, often without comprehensive consideration of the performance of the integrated system as a whole (the building or infrastructure as a STS). While this is arguably done to take advantage of subject matter expertise, the lack of holistic consideration of how performance goals should be described, defined, quantified (where possible), evaluated, and assured, can result in unintended consequences. Examples include: energy performance of buildings requirements for increased thermal insulation, which could result in the use of insulation materials with poor fire performance; desire for lightweight timber systems as sustainable materials, which are highly combustible and can contribute significantly to fire if not adequately protected; and, use of lightweight high-strength concrete, which can result in small structural members and less massive structural systems, but which can fail more significantly in fire due to spalling if not appropriately addressed by noncombustible fiber additives or other mitigation measures. [e.g., 161, 162].

If one considers the above eight steps, the first six can be used to envision a new framework for PBD for fire safety. This is reflected in Fig. 1.9. The entire

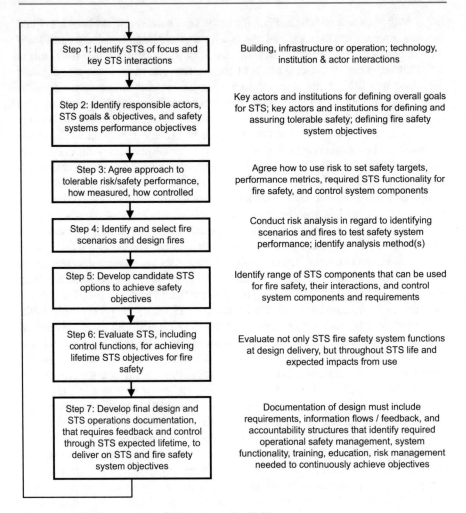

Step 1: Identify STS of focus and key STS interactions	Building, infrastructure or operation; technology, institution & actor interactions
Step 2: Identify responsible actors, STS goals & objectives, and safety systems performance objectives	Key actors and institutions for defining overall goals for STS; key actors and institutions for defining and assuring tolerable safety; defining fire safety system objectives
Step 3: Agree approach to tolerable risk/safety performance, how measured, how controlled	Agree how to use risk to set safety targets, performance metrics, required STS functionality for fire safety, and control system components
Step 4: Identify and select fire scenarios and design fires	Conduct risk analysis in regard to identifying scenarios and fires to test safety system performance; identify analysis method(s)
Step 5: Develop candidate STS options to achieve safety objectives	Identify range of STS components that can be used for fire safety, their interactions, and control system components and requirements
Step 6: Evaluate STS, including control functions, for achieving lifetime STS objectives for fire safety	Evaluate not only STS fire safety system functions at design delivery, but throughout STS life and expected impacts from use
Step 7: Develop final design and STS operations documentation, that requires feedback and control through STS expected lifetime, to deliver on STS and fire safety system objectives	Documentation of design must include requirements, information flows / feedback, and accountability structures that identify required operational safety management, system functionality, training, education, risk management needed to continuously achieve objectives

Fig. 1.9 A STS Framework for PBD for fire safety [164]

system is predicated on accepting buildings and infrastructure as STS, that this requires broader stakeholder participation (step 8), and that risks associated with inadequately prepared, unqualified and/or noncompetent actors must be considered.

1.5   Summary

While PBD for fire safety has been used for some 30 years, the approach has not changed significantly, and its acceptance remains low in some parts of the world. This does not reflect failures of the approach, per se, but failure to recognize that PBD for fire safety is applied within a complex sociotechnical systems environment and needs to account for sociotechnical system interactions and attributes to evolve to a place of greater acceptance and confidence in use.

As a means to advance the thinking about PBD for fire safety, it is proposed to modify the approach to one in which the objective of analysis—buildings, infrastructure, operations—are sociotechnical systems, and that the PBD for fire safety framework must appropriately consider and take advantage of sociotechnical interactions. This means changing the starting point from a focus on fire safety objectives as a unique property of buildings, infrastructure, or operations, and considering fire safety as one of several sociotechnical objectives. It also means focusing fire safety analysis and design on system attributes which can be controlled through design, and less on variables for which control is unlikely or not possible. As part of this, consideration of fire safety systems performance should be considered in terms of a "fail-safe" perspective, in which there is less focus on all possible events that could occur, and more on preventing those which could result in unacceptable performance. Evaluation of building fire safety as a sociotechnical systems problem would also need to consider the interactions of all components that contribute to safety over the lifetime of the system, including in-use safety system management and system performance over time.

In the end, while many aspects of fire safety analysis and design, as undertaken today, may not change, by considering better the environment within which the assessments are undertaken, the interconnections between sociotechnical system component interactions, the components of system safety that can be designed for, and the aim for lifetime performance assurance, can result in more robust, accepted, and better performance fire safety system designs.

Acknowledgments The author sincerely thanks Henrik Bjelland, Jonatan Gehandler and Nicholas Dembsey for their review and comments on this chapter and for their helpful comments. The author also gratefully acknowledges anonymous reviewer comments on the manuscript for [164], which reflects a shorter version of this chapter, and Springer Nature, for granting permission to republish [164] with enhancements and modifications as this chapter..

References

1. Lucht, D.A., Ed., Proceedings, *Conference on Firesafety Design in the 21st Century*, Worcester Polytechnic Institute, Worcester, MA, USA, 1991.
2. Meacham, B.J. and Custer, R.L.P., "Performance-Based Fire Safety Engineering: An Introduction of Basic Concepts," *Journal of Fire Protection Engineering*, Vol. 7, No. 2, 35-54, 1995.
3. Proceedings, *1st International Conference on Performance-Based Codes and Fire Safety Design Methods*, Society of Fire Protection Engineers, Boston, MA, USA, 1996.
4. Custer, R.L.P. and Meacham, B.J., *Introduction to Performance-Based Fire Safety*, NFPA, Quincy, MA, June 1997.
5. Meacham, B.J., *Assessment of the Technological Requirements for Realization of Performance-Based Fire Safety Design in the United States*, GCR 98-761, NIST, Gaithersburg, MD, 1998.

6. Hadjisophocleous, G.V., Benichou, N. and Tamim, A.S., "Literature Review of Performance-Based Fire Codes and Design Environment," *Journal of Fire Protection Engineering*, Volume 9, Issue: 1, pp. 12-40, doi:https://doi.org/10.1177/104239159800900102, 1998.

7. Meacham, B.J., "International Experience in the Development and Use of Performance-Based Fire Safety Design Methods: Evolution, Current Situation, and Thoughts for the Future," *Fire Safety Science*. 6: 59-76. doi:https://doi.org/10.3801/IAFSS.FSS.6-59, 2000.

8. Meacham, B.J., *The Evolution of Performance-Based Codes and Fire Safety Design Methods*, GCR 98-763, NIST, Gaithersburg, MD, 1998.

9. *Guidelines for the Introduction of Performance-Based Building Regulations (Discussion Paper)*, Inter-jurisdictional Regulatory Collaboration Committee (IRCC), Canberra, ACT, Australia, 1998 (https://ircc.info/Doc/Guidelines%20for%20the%20Introduction%20of%20Performance-Based%20Building%20Regulations%20[Discussion%20Paper]%20(1998).pdf, accessed 19 July 2021).

10. Meacham, B.J., Moore, A., Bowen, R. and Traw, J., "Performance-Based Building Regulation: Current Situation and Future Needs," *Building Research & Information*, 33, 1, 91-106, 2005.

11. Meacham, B.J., Ed., *Performance-Based Building Regulatory Systems: Principles and Experiences*, Inter-jurisdictional Regulatory Collaboration Committee (IRCC), Canberra, ACT, Australia, 2009 (https://ircc.info/Doc/A1163909.pdf, accessed 19 July 2021).

12. Kawagoe, K., *Fire Behaviour in Rooms*, Report No. 27, Building Research Institute of Japan, Tokyo, 1958.

13. Thomas, P. H., "Studies of Fires in Buildings Using Models, Part 1," *Research*, London, Vol. 13, No. 2, pp. 69–77, 1960.

14. Thomas, P. H. and Heselden, A. J. M., "Behaviour of Fully Developed Fire in an Enclosure," *Combustion and Flame*, Vol. 6, No. 3, pp. 133–135, 1962.

15. Kawagoe, K. and Sekine, T., "Estimation of Fire Temperature-Time Curves in Rooms," *Occasional Report No. 11*, Building Research Institute of Japan, Tokyo, 1963.

16. Salzberg, F. and Waterman, T. E., "Studies of Building Fires with Models," *Fire Technology*, Vol. 2, No. 3, 196–203, 1966.

17. Thomas, P. H., Heselden, A. J. M., and Law, M., "Fully Developed Compartment Fires — Two Kinds of Behaviour," *Fire Research Technical Paper No. 18*, H. M. Stationery Office, London, 1967.

18. Heselden, A.J.M., Thomas, P.H. & Law, M., "Burning rate of ventilation-controlled fires in compartments," *Fire Technology*, 6, 123–125, https://doi-org.ludwig.lub.lu.se/10.1007/BF02588898, 1970.

19. For a compilation of fundamental fire safety science concepts that underpin fire safety engineering, see for example Drysdale, D., *An Introduction to Fire Dynamics*, 2nd Edition, Wiley, London, 1999, and Hurley, M., Ed., *SFPE Handbook of Fire Protection Engineering*, 5th Edition, Springer, 2016.

20. Cornell., C.A., "Bounds on the Reliability of Structural Systems," American Society of Civil Engineers, *Journal of the Structural Division*, Volume 93, 171-200, 1967.

21. Moses, F. and Stevenson, J.D., "Reliability-Based Structural Design," American Society of Civil Engineers, *Journal of the Structural Division*, Volume 96, Issue 2, 1970.

22. Ang, A.H.S., "Structural Risk Analysis and Reliability-Based Design," American Society of Civil Engineers, *Journal of the Structural Division*, Volume 99 Issue 9, September 1973.

23. Ang, A.H.S and Cornell., C.A., "Reliability Bases of Structural Safety and Design," American Society of Civil Engineers, *Journal of the Structural Division*, Volume 100 Issue 9, September 1974.

24. Ellingwood, B., MacGregor, J.G., Galambos, T.V. and Cornell, C.A., "Probability-based load criteria: load factors and load combinations," American Society of Civil Engineers, *Journal of the Structural Division*, Volume 108. Issue 5, 978-997, 1982.

25. Farmer, F.R., "Siting Criteria—A New Approach," *IAEA Symposium on the Containment and Siting of Nuclear Power Reactors*, IAEA SM-89/34, Vienna, Austria, 1967.

26. Starr, C., "Societal Benefit vs. Technological Risk," *Science*, 165, pp. 1232–1238, 1969.
27. Rassmusen, N.C. et al., *Reactor safety study. An assessment of accident risks in U. S. commercial nuclear power plants.* WASH-1400 Report. U.S. Nuclear Regulatory Commission, USA, doi:https://doi.org/10.2172/7134131, 1975
28. Rowe, W.D., *Anatomy of Risk,* John Wiley and Sons, New York, 1977.
29. Rassmussen, N.C., "The Application of Probabilistic Risk Assessment Techniques to Energy Technologies," *Annual Review of Energy*, Vol. 6, pp. 123-138, 1981.
30. Kaplan, S. and Garrick, J.B., "On the Quantitative Definition of Risk," *Risk Analysis*, Vol. I, No. I, 1981.
31. IChemE. *Nomenclature for hazard and risk assessment in the process industries*, Institution of Chemical Engineers, UK, 1985.
32. Pettersson, O., Magnusson, S. E., & Thor, J., *Fire Engineering Design of Steel Structures*, Bulletin of Division of Structural Mechanics and Concrete Construction, Bulletin 52, Lund Institute of Technology, Sweden, 1975.
33. Lie, T.T., "Safety factors for fire loads," *Canadian Journal of Civil Engineering*, 6(4): 617-628, doi:https://doi.org/10.1139/l79-074, 1979.
34. Pettersson, O., *Reliability Based Design of Fire Exposed Concrete Structures.* LUTVDG/TVBB–3004–SE, vol. 3004, vol. 3004, Division of Building Fire Safety and Technology, Lund Institute of Technology, 1981.
35. Magnusson, S.E. and Pettersson, O., "Rational design methodology for fire exposed load bearing structures," Fire Safety Journal, Volume 3, Issue 4, Pages 227-241, doi:https://doi.org/10.1016/0379-7112(81)90046-1, 1981.
36. Harmathy, T and Mehaffey, J., "Design of Buildings for Prescribed Levels of Structural Fire Safety," in Fire Safety: Science and Engineering, T. Harmathy, Ed., ASTM International, West Conshohocken, PA, 160-175, doi:10.1520/STP35296S, 1985.
37. GSA, *Building Fire Safey Criteria, Appendix D: Interim Guide for Goal-Oriented Systems Approach to Building Fire Safety*, U.S. General Services Administration, Washington, DC, 1972.
38. Nelson, H.E., *Directions to Improve Applications of Systems Approach to Fire Protection Requirements for Buildings*, SFPE Technology Report 77-8, Society of Fire Protection Engineers, Boston, MA, USA, 1977.
39. Watts, J., *The Goal-Oriented Systems Approach*, NBS-GCR-77-103, National Bureau of Standards, Gaithersburg, MD, USA, 1977.
40. NFPA 550, *Guide to the Fire Safety Concepts Tree*, National Fire Protection Association, Qunicy, MA, USA, 1980.
41. Wakamatsu, T., "Fire Research in Japan - Development of a Design System for Building Fire Safety," *Proceedings of the 6th Joint Panel Meeting, UNJR Panel on Fire Research and Safety*, Tokyo, Japan, p. 882, 10-14 May 1982.
42. Fitzgerald, R.W., "An Engineering Method for Building Fire Safety Analysis," *Fire Safety Journal*, 9., 223-243, 1985.
43. Beard, A.N., "Towards a Systemic Approach to Fire Safety," *Proceedings, 1st International Symposium on Fire Safety Science*, Hemisphere Publishing co., New York, NY, USA, p 943, 1986.
44. Fire Safety and Engineering Project, *Project Report* and *Technical Papers, Books 1 and 2*, The Warren Centre for Advanced Engineering, the University of Sydney, Australia, 1989.
45. Rasbash, D. J., Ramachandran, G., Kandola, B, Watts, J. M., and Law, M., *Evaluation of Fire Safety*, John Wiley and Sons, London, 2004.
46. Emmons, H.W., "The prediction of fires in buildings," *Proc. Seventeenth Int. Symposium on Combustion*, The Combustion Institute, Pittsburgh, p. 1101, 1978.
47. Mitler, H.E., *The Physical Basis for the Harvard Computer Fire Code*, Home Fire Project Tech. Report. No. 34, Harvard University, 1978.
48. Yang, K.T. and Liu, V.K., *UNDSAFE-HA Computer Code for Buoyant Turbulent Flow in an Enclosure with Radiation*, Tech. Report TR79002-78-3, Dept. Aero. and Mech. Eng., Univ. of Notre Dame, 1978.

49. Zukoski, E.E. and Kubota, T., Two-layer modeling of smoke movement in building fires," *Fire Mater*. 4, 1980.
50. Tanaka, T., "A Model on Fire Spread in Small Scale Buildings," *BRI Research Paper 84*, Building Research Institute, Japan, 1980.
51. Quintiere, J.G., "An approach to modeling wall fire spread in a room," *Fire Safety Journal*, 3, p 201, 1981.
52. Buchanan, A., *Fire Engineering Design Guide*, Centre for Advanced Engineering, University of Canterbury, Christchurch, New Zealand, July 1994.
53. Fire Code Reform Centre, *Fire Engineering Guidelines*, Sydney, Australia, March, 1996.
54. *Fire Safety Engineering in Buildings*, DD 240: Parts 1 and 2: 1997, British Standards Institute, 1997.
55. ISO TR 13387, Part I: *The Application of Fire Performance Concepts to Design Objectives*, 1999.
56. Tanaka, T., "The Outline of a Performance-Based Fire Safety Design System of Buildings," *Proceedings of the 7th International Research and Training Seminar on Regional Development Planning for Disaster Prevention, Improved Firesafety Systems in Developing Countries*, United Nations Center for Regional Development, Tokyo, Japan, 1995
57. *SFPE Engineering Guide to Performance-Based Fire Protection: Analysis and Design of Buildings*, National Fire Protection Association, Quincy, MA, 2000.
58. Fitzgerald, R.W., *Building Fire Performance Analysis*, John Wiley & Sons, London, 2005.
59. Fitzgerald, R.W. and Meacham, B.J., *Fire Performance Analysis for Buildings*, John Wiley & Sons, London, 2017.
60. NFPA 550, *Guide to the Fire Safety Concepts Tree*, 2022 edition. Copyright© 2021, National Fire Protection Association. (For a full copy of NFPA 550, please go to www.nfpa.org)
61. SFPE *Engineering Guide to Performance-Based Fire Protection*, 2nd Edition, Society of Fire Protection Engineers and National Fire Protection Association, Quincy, MA, 2007.
62. *International Fire Engineering Guidelines*. National Research Council of Canada. International Code Council. New Zealand. Department of Building and Housing. Australian Building Codes Board. Canberra, ACT: Australian Building Codes Board, 2005.
63. *Leitfaden Ingenieurmethoden des Brandschutzes*, Technisch-Wissenschaftlicher Beirat (TWB) der Vereinigung zur Förderung des Deutschen Brandschutzes e.V. (vfdb), Altenberge, Deutschland, 2013.
64. *Verification Method C/VM2: Framework for Fire Safety Design*, Amendment 5, Ministry of Business, Innovation and Employment (MBIE), Wellington, New Zealand, 2017.
65. ISO 23932:2018, *Fire safety engineering — General Principles: Part 1 - General*, International Organization for Standardisation, Geneva, Switzerland, 2018.
66. BS7974:2019, *Application of fire safety engineering principles to the design of buildings. Code of practice*, British Standards Institution, London, 2019.
67. *Australian Fire Engineering Guidelines*, ©Commonwealth of Australia and States and Territories of Australia 2021, published by the Australian Building Codes Board, Canberra, ACT, Australia, 2021.
68. Wang, Y., Burgess, I., Wald, F. and Gillie, M., *Performance-Based Fire Engineering of Structures*, CRC Press, Boca Raton, FL, USA, 2013.
69. Purkiss, J.A. and Li, L.-Y., *Fire Safety Engineering Design of Structures*, CRC Press, Boca Raton, FL, USA, 2014.
70. Hurley, M.J. and Rosenbaum, E.R., *Performance-Based Fire Safety Design*, CRC Press, Boca Raton, FL, USA, 2015.
71. Hurley, M.J., Editor, *SFPE Handbook of Fire Protection Engineering*, Springer, 2016.
72. Buchanan, A.H. and Abu, A.K., *Structural Design for Fire Safety*, 2nd Edition, John Wiley & Sons, Chichester, England, 2017.
73. LaMalva, K., Editor, *Structural Fire Engineering*, Fire Protection Committee, American Society of Civil Engineers, Reston, VA, USA, https://ascelibrary.org/doi/book/10.1061/9780784415047, 2018.

74. Barry, T.F., *Risk-Informed, Performance-Based Industrial Fire Protection — An Alternative to Prescriptive Codes*, First Edition, TFBarry Publications, 704 pages, Publisher: Tennessee Valley Publishing, Knoxville, Tennessee, USA, 2002. ISBN 1-882194-09-8.
75. Lundin, J. Development of a Framework for Quality Assurance of Performance-Based Fire Safety Designs, *Journal of Fire Protection Engineering*, 15 (1): 14–19, doi:https://doi.org/10.1177/1042391505045581, 2005.
76. Johann, M. A., Albano, L. D., Fitzgerald, R. W., & Meacham, B. J., Performance-Based Structural Fire Safety, *Journal of Performance of Constructed Facilities*, 20(1), 45–53, doi:https://doi.org/10.1061/(ASCE)0887-3828(2006)20:1(45), 2006.
77. Meacham, B.J., "Chapter 2 - Extreme Event Mitigation Through Risk-Informed Performance-Based Analysis and Design," in *Extreme Event Mitigation in Buildings: Analysis and Design* (B.J. Meacham and M.A. Johann, eds.), IBSN-10:0877657432, NFPA, Quincy, MA, 2006.
78. Hamilton, S.R., *Performance-based fire engineering for steel framed structures: a probabilistic methodology*, Ph.D. Dissertation, Stanford University, Stanford, CA, USA, 2011.
79. Albrecht C. *A risk-informed and performance-based life safety concept in case of fire* [Ph.D. thesis]. Technical University of Braunschweig, Institute for Building Materials, Concrete Construction and Fire Protection (iBMB); URL: http://www.digibib.tu-bs.de/?docid=00043585. 2012.
80. Alvarez, A., Meacham, B.J., Dembsey, N.A. and Thomas, J.R., "20 Years of Performance-Based Fire Protection Design: Challenges Faced and a Look Ahead," *Journal of Fire Protection Engineering*, DOI: https://doi.org/10.1177/1042391513484911, Vol. 23, No. 4, 2013.
81. Van Hees, P., Validation and Verification of Fire Models for Fire Safety Engineering, *Procedia Engineering* 62 (January): 154–68, doi:https://doi.org/10.1016/j.proeng.2013.08.052, 2013.
82. Bjelland, H. and Borg, A.. "On the Use of Scenario Analysis in Combination with Prescriptive Fire Safety Design Requirements." *Environment Systems & Decisions* 33 (1): 33–42. doi:https://doi.org/10.1007/s10669-012-9425-2. 2013.
83. Alvarez, A., Meacham, B.J., Dembsey, N.A. and Thomas, J.R., "A Framework for Risk-Informed Performance-Based Fire Protection Design for The Built Environment," *Fire Technology*, DOI 10.1007/s10694-013-0366-1, Vol. 50, pp161-181, 2014.
84. Bjelland H, Njå O, Heskestad AW, Braut GS. The Concepts of Safety Level and Safety Margin: Framework for Fire Safety Design of Novel Buildings. *Fire Technology*. 51:409–441. 2015.
85. Park, H., Meacham, B.J., Dembsey, N.A. and Goulthorpe, M., Improved incorporation of fire safety performance into building design process, *Building Research and Information*, DOI:https://doi.org/10.1080/09613218.2014.913452, published on-line 16 May 2014, print January 2015.
86. Borg A, Njå O, Torero J., A Framework for Selecting Design Fires in Performance Based Fire Safety Engineering, *Fire Technology*, 51(4):995-1017. doi:https://doi.org/10.1007/s10694-014-0454-x. 2015.
87. Park, H., Meacham, B.J., Dembsey, N.A., and Goulthorpe, M., Conceptual Models for Holistic Building Fire Safety Performance Analysis, *Fire Technology*, DOI:10.1007/s10694-013-0374-1, Volume 51, Issue 1, pp 173–193, 2015.
88. Meacham, B.J. and Alvarez-Rodriguez, A. *Risk-Informed Performance-Based Design for Fire: Concepts and Framework, Final Report*, NIST GCR 15-1000, dx.doi.org/10.6028/NIST.GCR.15-1000, Gaithersburg, MD, March 2015.
89. Dai X, Welch S, Usmani A., A critical review of "travelling fire" scenarios for performance-based structural engineering. *Fire Safety Journal*. 91:568-578. doi:https://doi.org/10.1016/j.firesaf.2017.04.001. 2017.
90. Gehandler, J., The theoretical framework of fire safety design: Reflections and alternatives, *Fire Safety Journal*, Volume 91, Pages 973-981, https://doi.org/10.1016/j.firesaf.2017.03.034. 2017.

91. Shrivastava, M., Abu, A., Dhakal, R. and Moss, P., State-of-the-art of probabilistic performance based structural fire engineering, *Journal of Structural Fire Engineering*, 10(2):175-192. doi:https://doi.org/10.1108/JSFE-02-2018-0005. 2019.

92. Van Coile, R., Hopkin, D., Lange, D. et al. The Need for Hierarchies of Acceptance Criteria for Probabilistic Risk Assessments in Fire Engineering. *Fire Technology*. 55, 1111–1146. doi:https://doi.org/10.1007/s10694-018-0746-7. 2019.

93. Van Weyenberge, B., Deckers, X., Caspeele, R. et al. Development of an Integrated Risk Assessment Method to Quantify the Life Safety Risk in Buildings in Case of Fire. *Fire Technology*. 55, 1211–1242. https://doi-org.ludwig.lub.lu.se/10.1007/s10694-018-0763-6. 2019.

94. Gernay, T. and Elhami Khorasani, N., Recommendations for performance-based fire design of composite steel buildings using computational analysis, *Journal of Constructional Steel Research*, Volume 166, 105906, ISSN 0143-974X, doi:https://doi.org/10.1016/j.jcsr.2019.105906. 2020.

95. Kuehnen, R., Youssef, M. and El-Fitiany, S, Performance-based design of RC beams using an equivalent standard fire. *Journal of Structural Fire Engineering*, v. 12, n. 1, p. 98–109. DOI https://doi.org/10.1108/JSFE-02-2020-0008. 2020.

96. Vacca, P., Caballero, D., Pastor, E., and Planas, E., WUI fire risk mitigation in Europe: A performance-based design approach at home-owner level, *Journal of Safety Science and Resilience*, Volume 1, Issue 2, pp 97-105, ISSN 2666-4496, doi:10.1016/j.jnlssr.2020.08.001. 2020.

97. Mohan, A. T., Van Coile, R., Hopkin, D., Jomaas, G., & Caspeele, R. Risk Tolerability Limits for Fire Engineering Design: Methodology and Reference Case Study. *Fire Technology*, 57(5), 2235–2267. doi:https://doi.org/10.1007/s10694-021-01118-w. 2021.

98. Meacham, B.J., Accommodating Innovation in Building Regulation: Lessons and Challenges, *Building Research & Information*, Vol. 38, No. 6, 2010.

99. Meacham, B.J., Fire Safety Engineering at a Crossroad, *Case Studies in Fire Safety*, doi:https://doi.org/10.1016/j.csfs.2013.11.001, 2013.

100. Spinardi, G., Bisby, L. & Torero, J. A Review of Sociological Issues in Fire Safety Regulation. *Fire Technology*. 53, 1011–1037. doi:https://doi.org/10.1007/s10694-016-0615-1. 2017.

101. Lange, David, Jose L. Torero, Andres Osorio, Nate Lobel, Cristian Maluk, Juan P. Hidalgo, Peter Johnson, Marianne Foley, and Ashley Brinson. "Identifying the Attributes of a Profession in the Practice and Regulation of Fire Safety Engineering." *Fire Safety Journal*, 121 (May). doi:10.1016/j.firesaf.2021.103274. 2021.

102. Fleischmann, C.M. Is Prescription the Future of Performance-Based Design? *Proceedings - Fire Safety Science*. https://firesafetyscience.org/publications/fss/10/77/view/fss_10-77.pdf. 2011.

103. Trist, E. and Murray, H., Eds. *The Social Engagement of Social Science, Volume 2: A Tavistock Anthology-The Socio-Technical Perspective*, University of Pennsylvania Press. 1993.

104. Trist, E. "Introduction," in Trist, E. and Murray, H., Eds. T*he Social Engagement of Social Science, Volume 2: A Tavistock Anthology–The Socio-Technical Perspective*, University of Pennsylvania Press. 1993.

105. Meacham, B.J. and van Straalen, I., A Socio-Technical System Framework for Risk-Informed Performance-Based Building Regulation, *Building Research & Information*, DOI https://doi.org/10.1080/09613218.2017.1299525, 2017.

106. Rasmussen, J. Risk Management in a Dynamic Society: A Modelling Problem. *Safety Science*. 1997:27(2/3):183-213.

107. Rasmussen, J. and Svedung, I. *Proactive Risk Management in a Dynamic Society*. Swedish Rescue Services Agency, Stockholm, 2000.

108. Leveson, N. A new accident model for engineering safer systems. *Safety Science*. 2004:42, 237-270.

109. Petak, W. Earthquake Resilience through Mitigation: A System Approach. Lecture paper. International Institute for Applied Systems Analysis, January 2002.

110. Meacham, B.J., Stromgren, M. and van Hees, P. "A Holistic Framework for Development and Assessment of Risk-Informed Performance-Based Building Regulation," *Fire and Materials*, DOI:https://doi.org/10.1002/fam.2930, 2020.
111. Rohracher, H. Managing the Technological Transition to Sustainable Construction of Buildings: A Socio-Technical Perspective. *Technology Analysis & Strategic Management.* 2001:13(1)137-150. 2001.
112. Harty, C. Innovation in construction: a sociology of technology approach. Building Research & Information. 2005:33:6, 512-522, DOI: 10.1080/09613210500288605. 2005.
113. Schweber, L. and Harty, C. Actors and Objects: a socio-technical networks approach to technology uptake in the construction sector. *Construction Management and Economics.* 2010:28(6):657-674, DOI: 10.1080/01446191003702468. 2010.
114. Guy, S., Marvin, S., Medd, W. and Moss, T. *Shaping Urban Infrastructures: Intermediaries and the Governance of Socio-technical Networks*, Earthscan, London. 2011.
115. Edwards, P. N. *Infrastructure and modernity: Force, time, and social organization in the history of sociotechnical systems.* Modernity and Technology, Massachusetts Institute of Technology, Cambridge, MA, USA. 2003:185-225. 2003.
116. Hansman, R. J., Magee, C., De Neufville, R., & Robins, R.. Research agenda for an integrated approach to infrastructure planning, design and management. International Journal of Critical Infrastructures. 2006:2(2):146-159. 2006.
117. Ottens, M., Franssen, M., Kroes, P., & Van De Poel, I. Modelling infrastructures as socio-technical systems. International Journal of Critical Infrastructures. 2006:2(2):133-145. 2006.
118. Kroes, P., Franssen, M., Poel, I. V. D., & Ottens, M. Treating socio-technical systems as engineering systems: some conceptual problems. Systems Research and Behavioural Science. 2006: 23(6):803-814. 2006.
119. Jönsson, H., Johansson, J., & Johansson, H. Identifying critical components in technical infrastructure networks. Journal of Risk and Reliability. 2008:222(2):235-243. 2008.
120. Meacham, B.J. and Stromgren, M. *A Review of the English and Swedish Building Regulatory Systems for Fire Safety using a Socio-Technical System (STS) Based Methodology*, HOLIFAS Project WP 3 Report, Briab Brand & Riskingenjörerna AB (Sweden) and Meacham Associates (USA) Research Report 2019:01. https://doi.org/10.13140/RG.2.2.34702.72001. 2019.
121. Rasmussen, J. The role of hierarchical knowledge representation in decision making and system management. *IEEE Transactions on Systems, Man and Cybernetics*, SMC-15(2), 234-243. https://doi.org/10.1109/TSMC.1985.6313353. 1985.
122. Rasmussen, J., Vicente, K. Ecological interface design: theoretical foundations. *IEEE Transactions on Systems, Man and Cybernetics.* 22 (4) (July/August). 1992.
123. Leveson, N.G. *Engineering a Safer World.* MIT Press, Cambridge, MA. 2012.
124. Leveson, N.G. *Safety III: A Systems Approach to Safety and Resilience.* MIT. Cambridge, MA. http://sunnyday.mit.edu/safety-3.pdf
125. Cherns, A.B. The principles of sociotechnical design. *Human Relations.* 1976:29:783–792.
126. Torero, J., Lange, D., Horasan, M., Osorio, A., Maluk, C., Hidalgo, J. and Johnson, P. *Current Status of Education, Training and Stated Competencies for Fire Safety Engineers.* The Warren Centre for Advanced Engineering. University of Sydney. Australia. DOI:10.25910/a75g-gn88. https://hdl.handle.net/2123/23469. 2019.
127. Bjelland, H. *Engineering Safety with Applications to Fire Safety Design of Buildings and Road Tunnels*, Faculty of Science and Technology, University of Stavanger, Norway, Stavanger, 2013.
128. Gehandler, J. *Fire safety design of road tunnels.* Lund University. Department of Fire Safety Engineering. Lund. Sweden. 2020.
129. Checkland, P. Systems Thinking, Systems Practice, Chichester, UK: Wiley. 1981.
130. Checkland, P. and Poulter, J. Soft Systems Methodology, Chapter 5, in M. Reynolds and S. Holwell (eds.), *Systems Approaches to Managing Change: A Practical Guide*, DOI 10.1007/978-1-84882-809-4_5, © The Open University 2010. Published in Association with Springer-Verlag London Limited.

131. Checkland, P. Four Conditions for Serious Systems Thinking and Action, Systems Research and Behavioral Science Syst. Res. 29, 465–469, DOI: https://doi.org/10.1002/sres.2158, 2012.
132. https://www.nrc.gov/about-nrc/regulatory/risk-informed/history/2007-present.html
133. NFPA 805. *Performance-Based Standard for Fire Protection for Light Water Reactor Electric Generating Plants*. National fire Protection Association. Quincy. MA. 2020.
134. Perez, C. *Technological Revolutions and Financial Capital*. Cheltenham, UK: Edward Elgar. 2002.
135. https://www.gartner.com/en/documents/3887767/understanding-gartner-s-hype-cycles
136. Gartner Research's Hype Cycle Diagram, Source: Jeremey Kemp, https://commons.wikimedia.org/wiki/File:Gartner_Hype_Cycle.svg (CC BY-SA 3.0, https://creativecommons.org/licenses/by-sa/3.0/deed.en)
137. Steinert, M. and Leifer, L. Scrutinizing Gartner's Hype Cycle Approach. *PICMET 2010 Proceedings (2010 Portland International Conference on Management of Engineering & Technology)*, IEEE, ISBN: 978-1-4244-8203-0, pp 254-265.
138. *SFPE Handbook of Fire Protection Engineering, 1st Edition*. DiNenno, P., Ed., Society of Fire Protection Engineers, Boston, MA. 1983.
139. Hamburger, R.O., Court, A.B. and Soulages, J.R. Vision 2000: A Framework for Performance-Based Engineering of Buildings. *Proceedings of the 64th Annual Convention. Structural Engineers Association of California*. Pages 127-146. 19-21 October 1995.
140. King, J. and Perry, C., *Smart Buildings: Using Smart Technology to Save Energy in Existing Buildings*, Report A1701, American Council for an Energy-Efficient Economy, Washington, DC, 2017.
141. Lea P. *Internet of Things for Architects: Architecting IoT solutions by implementing sensors, communication infrastructure, edge computing, analytics, and security*. Packt Publishing Ltd. 2018
142. US DOE. *Innovations in Sensors and Controls for Building Energy Management: Research and Development Opportunities Report for Emerging Technologies*, U.S. Department of Energy, Office of Energy Efficiency and Renewable Energy, Building Technologies Office, Washington, DC, (https://www1.eere.energy.gov/buildings/pdfs/75601.pdf). 2020.
143. Cowlard A, Jahn W, Abecassis-Empis C, Rein G, Torero JL. Sensor Assisted Fire Fighting. Fire Technology. 2008;46:719-41. doi:https://doi.org/10.1007/s10694-008-0069-1. 2008.
144. Han, L., Potter, S., Beckett, G., Pringle, G., Welch, S., Koo, S-H., Wickler, G., Usmani, A., Torero, J. and Tate, A. (2010) FireGrid: An e-Infrastructure for Next-Generation Emergency Response Support, Journal of Parallel and Distributed Computing, 70 (2010) 1128-1141. 2010.
145. Su, L. C., Wu, X., Zhang, X., & Huang, X. Smart performance-based design for building fire safety: Prediction of smoke motion via AI. *Journal of Building Engineering*, 43, [102529]. https://doi.org/10.1016/j.jobe.2021.102529. 2021.
146. Wang, H., Dembsey, N.A., Meacham, B.J., Liu, S. and Simeoni, A. "Conceptual Design of a Building Fire Performance Monitoring Process," *Fire Technology*, Manuscript FIRE-D-20-00271 (in review).
147. Wang, H., Dembsey, N.A., Meacham, B.J., Liu, S. and Simeoni, A. "A Sensitivity Matrix Method to Understand the Building Fire Egress Performance Gap," *Fire Safety Journal*, Manuscript, (in review).
148. Meacham, B.J., Understanding Risk: Quantification, Perception and Characterization, *Journal of Fire Protection Engineering*, Vol. 14, No. 3, pp.199-228, 2004.
149. Meacham, B.J., Johnson, P.J., Charters, D. and Salisbury, M., Building Fire Risk Analysis, Chapter 75, *SFPE Handbook of Fire Protection Engineering*, 5th Ed., Springer, USA, 2015.
150. Mossberg, A., Nilsson, D. & Andrée, K. Unannounced Evacuation Experiment in a High-Rise Hotel Building with Evacuation Elevators: A Study of Evacuation Behaviour Using Eye-Tracking. Fire Technol 57, 1259–1281. https://doi-org.ludwig.lub.lu.se/10.1007/s10694-020-01046-1. 2021.

151. Galea, E.R., Xie, H., Deere, S., Cooney, D. and Filippidis, L. Evaluating the effectiveness of an improved active dynamic signage system using full scale evacuation trials. *Fire Safety Journal*, 91 (2017), pp. 908-917, https://doi.org/10.1016/j.firesaf.2017.03.022. 2017.
152. Fu, M. and Liu, R. BIM-based automated determination of exit sign direction for intelligent building sign systems, *Automation in Construction*, Volume 120, 2020, 103353, ISSN 0926-5805, https://doi.org/10.1016/j.autcon.2020.103353. 2020.
153. Naser, M.Z. Autonomous and resilient infrastructure with cognitive and self-deployable load-bearing structural components, *Automation in Construction*, Volume 99, Pages 59-67, ISSN 0926-5805, https://doi.org/10.1016/j.autcon.2018.11.032. 2019.
154. Naser, M.Z. Autonomous Fire Resistance Evaluation. *Journal of Structural Engineering*. Vol. 146. Issue 6. American Society of Civil Engineers. DOI: 10.1061/(ASCE)ST.1943-541X.0002641. 2020.
155. *Building a Safer Future - Independent Review of Building Regulations and Fire Safety: Interim Report.* Secretary of State for (Housing) Communities and Local Government, England. December 2017. (note – Housing was added in 2018, but used here for consistent style).
156. *Building a Safer Future - Independent Review of Building Regulations and Fire Safety: Final Report.* Secretary of State for Housing Communities and Local Government, England. May 2018.
157. *Building a Safer Future – An Implementation Plan.* Secretary of State for Housing, Communities and Local Government, England. December 2018.
158. Genco, G. *Lacrosse Building Fire.* Report. City of Melbourne, Victoria, Australia. 2015. https://www.melbourne.vic.gov.au/sitecollectiondocuments/mbs-report-lacrosse-fire.pdf
159. Shergold, P. and Weir, B. *Building Confidence: Improving the effectiveness of compliance and enforcement systems for the building and construction industry across Australia.* Ministers Forum, Canberra, ACT, Australia. 2018. https://www.industry.gov.au/sites/default/files/July%202018/document/pdf/building_ministers_forum_expert_assessment_-_building_confidence.pdf
160. Cheng, L. Judge finds architect proportionately liable for Lacrosse fire damages. *ArchitectureAU.* 2019. https://architectureau.com/articles/judge-finds-architect-proportionately-liable-for-lacrosse-fire-damages/
161. Meacham, B.J. and McNamee, M. F*ire Safety Challenges of 'Green' Buildings and Attributes,* Fire Protection Research Foundation, Quincy, MA November 2020 (https://www.nfpa.org/~/media/Files/News%20and%20Research/Fire%20statistics%20and%20reports/Building%20and%20life%20safety/RFGreenBuildings2020.pdf, last accessed December 2020)
162. Meacham, B.J. and McNamee, M. *Handbook of Fire and the Environment: Impacts and Mitigation,* Springer (April 2022).
163. Cornell, C.A., "Structural Safety: Some Historical Evidence that it is a Healthy Adolescent," *Proceedings of the 3rd International Conference on Structural Safety and Reliability,* Trondheim, Norway, June 1981, pp. 19-29.
164. Meacham, B.J. A Sociotechnical Systems Framework for Performance-Based Design for Fire Safety. *Fire Technol* (2022). doi:https://doi.org/10.1007/s10694-022-01219-0

A Twenty-First Century Approach to Fire Resistance

Vytenis Babrauskas

2.1 History of Fire Resistance Concepts

Most construction types are adversely affected by fire, if it occurs. If the construction is combustible, then the structure may burn down. But even if it is not combustible, a serious fire, especially a post-flashover fire, may result in major damage or even collapse. This is because few materials are available which can stand prolonged application of high temperatures without degradation or failure.

Up through the first half of the nineteenth century, serious fires, conflagrations, and entire towns burning down used to be considered as unavoidable disasters [1]. But in the second half of the nineteenth century, tall buildings started being built in large cities. This focused attention on the fact that major fires involving tall buildings should be considered as a solvable engineering problem. That is, techniques should be developed to limit the damage sustained from fires. Specifically, two primary social objectives became at least implicitly recognized: (1) Efforts should be taken to prevent city-wide conflagrations and (2) Fire safety measures should be adopted to reduce the likelihood of life loss during structure fires. The first objective was generally accomplished legislatively, without requiring overt engineering measures. This typically involved restriction of the use of combustible materials in construction, measures to provide streets of adequate width, and measures to restrict use of the more combustible types of roofings.

But the second objective required development of engineering solutions. Measures to reduce flame spread or to minimize ignition potential had to wait until much later in the twentieth century. What was possible in the nineteenth century was to provide what later became known as *fire resistance*. During the nineteenth century, however, this was known as designing of "fireproof" buildings [2]. The term

V. Babrauskas (✉)
John Jay College of Criminal Justice, and Fire Science and Technology Inc., Clarkdale, AZ, USA
e-mail: vytob@doctorfire.com

© Springer Nature Switzerland AG 2022
M. Z. Naser, G. Corbett (eds.), *Handbook of Cognitive and Autonomous Systems for Fire Resilient Infrastructures*, https://doi.org/10.1007/978-3-030-98685-8_2

"fireproof" was later deprecated, since it was clearly incorrect—no building can be practicably designed where fires can be staged with impunity and without damage to the building. Instead, the term was replaced with the term "fire resistance," which correctly implied that the building had the ability to withstand some effects of fire, for some time. It was early on understood that fire resistance had two quite different aspects associated with it: (1) When threatened by fire, the building should resist collapse for some agreed-upon length of time and (2) a fire originating in one room of the structure should not readily propagate into other rooms or other floors. The latter requirement implies that a role for walls and floors is to prevent fire from transgressing this barrier for a certain length of time. In modern usage, the first objective is termed *stability*, while the latter one is separated into *integrity* and *insulation*. Integrity indicates that holes should not open up in a barrier allowing flames to move through them, while insulation refers to the ability to prevent fire propagation by means of heat conduction. A steel plate may not collapse and may not open up holes through it, but if no insulation is used, combustible materials on the unexposed face would be likely to ignite quickly.

As we will see later, in the twentieth century it became possible to use engineering design methods to provide fire resistance. But for the first century (1860s–1870s onward), fire resistance could only be demonstrated by testing. Thus, the early history involves solely efforts to intuitively design fire resistant floors, walls, and other building elements, along with the development of testing methods to demonstrate that the expected fire resistance was indeed achieved. The detailed history of how fire resistance testing evolved has been described by Babrauskas [3], by Babrauskas and Williamson [4], and more recently by Gales et al. [5]. Here, we shall summarize just some major highlights below.

During the crucial decades of 1890 to 1910, two individuals were mostly responsible for establishing the concepts of fire resistance testing that have remained in use for over 100 years. In the US, the effort was led by Ira Woolson, who was affiliated with both the National Board of Fire Underwriters and Columbia University. The latter was the first institution in the US to establish a fire testing station. In the UK, the effort was led by Sir Edwin Sachs, who was an architect who founded the British Fire Prevention Committee, BFPC and proceeded to build a fire resistance testing station (Fig. 2.1). In 1903, Sachs proposed, in very imprecise terms, how a fire resistance testing standard might be configured [6]. Sachs and the BFPC became prolific testing engineers in London, publishing numerous fire resistance tests as a series of "Red Books." This inspired later researchers, but did not directly lead to the later—published—standards.

The work of Woolson, Rudolph Miller from the Dept. of Buildings of New York City and other researchers, mostly based in New York City, eventually did result in the first edition (1918) of a published standard, ASTM C 19 [7], later renumbered E119 [8]. The parallel UK standard, BS 476 [9], was not issued until 1932, while the international standard, ISO 834 [10] only emerged in 1975. It may be interesting to note that C 19 had a rather grandiose title of "Standard Specifications for Fire Tests of Materials and Construction," implying that this constituted the only type of fire testing that should be needed. Later tests for different fire properties, however,

Fig. 2.1 The test huts at the first UK fire testing station, built in London in 1899 by Sir Edwin Sachs

were issued as separate ASTM standards. In the UK, however, BS 476 eventually encompassed a hodge-podge of unrelated testing methods. Although even there eventually additional fire test types were issued under separate numberings. The opposite was true of ISO, where fire resistance testing concepts in later editions of ISO 834 devolved into diverse "Parts," published as separate documents.

2.2 Principles of Standardized Fire Resistance Tests

Despite differences in detail, fire resistance tests in all countries evolved in a fairly similar manner. Unlike tests of small products, it is clearly impractical to arrange for true full-scale testing of buildings, since this would involve erecting and burning down of an exemplar building. Instead, two concepts arose:

1. That buildings can be subdivided into a small number of discrete components, e.g., walls, floors, beams, columns, etc.
2. That realistic testing results can be obtained by testing components of large, but not excessive scale, generally somewhere in the range of 8 ft. (2.4 m) to 15 ft. (4.6 m).

It may be noted what this philosophy does *not* encompass:

(a) The performance of joints or connections;
(b) Frame action and load redistribution.

Real building members typically do not involve what the structural engineers refer to as "pinned" connections. In other words, the ends of the member can resist moments and not just axial loads. Much later in the development of fire resistance testing, the concept of "restrained" ratings for floors was adopted [11]. This testing strategy endeavored to simulate a certain amount of moment resistance at the ends of the test specimen. This was a half-way measure since it still does not simulate what would be happening in a real building fire. This concept was added in the 1960s, and since then, fire resistance testing has changed only in minor details.

2.3　　Simulation of Fires and Control of Fire Test Furnaces

While the concept of flashover did not originate until the 1950s [12], today we understand that the fire resistance test is a test which examines the post-flashover behavior of building elements. In other words, time $t = 0$ represents flashover, or at least to the extent practicable. But how did the testing concepts originate in the 1890–1910 time period? A modern engineer might consider quantifying heat release rate, heat flux, or possibly several other variables, but, as the adage goes, "When all you have is a hammer, everything looks like a nail" [13]. The "hammer" in 1900 was a temperature-measuring device, typically a thermocouple, which was already invented in the 1820s. There was no other way of characterizing fires in 1900 except by a temperature measurement. What is interesting in hindsight, is that the early developers of fire resistance testing concepts did not go about characterizing real building fires by their temperatures. Instead, they made the tacit assumption that real fires will not be hotter than the hottest fires that they can create in the laboratory, the latter being done by either stoking the furnaces with wood or firing them with gas burners. Here, we can add that the very earliest test furnaces did not much resemble the dedicated fire testing furnaces of today. Instead, they were often ad hoc built huts, where the specimen formed the to-be-tested portion of the hut, e.g., a wall.

In a modern view, the thermal attack upon a building element would more commonly be represented by a heat flux, rather than by a temperature. Due to the importance of radiant heat transfer, the heat fluxes are scale-dependent and would be distinctly lower if the same temperature were maintained, but a burning room would be reduced to a small-scale model. However, there is very little scale effect [14] once the size exceeds 2 or 3 m, thus the instinctive understanding of the early researchers that fire resistance testing should not be done in small scale proved to be prescient.

What may be striking, however, is that a decade had to elapse after the 1918 publication of ASTM C 19 before somebody became curious enough to study the temperatures in real fires. Since the thermocouple was invented in the 1820s, this means that a century elapsed before the profession started learning what fire temperatures really are. The curious researcher was Simon Ingberg, who worked at the US Bureau of Standards, the institution which today is known as NIST, the National Institute of Standards and Technology. In 1928, Ingberg reported on an extensive series of full-scale fire experiments [15], conducted in some buildings in Washington DC. Ingberg extended this original study by additional tests conducted in 1939, but the results were not published until 1967 [16].

2.4   Modern Data on Room Fire Temperatures

These two Ingberg test series are now mainly of historical interest, but modern data on room fire temperatures were published by Fang and Breese [17], working for the same institution as did Ingberg. Figure 2.2 shows some of their results, obtained by testing modern furniture (substantive use of plastics, instead of cellulosic materials). Two important observations can be made from this figure:

1. During the early part of the test (0 – 30 min), recorded gas temperatures are substantially higher than the standard ASTM time/temperature curve.
2. But when averaged over a 60 min interval, average temperatures are similar to, or lower than the standard time-temperature curve. Fire resistance test rating periods for walls, floors, etc. in the US are not shorter than 60 min, but may be 2 h, 3 h, or 4 h. Thus, it can be concluded that, when averaged over a 1 h, or longer, time period, the ASTM E119 standard time/temperature curve is not unconservative.[1]

The comparison is not straightforward, however, due to differences in temperature measurement technology. Unlike ISO 834 or BS 476, the US standard uses very peculiar thermocouples, which are enclosed in a heavy pipe ("thermowell") [5]. The temperatures registered in fires by thermocouples depend significantly on the physical characteristics of the thermocouples. Notably, increasing diameter leads to lower observed temperatures [18]. But the thermocouples in the ASTM standard are not only of a large size (1.02–1.63 mm wire diameter, but they are enclosed in an 21.3 mm O.D. Inconel (earlier, iron) thermowell. In the early days of the twentieth century, such a protective thermowell was seen as necessary to ensure reliability and longevity for the thermocouples. This was achieved, but at a serious cost of accuracy. Figure 2.3 shows that after about 20 min, there is little difference in temperature readings between small-diameter, bare thermocouples (such as are normally used to measure temperatures in test room fires), and the ASTM thermocouples. But early in the test, recorded temperatures from ASTM thermocouples are up to 550 °C lower than obtained from a more realistic temperature measurement technology (bare-wire thermocouples). This effect was *not* taken into account by Fang and Breese in Fig. 2.2. If this effect were properly taken into account, it would be seen that the

---

[1] There are other countries which use 20 or 30 min fire resistance ratings. It is understood that the purpose of such ratings is to allow for a minimum time period during which occupants can make a safe escape and fire services can complete their search. Such short ratings do not indicate that fires are expected to be only 20 or 30 min in duration.

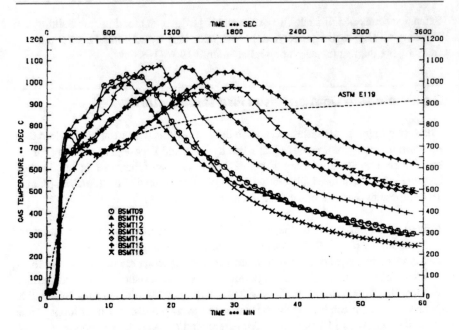

Fig. 2.2 Some example room fire temperature data reported by Fang and Breese [14]

ASTM standard time/temperature curve is notably conservative with respect to real fires in buildings.

Based on the above considerations, the data of Fang and Breese can be compared to "True ASTM" temperatures, with the latter being defined as nominal ASTM temperatures plus the error (difference between bare thermocouples and ASTM thermocouples) presented by Babrauskas (Fig. 2.3). Thus, the "True ASTM" temperatures represent values which would be measured by bare, 0.81 mm thermocouples while the ASTM thermocouples are following the standard ASTM time/temperature curve. It can be clearly seen that, for every test, there are significantly longer times for which the test temperatures fall below the "True ASTM" value, compared to intervals where they exceed this curve (Fig. 2.4).

2.5   Multiple Time/Temperature Curves?

It is very common to find suggestions that the standard time/temperature curve is not right for some purposes, and that different testing curves are needed. One of the earliest such suggestions was by Prof. Boris Bresler [19] in 1972. He observed that by 1972, plastics, and especially foam plastics, were being used in furniture, and that these products, when ignited, could spread fire rapidly and quickly show high peak rates of burning. Thus, he proposed that a Short Duration High Intensity (SDHI) curve be an alternative time/temperature curve for characterizing post-flashover

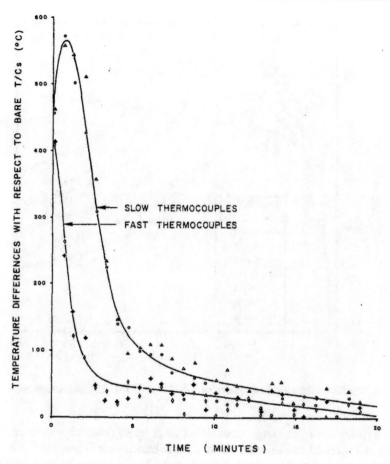

Fig. 2.3 Temperature difference in the ASTM E119 test furnace between bare-wire thermocouples (bare 0.81 mm wires), as compared to standard ASTM furnace thermocouples ("slow thermocouples"), and fast thermocouples (sheathed thermocouples with a 6.35 mm O.D. of sheath); from Babrauskas [1]

fires. Figure 2.5 shows Bresler's SDHI curve, in direct comparison to the nominal ASTM E119 curve. However, it is important to appreciate the context of Bresler's recommendation—he was interested in *modeling, not testing* building elements under post-flashover conditions [20]. For modeling purposes, certainly it is just as easy to use one curve, as it is to use another, or an alternative. We further discuss Bresler's work and modeling approaches below. But here, we wish to consider *testing* paradigms, especially since engineers might consider the presentation of such SDHI curves as a suggestion that the ASTM E119 test curve is inadequate, and should be either supplemented or supplanted.

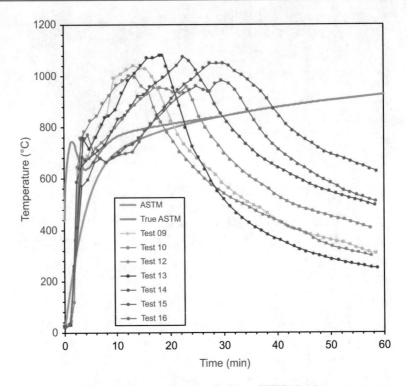

Fig. 2.4 The data of Fang and Breese along with the "True ASTM" time/temperature curve

In a narrow sense, one may consider that the standard time/temperature curve is, in fact, a family of curves, rather than a single curve. This is because the ASTM curve is defined over the interval of 0–8 h, and a product might be tested for 1, 2, 3, or 4 h.[2] Whether this is considered to be multiple curves, or not, is a matter of perspective.

What is important to appreciate in Fig. 2.4 is the area under the curve, which effectively represents the integrated value of the thermal attack from the fire upon the test specimen. It is clear that the area under the SDHI curve is much smaller than the area under the ASTM E119 curve. Thus, if the more appropriate test exposure was considered to be the SDHI curve, and the test was conducted by following the standard ASTM E119 prescription, the results would be conservative.[3]

The main reason why multiple time/temperature curves are not used is practical. Multiple tests of a product would be needed to provide results under the various

---

[2] No fire tests have been used or reported for durations over 4 h in the modern era, although some very long fire tests were being conducted in the 1890s and early 1900s.

[3] It might be thought that this is not necessarily true, if the response of the test specimen to fire is highly non-linear, in that destruction is disproportionately higher at temperatures which exceed the ASTM E119 curve. However, in practice, no such materials have been identified.

Fig. 2.5 The SDHI
time/temperature curve, as
envisioned by Bresler [17]

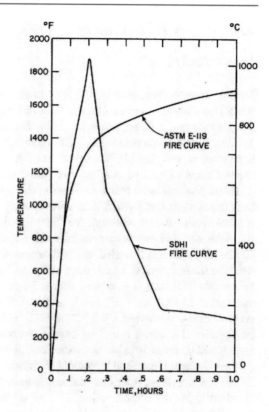

test curves. Fire resistance testing is very different from small-scale reaction-to-fire tests. It is common for the latter to be run at several test conditions, since the tests are relatively inexpensive and quick to run. But fire resistance tests are more expensive by several orders of magnitude, thus, there would have to be an enormous societal benefit for this type of testing, and such benefit has not been seen.

As shown above, for residential occupancies, the behavior of typical fuel loads is such that testing under the standard time/temperature curve is conservative and acceptable. The main exception would be libraries and storage facilities [21], where nearly-unlimited fuel loads may sometimes be encountered. In such cases, fires may indeed burn for many hours, or even days, and building codes do not consider that commensurate fire resistance should be provided. In other words, after a facility has burned for several hours, it is highly unlikely that there are still some unevacuated occupants; meanwhile, the costs of providing such long-term fire resistance would not be economical.

2.6   Petrochemical Industry Tests

2.6.1   Pool Fires

Burning hydrocarbon liquids will tend to show a much higher heat release rate (HRR) than wood materials. But this does not imply higher peak temperatures. The temperature that a flame would exhibit under conditions of no heat losses is the *adiabatic flame temperature*, T_{ad} [22]. Values of T_{ad} for wood and for hydrocarbon liquids are very similar [3]. Yet, operators of petrochemical facilities noted that high thermal assault from hydrocarbon fires can often be expected.

It has been claimed that fires from burning hydrocarbon liquids develop high heat fluxes very rapidly and that, consequently, the ASTM E119 curve, with its gradual rise does not represent the "thermal shock" from those conditions. The importance of this was never demonstrated, nonetheless, ASTM published standard E1529 [23] which is intended to simulate thermal attack from a hydrocarbon pool fire. The furnace control here is done in a peculiar manner, with the primary control being cold-wall heat flux, which, after a 5-min warm-up period, is required to be constant at 158 kW m^{-2} $\pm$ 25%. In addition, the furnace temperature is also to be controlled, being between 1010 °C and 1180 °C after the first 5 min. There has been no significant research justifying such a test method, although some of the ideas were based on an early paper by Castle [24], and unpublished testing was later done by the US Coast Guard and by Sandia National Laboratories.

UL has published a similar, but not identical test as UL 1709 [25]. This uses standard thermal instrumentation, in contrast to the ASTM test, which requires unique instrumentation, generally not used elsewhere. Consequently, this test can be considered to be preferable to the ASTM version. But again, there is no known research detailing how it was developed. Some modeling results using ASTM E119 and UL 1709 thermal exposures do not suggest major differences [26]. IEEE 1717 [27] is an offshoot of UL 1709, intended for testing cables instead of structural members. And, again, background research is nonexistent.

In the EU, a similar concept is defined in EN 1991-1-2 [28]. But here, the hydrocarbon pool fire exposure is defined as a temperature *curve* (the "hydrocarbon curve"), similar to the standard time/temperature curve, but showing more rapid rise.

The main conclusion is that there has been no credible demonstration that "thermal shock" is an important variable in establishing the fire endurance of petrochemical, nor of building, products. In the absence of such demonstration, a slightly greater thermal attack can always be presented in the context of the ASTM E119 standard furnace exposure by providing for a slightly longer required exposure. Conversely, without relevant research, the possibility cannot be precluded, but a possibility alone would not seem to be the ideal way to justify a testing paradigm. If focused research were to establish the need for "thermal shock" testing, then, to be useful, it would also need to identify the categories of materials and the types of circumstances where such extra challenging testing needs to be used.

2.6.2 Jet Fires

Operators of petrochemical facilities observed that fires due to broken piping producing jets of burning hydrocarbon liquids can show exceedingly damaging effects. This is due to the fact that the typical failure incident is likely to produce a jet flame with very high flow velocity, and this jet may impinge on structural components. The heat fluxes created due to high jet velocities and an impinging flow geometry can be very high [29]. Heat fluxes from such heating will not only be greater than what is expected in buildings of other occupancies, but also greater than the thermal attack from pool fires. Thus, industry considered that a separate, specialized test is required.

Unlike for pool fires, the research leading to a jet fire test has been documented at significant length. Parker [30], Roberts et al. [31], and Mather and Smart [32] provide some overviews of the problem and the test development. A large number of detailed research studies have also been published. The basic testing details are standardized in ISO 22899-1 [33], although industry testing in practice often involves deviations [34]. The test is generally intended for testing of fireproofing materials applied to steel piping or equipment products, rather than assemblies from buildings or building frames. Much of the development occurred at the UK Health & Safety Executive, who published the preliminary version of the testing procedures, in cooperation with Shell Research, British Gas, SINTEF NBL, and several other institutions [35]. The scale of these tests is usually 1 m, or less, in contrast to tests of building elements, which are several-fold greater in size.

2.7 What Is the Basis for the Required Fire Resistance Rating?

Some early nineteenth century thinking was based on the idea (but not clearly delineated) that the required fire resistance rating be such that the structure withstand a full burnout. Even in the twentieth century, some authors argued for this concept [15, 36]. An essentially equivalent formulation is that the required fire resistance rating increases proportionately to the fuel load present. However, this notion was never accepted by US building codes. Instead, the building codes effectively espoused a *risk* concept, although not labeling it as such. Within a risk framework, more conservative designs need to be provided, if the consequences of failure are more severe. Single-family homes have few occupants, are short in height, and normally easy to escape from. Thus, in most cases, US codes have not laid any fire resistance requirements on them. Commercial buildings may be taller, and may hold many more persons. Thus, depending on details, 1- or 2-h ratings are typically required. Structural frames and bearing walls may, in the most stringent applications be required to have 3-h ratings [37]. During much of the twentieth century, however, some situations required up to 4-h ratings. Also, within commercial buildings, more important structural components (e.g., columns) require greater ratings than less important ones (e.g., non-bearing partition walls). These are clearly risk concepts, even though the codes do not label them as such.

2.8   Design Practice

During the late 1970s, the fire safety engineering profession considered that standardized tests, of the type represented by ASTM E119, BSI 476, or ISO 834, would be shortly obsolete, due to advances in fire modeling. Babrauskas published the first computer model, COMPF, for predicting post-flashover room temperatures in 1975 [38]. Meanwhile, at the same institution, University of California Berkeley, Prof. Boris Bresler took the next step. He considered that, once the fire temperatures are known, the fire resistance design of a building can be achieved if two more models are available: [1] a model to predict the thermal response of a building's structural elements and [2] the mechanical (thermostructural) response of the structural elements to these temperatures. In short order, he and his graduate students produced two computer models, FIRES-T [39] and FIRES-RC [40], for accomplishing task #1 and task #2, respectively. These were limited to analyzing buildings with concrete frames, since he considered this to be the first priority. These were shortly (1977) followed by expanded versions, FIRES-T3 [41] and FIRES-RC II [42].

There is little evidence that any of Prof. Bresler's models received any significant use. Instead, for the next 20 years, nothing changed within the profession. Eventually, during the late 1990s, the profession rediscovered the potential applications for providing fire resistance to buildings by use of fire modeling and thermostructural modeling.

Fast-forwarding to today, a large or expensive structure is likely to have its fire resistance protection provided by modeling. Fire modeling is generally done by using the FDS [43] program of NIST. This is vastly more capable than COMPF, in that it can treat multiple rooms and encompasses both pre- and post-flashover modeling. Thus, it finds use in other applications, e.g., smoke management, not just as a tool for fire resistance design. There is no single dominant thermostructural model, but common commercial packages are typically used, especially ANSYS and ABAQUS, although there are numerous others also. Some useful references describing useful techniques for providing fire resistance by means of thermostructural modeling include those by Buchanan [44], ASCE [45], and Wang et al. [46] A brief overview of the subject has been published by the (UK) Institution of Structural Engineers [47]. However, most of the books on this topic are written from an academic point of view, rather than that of the practicing design engineer or architect. In the EU, a series of Eurocodes prescribe requirements for structural design, including thermostructural design. Franssen and Vila Real [48] and Narayanan and Beeby [49] have published designers' guides to steel and concrete structural fire design, respectively.

2.9   Hose Stream Testing

The United States and Canada are deviant with respect to the rest of the world, in that a hose stream test [50] is included as an integral part of fire resistance testing (Fig. 2.6). This means that, after the fire test is concluded, water from a hose stream is applied to the specimen according to certain specifications. The test is passed if the specimen does not collapse, and the hose stream does not penetrate the far side of the assembly. The test used to be required for all types of assemblies, but in 1955 it was removed as a requirement for floor assemblies. The reason had nothing to do with appropriateness of the test. Instead, the requirement was deleted due to excessive damage to floor furnaces. Specimens are not handled identically in floor and in wall furnaces. In a wall furnace, apart from early testing activities, the procedure has been to provide a specimen frame into which the specimen is constructed. This gets wheeled (typically by an overhead crane-type device) into the furnace at the start of the test, and wheeled out afterwards, where the hose stream test can be conducted some distance away from the furnace. Specimen frames for floor furnaces, on the other hand, are typically much more massive, and lack provisions for rapidly wheeling out from the furnace at the completion of the fire test. Thus, hose stream testing used to be conducted as shown in Fig. 2.6, leading to significant thermal shock damage to the furnace as a result.

The origin of the test was in the fact that during the 1840s through the 1880s, cast iron used to be a popular material for constructing the facades of commercial

Fig. 2.6 Hose stream testing at the Columbia University Testing Station, ca. 1913

buildings in New York and some other large cities. Cast iron is a very brittle material, and if hot cast iron is hit by a stream of cold water, it is likely to shatter precipitously. Obviously, this would create an unsafe situation for firefighters, since the façade would be likely to tumble down upon them, if they are standing below in the street. Since fire resistance testing was primarily developed in the 1880s and 1890s, it made good sense to establish specific provisions that this would not happen, at least for buildings intended to be fire resistive.

What makes much less sense is that the first edition of the ASTM standard on fire resistance testing did not appear until 1918. By that time, cast iron architecture was obsolete and not used for new building constructions. Furthermore, fire resistance testing has always been seen as a test for the materials or products to be put into new buildings, not as a means of examining the performance of historical buildings. Thus, putting the hose stream provision into the standard made no sense, but put in it was. The hose stream test also appeared in the original 1932 edition of the British standard BS 476, although it was sensibly removed from the next (1953) edition.

So why does the hose stream test still exist as a requirement for wall tests in ASTM E119? The answer is, due to industry influence. The building products industry is extremely loathe to make any changes in the standard for two reasons:

(a) It might invalidate the massive data bank of past tests; and
(b) It might change the marketplace, or allow entry for new competitors.

If the hose stream test were not present, more lightweight wall assemblies could pass the test, leading to lower prices for the product category. This is not advantageous to the existing producers. Fortunately, the rest of the world is not saddled with this unsatisfactory history.

2.10 Additional Issues

The combination of E119 testing and design-by-modeling currently addresses the majority of fire resistance problems for the design profession. Yet there are certain areas that fall outside the scope of these two types compliance mechanisms.

Specialized Fire Resistance Tests ASTM E119 testing encompasses walls (partitions), floor/ceiling assemblies, beams, and columns. But in more recent years, some additional tests have been established by ASTM for certain building components which are sufficiently different from E119 so that they are described in different standards. The hydrocarbon pool-fire test has already been discussed above. The oldest of the specialized fire resistance tests is ASTM E814 [51], a test for firestopping of penetrations ("poke-through") in wall or floor assemblies, originally published in 1981. Such products are small scale, thus, they do not require a full-scale test furnace. As a result, this standard notably differs from ASTM E119 in that a small (typically, 1 m cube) furnace is used. The standard time/temperature curve is followed, using ASTM E119 thermocouples.

Another specialized fire resistance test is ASTM E1966 [52], a test method for seismic or expansion joints, established in 1998. This appears to be a solution in search of a problem. Theoretically, fire could spread in a building by burning through lightweight or flimsy joints. However, no such case incidents have been identified. It can also be noted that the standard is bereft of any references to the scientific literature.

Finally, ASTM E2307 [53] is described as a fire resistance test for perimeter fire barriers. One may begin to view this situation by noting that US building codes and US design practice is such that for most buildings, there is no fire resistance requirement of the façade walls. This can readily be verified by noting that the overwhelming majority of buildings are fitted with windows on their outside walls, and these are made of ordinary window glass, a product which has no fire resistance rating. What the test method actually tests are firestopping products which go between the end of the floor slab and the exterior wall. Now, one might ask, if the exterior wall is not required to have any fire resistance, what benefit is there of making sure that the few centimeters spanning between the end of the floor slab and the wall have a fire resistance rating? Proponents will claim that the 1970 fire in the One New York Plaza building [54] is a good example. In that event, light-gauge aluminum panels were used to span a void space between the ends of the floor slab and the curtain wall, and fire spread upwards by melting through these flimsy barriers. But is that really a robust justification? Fire spread from storey to storey via the façade is obviously a very dangerous situation. This has occurred numerous times in high-rise buildings outside of North America, typically where combustible ceilings are used (we are assuming here that the façade itself is not combustible). The simplest way to guard against this mode of fire propagation is to prohibit combustible ceilings from being used in high-rise buildings. Theoretically, progressive upward fire propagation along a non-combustible façade could also be precluded by using non-combustible spandrel panels. However, to serve this purpose, the spandrel panels would have to be unreasonably high [55], much higher than 3 ft. or 1 m. Thus, it is not at all clear that the One New York Plaza fire could have been prevented by fire resistive blocking of the end-of-floor-slab gap; fire might well have propagated upwards by going directly through the outside, via the glazing. There may indeed be a practical value from fire testing of these barriers, but the situation is unproven, based on currently available research.

Unexplored Areas It may be noted that E1966 and E2307 are the only tests where some form of joint or intersection between structural members is assessed. Both of these are intended for testing firestopping-type materials, i.e., non-structural materials used to fill in some gap and to thereby provide fire resistance. But the actual structural joints are not subjected to testing. These might be inadequately designed, so that they fail prematurely or precipitously. But tests do not exist to examine for this.

Another area of interest that has received very little attention is frame action. Significant buildings are typically designed with a plurality of load-bearing (and moment-resisting) joints. If a fire occurs in one, or a few compartments, heating of one portion of the frame will cause load redistribution around that area, and possibly throughout the whole structure. The research of Prof. Bresler in the 1970s focused on that type of behavior. But in more recent times, there has been little interest in pursuing research on this topic.

2.11   Conclusions

The provision of fire resistance to buildings, structures, or equipment can be done by testing, as has been done for over 140 years now. Conversely, it can be done by computer modeling in the modern era. The primary "driving force" in a fire resistance test or model is the fire temperature. Since this can vary over time, it is usually referred to as the time/temperature curve. In this chapter, we have undertaken to examine how the exposure temperature definition originated, and why there is only one standard time/temperature curve, instead of a family of curves.

The ASTM E119 test is now over 100 years old, yet it has changed surprisingly little over its exceedingly long history. Since the test requirements are driven by industry, this is to be expected. The main reason is that fire resistance tests are hugely expensive and industry would not consider itself to be well-served by any efforts to abrogate the validity of the exceptionally expensive investment in the database of product tests. This is the overriding reason why a family of testing curves has not been found to be practical.

Nonetheless, testing alternatives exist, and some have flourished. The petrochemical industry is substantially different from the building products industry, and they have elected to develop different tests, notably comprising pool-fire and jet-fire arrangements. The ASTM test for pool-fire exposures, ASTM E1529, requires some quixotic instrumentation whose necessity has never been justified. Consequently, the later UL 1709 test is preferred nowadays, since it utilizes standard thermal instrumentation. Jet fire testing is typically done under guidance from the ISO test, although, due to the nature of the industry, tests often require deviations.

When fire resistance is provided by means of computer modeling, there is of course no restriction as to the nature of the thermal attack to be simulated. Typically, in such cases, no standard time/temperature curve is employed. Instead, designers model fire temperatures, and then use these data to compute the thermal and mechanical performance of the building assemblies being studied. Despite great progress in both computer hardware and computer fire models, such an approach is still time-consuming and expensive. Thus, this approach has been primarily used for the design of unusual or high-value projects, such as airport terminals, sports arenas, and train stations.

It is unlikely that computer modeling approaches will start being used for low-cost, mundane projects anywhere in the near future. But what is likely to happen, is that the use of this approach will gradually spread downward into mid-cost projects.

This may well entail some automation or simplification of the computer modeling design process.

Finally, in view of the enormous costs of providing fire resistance for buildings, one might surmise that there exist good benefit/cost analyses. One would be wrong. In general, there have been no significant studies on case histories of failures of fire resistance. Without good knowledge of failures, it is impossible to rationally assess how much expenditure is economically justifiable to avert such failures. It should be self-evident that providing fire safety features should not be mandated, if they are not cost effective. Yet this topic is painfully neglected in the codes, standards, and regulations [56]. Instead, both industry and regulatory officials tend to pursue the philosophy that more safety is better, and cost is irrelevant. This is not societally responsible regulation.

References

1. Lyons, P. R., **Fire in America!** NFPA, Boston (1976).
2. Wermiel, S. E., **The Fireproof Building: Technology and Public Safety in the Nineteenth-Century American City**, The Johns Hopkins Univ. Press, Baltimore (2000).
3. Babrauskas, V., Fire Endurance in Buildings (Ph.D. dissertation), Univ. California, Berkeley (1976).
4. Babrauskas, V., and Williamson, R. B., The Historical Basis of Fire Resistance Testing, *Fire Technology***14**, 184-194, 205, 304-316 (1978).
5. Gales, J., Chorlton, B., and Jeanneret, C., The Historical Narrative of the Standard Temperature–Time Heating Curve for Structures, *Fire Technology***57**, 529-558 (2021).
6. Sachs, E. O., et al., **First International Fire Prevention Congress, The Official Congress Report**. British Fire Prevention Committee and "The Public Health Engineer," London (1903).
7. Standard Specifications for Fire Tests of Materials and Construction (C 19), ASTM, Philadelphia (1918).
8. Standard Test Methods for Fire Tests of Building Construction and Materials (ASTM E119), ASTM.
9. British Standard Definitions for Fire-Resistance, Incombustibility and Non-Inflammability of Building Materials and Structures (Including Methods of Test), BS 476, British Standards Institution, London (1932).
10. Fire Resistance Tests — Elements of Building Construction (ISO 834), International Organization for Standardization, Geneva (1975).
11. Bletzacker, R. W., Fire Resistance of Protected Steel Beam Floor and Roof Assemblies as Affected by Structural Restraint, pp. 63-90 in *Fire Test Methods—Restraint and Smoke* (STP 422), ASTM, Philadelphia (1966).
12. Hird, D., and Fischl, C. F., Fires in Model Rooms (FR Note 12), Joint Fire Research Organization, Borehamwood, UK (1952).
13. As paraphrased from Maslow, A. H., **The Psychology of Science: A Reconnaissance**, Harper & Row, New York (1966).
14. Babrauskas, V., Estimating Large Pool Fire Burning Rates, *Fire Technology***19**, 251-261 (1983).
15. Ingberg, S. H., Tests of the Severity of Building Fires, *Q. NFPA***22**:3, 43-61 (July 1928).
16. Rodak, S., and Ingberg, S. H., Full-scale Residential Occupancy Fire Tests of 1939 (Report 9527), NBS, Washington (1967).
17. Fang, J. B., and Breese, J. N., Fire Development in Residential Basement Rooms (NBSIR 80-2120), [U.S.] Natl. Bur. Stand., Gaithersburg MD (1980).

18. Brundage, A. L., et al., Thermocouple Response in Fires, Part 1: Considerations in Flame Temperature Measurements by a Thermocouple, *J. Fire Sciences* **29**, 195-211 (2011).
19. Bresler, B., Reinforced Concrete Column Response to Fire. Discussion No. 9, pp. 723-726 in *Intl. Conf. on Planning and Design of Tall Buildings*, Vol. 1B, Bethlehem PA (1972).
20. Bresler, B., Fire Protection of Modern Buildings: Engineering Response to New Problems (Report UCB FRG 77-3), Univ. California, Berkeley (1977).
21. Sharry, J. L., and Walker, E., Military Personnel Records Center Fire, Overland, Missouri, *Fire J.* **68**:3, 5-9 (May 1974); **68**:4, 65-70 (July 1974).
22. Glassman, I., and Yetter, R. A., **Combustion**, 4th ed., Elsevier, Amsterdam (2008).
23. Standard Test Methods for Determining Effects of Large Hydrocarbon Pool Fires on Structural Members and Assemblies (ASTM E1529), ASTM.
24. Castle, G. K., The Nature of Various Fire Environments and the Application of Modern Material Approaches for Fire Protection of Exterior Structural Steel in Them, *J. Fire & Flammability* **5**, 203-222 (1974).
25. Rapid Rise Fire Tests of Protection Materials for Structural Steel (UL 1709), UL, Northbrook IL
26. McGrattan, K. B., Numerical Simulation of the Caldecott Tunnel Fire, April 1982 (NISIR 7231), NIST, Gaithersburg MD (2005).
27. IEEE Standard for Testing Circuit Integrity Cables Using a Hydrocarbon Pool Fire Test Protocol (IEEE 1717), IEEE, New York (2012).
28. Eurocode 1: Actions on Structures. Part 1-2: General Actions – Actions on Structures Exposed to Fire (EN 1991-1-2), CEN, Brussels (2002).
29. Baukal, C. E. jr, and Gebhart, B., A Review of Empirical Flame Impingement Heat Transfer Correlations, *Int. J. Heat and Fluid Flow* **17**, 386-396 (1996).
30. Parker, A. J., Evaluating High-Temperature Intumescent Insulation Materials under Fire and Blast Conditions, pp. 498-509 in **Insulation Materials: Testing and Applications**, vol. 3 (ASTM STP 1320), ASTM, West Conshohocken PA (1997).
31. Roberts, T. A., Brown, D., Beckett, H., and Buckland, I., Comparison of the Effects of Different Fire Test Regimes on Passive Fire Protection Materials, pp. 253-266 in **Major Hazards Onshore & Offshore** (IChemE Symp. Series No. 139), Institution of Chemical Engineers, Rugby UK (1995).
32. Mather, P., and Smart, R. D., Report on Comparable Testing Using Large Scale and Medium Scale Jet Fire Tests (OTO 97079), UK Health & Safety Executive, Buxton, UK (1998).
33. Determination of the Resistance to Jet Fires of Passive Fire Protection Materials. Part 1: General Requirements (ISO 28899-1), ISO, Geneva.
34. Stølen, R., and Reitan, N. K., High Heat Flux Jet Fire Testing at SP Fire Research, Norway (SPFR INFO 2014-06), SP Fire Research AS, Trondheim, Norway (2014).
35. The Jet Fire Test Working Group, Jet-Fire Resistance Test of Passive Fire Protection Materials (OTI 95634), UK Health & Safety Executive, Buxton, UK (1996).
36. Robertson, A. F., Gross, D. Fire Load, Fire Severity and Fire Endurance, pp. 3-29 in *Fire Test Performance* (ASTM STP 464), ASTM, Philadelphia (1970).
37. International Building Code, International Code Council, Country Club Hills IL (2021).
38. Babrauskas, V., COMPF: A Program for Calculating Post-flashover Fire Temperatures (UCB FRG 75-2), Fire Research Group, Univ. California, Berkeley (1975).
39. Becker, J., Bizri, H., and Bresler, B., FIRES-T: A Computer Program for the Fire Response of Structures—Thermal (UCB FRG-74-1), Univ. California, Berkeley (1974).
40. Becker, J. M., and Bresler, B., FIRES-RC – A Computer Program for the Fire REsponse of Structures - Reinforced Concrete Frames (UCB FRG 74-3), University of California, Berkeley (1974).
41. Iding, R., Bresler, B., and Nizamuddin, Z., FIRES-T3: A Computer Program for the Fire Response of Structures—Thermal (Fire Research Group Report UCB FRG 77-15), Univ. California, Berkeley (1977).

42. Iding, R. H., Bresler, B., and Nizamuddin, Z., FIRES-RC II: Computer Program for the FIre REsponse of Structures—Reinforced Concrete Frames. Second Revised Version (UCB FRG-77-8), Univ. California, Berkeley (1977).
43. McGrattan, K., Hostikka, S., Floyd, J., McDermott, R., and Vanella, M., Fire Dynamics Simulator User's Guide, 6th ed. (SP 1019), NIST, Gaithersburg MD (2020).
44. Buchanan, A. H., and Abu, A. K., **Structural Design for Fire Safety**, 2nd ed., Wiley, Chichester (2017).
45. **Performance-Based Structural Fire Design**, ASCE, Reston VA (2020).
46. Wang, Y., Burgess, I., Wald, F., and Gillie, M., **Performance-Based Fire Engineering of Structures**, CRC Press, Boca Raton FL (2013).
47. **Introduction to the Fire Safety Engineering of Structures**, The Institution of Structural Engineers, London (2003).
48. Franssen, J.-M., and Vila Real, P., **Fire Design of Steel Structures**, 2nd ed., ECCS/Wilhelm Ernst & Sohn, Berlin (2015).
49. Narayanan, R. S., and Beeby, A., Designers' Guide to EN1992-1-1 and EN1992-1-2 Eurocode 2: Design of Concrete Structures. General Rules and rules for buildings and structural fire design, Thomas Telford, London (2009).
50. Practice for Application of Hose Stream (ASTM E2226), ASTM.
51. Standard Test Method for Fire Tests of Through-Penetration Fire Stops (ASTM E814), ASTM Intl., West Conshohocken PA.
52. Standard Test Method for Fire-Resistive Joint Systems (ASTM E1966), ASTM.
53. Test Method for Determining Fire Resistance of Perimeter Fire Barriers Using Intermediate-Scale, Multistory Test Apparatus (ASTM E2307), ASTM
54. Powers, W. R., One New York Plaza Fire, New York, N.Y., August 5, 1970, The New York Board of Fire Underwriters, NY (1970).
55. Yokoi, S., Study on the Prevention of Fire-Spread Caused by Hot Upward Current (Report of the Building Research Institute No. 34), Building Research Institute, Tsukuba, Japan (1960).
56. Babrauskas, V., Some Neglected Areas in Fire Safety Engineering, *Fire Science & Technology (Tokyo)***32**:1, 35-48 (2013).

Integrating Modern Technologies to Realize Fire-Resistant Infrastructures

3

Casey Grant

3.1 Introduction and Background

Fire has long been both a friend and foe to the built infrastructure. Unwanted fire has been a major hazard since the earliest civilization, and today it continues to be a major aspect in societal planning worldwide.

Fire loss is a staggering burden. In the USA, for example, public fire departments responded to 1,318,500 fires in 2018, that had an overall result of 3655 civilian fire fatalities and 15,200 civilian fire injuries [1]. Just as disconcerting as the fire loss are the resources expended to address these losses. For example, in the year 2014 in the U.S. the total overall cost was $328.5 billion, which was 1.9% of the U.S. Gross Domestic Product. This includes fire protection expenditures of $273.1 billion (83.1% of total) and losses of $55.4 billion (16.9% of total) [2].

Today's fire-resistant infrastructure can be significantly enhanced by exploiting new opportunities in technology development. We are stepping into the new age of Cyber Physical Systems, with dramatic technological advances that are opening new doors for the safety infrastructure.

These advances are on the backbone of emerging conceptual fields of understanding, such as artificial intelligence and machine learning, and are being enabled by sweeping evolutionary changes such as the Internet of Things (IoT). All of this is fostering significant advances that have great potential for an improved safety infrastructure. Not surprisingly, these same advances are also introducing new vulnerabilities, such as cybersecurity, and these vulnerabilities are requiring completely new focus points to assure reliable safety [3].

Integrating modern technologies to realize fire-resistant infrastructures requires an understanding of the world of fire protection and emergency management, as

C. Grant (✉)
DSRAE, LLC, Belmont, MA, USA
e-mail: cgrant@dsrae.net

© Springer Nature Switzerland AG 2022
M. Z. Naser, G. Corbett (eds.), *Handbook of Cognitive and Autonomous Systems for Fire Resilient Infrastructures*, https://doi.org/10.1007/978-3-030-98685-8_3

well as of Cyber Physical Systems, and the integration thereof of these two spheres of knowledge. Today's fire protection and emergency management infrastructure is evolving into an all-hazards focus, with specific attention to pre-event, during-event, and post-event [4]. The stakes are often very high, and regularly involve life and death consequences.

Cyber Physical Systems are based on the root concept of collecting data, processing data, and delivering data [5]. These systems are understood to be the fusion of emerging sensor and computing technologies with building control systems, fire fighting equipment, and so on. Cyber Physical Systems will revolutionize fire protection and emergency response and have significant potential to provide effective and efficient infrastructure improvements.

Today we are fortunate to witness a merging of the worlds of fire protection and emergency management, with Cyber Physical Systems. All of this is evolving through systems integration, robotics, new lexicons, and other features that are adding to our body of knowledge and understanding. This includes a range of approaches that at their core involves collecting data globally, processing the information centrally, and distributing the results locally [6].

Multiple case study examples are included that provide clarifying base scenarios. This includes deeper review of certain specific notable events that involve a complex fabric of dynamically changing details involving all branches of emergency response. These highlight meaningful technological innovation and advances, amidst the daunting and dynamic challenges faced during an emergency [7].

As we turn the pages into tomorrow, we are learning that some of our greatest challenges are less technical, and instead are dwarfed by other aspects of the overall solution approach such as political, legal, ethical, and so on. A stark example is data privacy and confidentiality.

Technological development is already on a fast-track, with innovation occurring everywhere we look and rapidly emerging in other forums and parallel sectors. As a result, applied researchers are discovering effective alternative approaches to leverage this technological innovation, such as research competitions and proof-of-concept type showcases.

The path forward into the future is promising. We are on the cusp of profound changes with the safety infrastructure, and if we stay focused, we can guide this in a positive and appropriate direction. A world absent of the scourge of fire and related hazards is a worthy and noble goal.

3.2 Fundamentals of Fire Protection and Emergency Management

In its simplest form, the safety infrastructure comprises all activities that keep us safe from harm. In terms of resistance from fire and similar emergencies, the classic approach falls into multiple pathways: either preventing a fire, or managing fire impact, or both [8].

Consistent with these approaches are the following. First is fire resiliency built into a particular design of a building, facility, operation or application. Second, the efforts to deal with a fire as it occurs, or after it occurs, through mitigation measures such as fire fighting and occupant egress.

3.2.1 Event Time Spectrum

Both of these approaches follow the event spectrum of either being pre-event, during-event, or post-event. Each approach has aspects that fall within all parts on the overall time spectrum [9]. From a technology perspective, and with a focus on achieving the mission of any given strategy, this time spectrum becomes quite important on the overall processing of technology-handled-data.

The time spectrum importance becomes obvious when realizing that the during-event time frame can be extremely precious with normally very limited available time to make decisions and take action. Pre-event and post-event, on the other hand, do not have these same restrictions and generally have ample time. During an emergency the significant challenge is to provide near real-time processing actions, when seconds count and make the difference between successful outcomes and possible catastrophic consequences (e.g., distinct life and death situations).

For the classical approach of fire prevention measures, passive and active fire protection systems for example are heavily focused on relatively extensive pre-event preparations to minimize the potential emergency or prevent it from happening all together. Thus, their time commitments in the other parts of the event spectrum are not as extensive, such as with during-event and post-event.

In contrast, measures to manage fire impact are more heavily centralized on the time frame of during-event. Nevertheless, they still require significant pre-event preparations that scale with the magnitude of a particular emergency (e.g., equipment, training, incident command, pre-planning, etc.).

3.2.2 Passive and Active Approaches

The approach to provide and support fire resistance is dependent on measures of built-in fire protection and techniques supporting emergency management. With both, there are passive and active approaches, though these are more obvious with built-in fire protection [10].

Examples of passive measures include zoning distances between buildings to minimize fire spread. These are fully embedded and require no subsequent specific control action. To some extent, passive safety measures are designed into virtually every product and material available in today's consumer marketplace.

In contrast, an active system takes a defined action in response to an emergency either automatically or through manual activation. An example of an active approach is a water-based automatic sprinkler system that automatically activates in direct response to an unwanted fire.

3.2.3 Fire Safety Goals

It is important to remain focused on the overall goals that are the genesis of fire resistance. What, for example, are we really trying to achieve when we seek to prevent or control unwanted fire? The classic primary underlying fire protection goals can be generally described as the following: [11].

1. Life safety;
2. Property protection;
3. Continuity of operations;
4. Environmental protection; and
5. Heritage conservation.

These underlying goals are sometimes described in different ways but are generally the same. The first three have long been the underpinnings of fire protection through the centuries, and in the modern era we have become much more cognizant of environmental protection as inhabitants of a planet whose resources are no longer seemingly endless, and heritage conservation with cultural resources that are virtually priceless and irreplaceable from a property protection standpoint.

Normally, but not always, life safety is the preeminent goal, seeking to prevent or minimize death and physical injury to occupants, responders, and others. As an example of this competing with other goals as the overall top priority, a naval ship during battle must have continuity of operations and mission, despite the sacrifices of individual life along with other goals to achieve its primary goal.

These concepts are summarized together in Fig. 3.1. In the quest to integrate modern technologies to realize fire-resistant infrastructures, it is important to remember and respect these fundamental principles. This includes time-dependency, which directly translates on the challenges we face meeting these goals.

3.2.4 Fire Protection Measures

The need to integrate modern technologies into our safety infrastructure requires a fundamental understanding of this infrastructure. Furthermore, if the technologies are to be realistically applied for some direct value-added gain, then the various components must be defined and understood. A deeper review is offered for the two primary approaches, first with built-in fire protection measures, and second with manual intervention techniques.

First, built-in fire protection measures encompass a large collection of fire safety efforts for the built environment. It is important to understand the depth of the knowledge base required for these built infrastructure safety measures. These are much more than simply passive and active inherent and automatic fire fighting systems composed of interoperable hardware and other components that serve a specific safety mission.

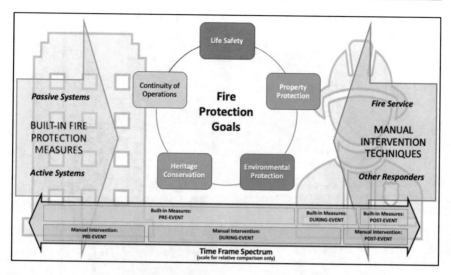

Fig. 3.1 Overview of fire protection goals

More specifically, the built infrastructure knowledge base addressing safety measures is extensive. For fire resistance, one way to illustrate this knowledge base is by clarifying the role of a fire protection engineer (FPE) who address these issues as a profession. Figure 3.2 illustrates the FPE core competencies, that include fire science, fire protection analysis, human behavior, and fire protection systems. Each of these has deep areas of professional activity and subject matter expertise [12].

A deeper discussion of these core competencies if further revealing. Each of these four have knowledge areas associated with them. Table 3.1 provides a compilation of topics that further illustrate the deeper areas addressed by FPE professionals [13]. These exemplify the depth of the subject matter expertise needed to address the built infrastructure when considering safety and overall fire protection.

From an outside perspective, passive and active fire protection systems are the obvious tools in our front-line arsenal to assure fire safety. But from the perspective of the built infrastructure, all the measures within a building are important during an emergency. If a fire or other emergency occurs, it is of course important how the fire alarm system, sprinkler system, or other dedicated safety systems perform and handle the event. However, all the other building systems also rise to new levels of importance, and to varying degrees can serve critical roles for occupants and emergency responders.

For example, during a fire the incident commander for the responding fire service units will be seeking information and control of the building air handling and ventilation systems, elevators, electric utilities, and so on. All information becomes important during an emergency, and the emergency responders need to work with facility managers to coordinate this data near real-time. These are the root ingredients for smart data exchange for enhanced fire safety.

Fig. 3.2 Fire protection engineering core competencies [12]

Table 3.1 Summary of recommended minimum knowledge areas [13]

Core competencies	Knowledge areas
Human behavior and evacuation	• Egress and life safety design concept • Human behavior and physiological response to fire
Fire protection analysis	• Building and fire regulations andand standards • Evacuation analysis • Numerical methods and computer fire modeling • Performance-based design • Risk management • Smoke management • Structural fire protection
Fire protection systems	• Active systems • Fire detection and alarm • Fire suppression • Passive systems
Fire science	• Fire chemistry • Fire dynamic • Heat transfer

3.2.5   Fire Protection Enforcement

The built infrastructure and its built-in fire protection measures have evolved with checks and balances to assure that restrictions in the name of safety, on any one person or organization, is fair and appropriate relative to the common good. In most countries' safety is enforced without infringing on individual rights and similarly the common good.

As such, new or altered buildings or facilities are constructed so that they will not compromise safety for the occupants, neighbors, the community, and others. Personal freedom is a fundamental basic right of today's society, and it is constantly balanced against the need to protect public health, safety, and welfare [14].

Who enforces the regulatory requirements in any particular jurisdiction of the built infrastructure? With many governments, especially those smaller in geographic size or population, this is handled at a national or federal level. In larger countries this is de-centralized, and the safety infrastructure is handled and enforced based on regional or local levels. This is the case, for example, in the USA, where the fundamental powers to enforce building regulations are delegated to the individual States based on the Tenth Amendment to the U.S. Constitution [15].

The individual or organization that enforces a statute in a particular jurisdiction is referred to as the "Authority Having Jurisdiction" or the "AHJ." Specifically, for the built infrastructure the AHJ is defined as the "organization, office, or individual responsible for enforcing the requirements of a code or standard, or for approving equipment, materials, an installation, or a procedure" [16]. Understandably, with a centralized national top-down enforcement infrastructure this is more straight-forward, but likewise it is less flexible than a de-centralized approach.

Thus, in countries like the U.S., the requirements will differ from one jurisdiction to the next (despite similarities) and can be more stringent in some areas and less in other. Furthermore, some AHJs are enforcing legislative code requirements with civil or criminal penalties, while others are enforcing requirements through other indirect means, such as, for example, insurers who will ultimately raise insurance premiums. Often there are multiple AHJs for a particular application in a specific jurisdiction. This is illustrated in Fig. 3.3, which provides a case study example of the multitude of possible AHJs for a specific building, facility or other applications [17].

3.2.6   Manual Intervention Techniques

The second of the primary approaches to achieve our overall fire safety challenges are the manual intervention techniques that we use to address and mitigate the impact of an unwanted fire or similar emergency. Responding to and addressing an emergency has been an underpinning for civilization since its earliest beginnings.

Throughout the world, a person can pick up a phone and call for help if an emergency occurs (assuming at least a partially stable government), even in remote

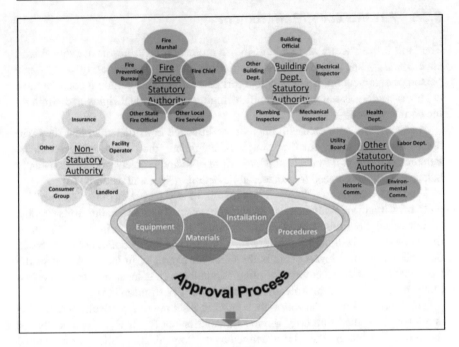

Fig. 3.3 Overview of enforcement infrastructure [17]

wilderness areas. The emergency may be an unwanted fire, medical episode, rescue situation, security incident, or a multitude of other possible scenarios.

3.2.7   Emergency Responders

From a classical perspective, emergency responders are typically the local Fire, EMS (emergency medical service), or law enforcement (Police) department. Sometimes these are the same organization, and they are generally referred to as "emergency 1st responders" since they are expected to be the first to arrive. Because these three traditional emergency response organizations will respond to any and all emergencies (i.e., Fire, EMS, and Police), the concept has evolved today that they are "all-hazards" responders (and not just unwanted fire).

There are multiple responding individuals and organizations required with any particular emergency event. After the initial first responders arrive, they may call additional external responders with different capabilities and special expertise, often referred to as emergency second responders.

Examples include the electric or gas utilities to secure a fire scene, the medical examiner with clearly deceased victims, or social workers for a law enforcement incident. Furthermore, there are additional individuals and organizations in the emergency response chain who serve as emergency receivers and are directly

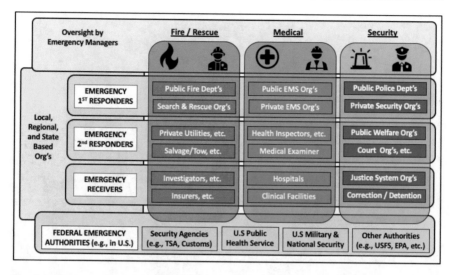

Fig. 3.4 Emergency response framework [18]

responsible for post-event efforts, such as hospitals and correction/detention facilities. They too are part of the overall support and services addressing an emergency event.

These concepts are illustrated in Fig. 3.4 [18]. For large disasters and emergencies that have broad impact, the emergency managers provide overall coordination of the services provided by Fire, EMS, and Police organizations. Much of the resources to deal with an emergency are widely distributed locally to provide rapid deployment, with the expectation that if called they will be available and on-site within minutes.

As emergencies escalate, or for certain specific types of emergencies, federal or national organizations are involved and are the key authority having jurisdiction. Examples include customs/immigration to control national boundaries, forest service for wildland stewardship, environmental protection agencies during an oil spill crisis, or public health services to handle a major health crisis like a pandemic. Each of the entities in the emergency response framework has its own specific subworld, and each can be relatively deep in terms of organization structure, application approach, and other applicable details.

3.2.8 Fire Service

For a fire-resistant infrastructure, the fire service is a key activity in the emergency responder portfolio. For addressing unwanted fires, fire fighters operate as part of a bigger unit, team or entity generally referred to as a "Fire Department" (or "Fire

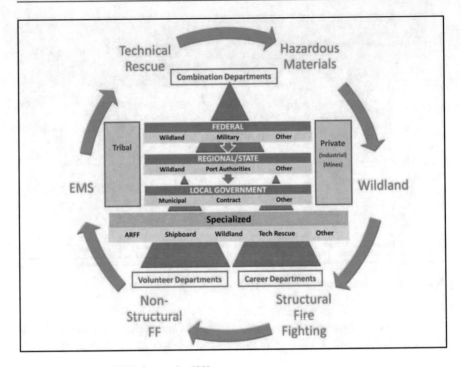

Fig. 3.5 Overview of U.S. fire service [20]

Brigade" in some countries), this being the basic organized group constituting fire service organizations.

In the U.S. these are mostly municipal agencies that are part of the local government, operating side-by-side with law enforcement, EMS, and other services. In 2018, there were almost 30,000 Fire Departments in the U.S., with an estimated 1,115,000 career and volunteer fire fighters. Of the total number of fire fighters, 370,000 (33%) were career fire fighters and 745,000 (67%) were volunteer fire fighters. Of the approximate 30,000 fire departments, 18% were all career or mostly career departments and protected 68% of the US population [19].

Fire departments in the U.S. might therefore be all career, all volunteer, or a mix of both known as combination departments. While these are typically municipal departments with the local government, some are with special entities like Port Authorities, contracted private organizations, federal operations like a military base, or have some other special arrangement. This is especially true with certain fire fighting activities such as wildland, aircraft rescue, and fire fighting (ARFF) at airports, shipboard at ports, etc. An overview of these concepts is provided by Fig. 3.5 [20].

Going forward, we will continue to evolve to deeper levels of understanding as to how we define and utilize the relatively sophisticated levels of what we are seeking to accomplish with built-in fire protection measures and manual intervention techniques.

In summary, it is imperative that we understand the detail of our safety infrastructure as we seek to integrate modern technologies to improve efficiencies and effectiveness. As we proceed to cultivate advances, we will further define and refine all the detailed sub-areas and components, their time-dependency, their interaction, and other critical features and characteristics.

3.3 Fundamentals of Cyber Physical Systems

In its simplest form, Cyber Physical Systems (CPS) are integrations of computation and physical processes [21]. These systems are composed of interacting digital, analog, physical, and human components engineered for function through integrated physics and logic [22].

3.3.1 The Era of Cyber Physical Systems

From a grand perspective, we are in the midst of a major technology revolution. In the last two centuries history has demonstrated major advances in physical technologies and systems in what we know as the "Industrial Revolution." In more recent decades we have been in the digital age and now deeply within the Internet Revolution. As the cyber-oriented hardware, software and systems of today further mesh and combine with our physical world, we find ourselves emerging into an entire new era, that of "Cyber Physical Systems," a.k.a. "CPS" [23].

CPS is evolving as the underpinning for everything we hear today as "smart" in the mainstream media and in street slang. We hear it everywhere: Smart Homes, Smart Buildings, Smart Grid, Smart Cities, Smart Transportation, Smart Healthcare, Smart Infrastructure, Smart Agriculture, Smart Warfare, and so on. CPS is the crossroads of all things both electronic and physical, and it is transforming the way we interact with engineered systems, in the same way the internet has already transformed how we handle information. Swept into this movement are advances and concepts that are blazing new pathways such as the Internet of Things (IoT), machine learning, artificial intelligence, machine-to-machine communication, augmented reality, analytical software, and so on.

3.3.2 Data, Data, in a Sea of Data

Data, data, and more data: in the modern world we are in a sea of data. It is sweeping all corners of our civilization, riding on the wave of the Internet of Things (IoT), and inundating us with massive amounts that only keep increasing, flooding us from all directions.

Data is the key to the future; it is the bloodline of the information age. But if data is key, analyzing data is an even bigger key. How we collect, analyze, and process it is really what produces genuine, highly impactful value.

A useful analogy is to compare "data" with "oil." In 2006 this oft-quoted concept found its way into our mainstream dialogue through the words of Clive Humby, who stated: "Data is the new oil. It's valuable, but if unrefined it cannot really be used. It has to be changed into gas, plastic, chemicals, etc., to create a valuable entity that drives profitable activity; so must data be broken down, analyzed for it to have value." [24].

This analogy is convenient and informative. The existing commodity of oil and the new commodity of data have striking parallels. Today's world has been built of the commodity of oil as a fundamental building block. It's not the barrels of raw crude that we see in everyday life, but instead endless things oil is refined into, with countless everyday products and processes all around us, e.g., fuels, building products, highway pavement, pharmaceuticals, space-age products, etc.

Oil, and now data, are fundamental building blocks of modern civilization. Both are essential and require safe and proper handling from cradle to grave, including identification, collection, containment, management, transport, use, disposal, and destruction. From the perspective of the safety infrastructure, the safe protection of these commodities is also paramount. Just as we provide extensive fire protection and emergency response for oil pipelines, tankers, refineries etc., so to we are awakening to the new frontier of data safety, i.e., cybersecurity, etc.

3.3.3 The Three Realms of Cyber Physical Systems

For fire resiliency, Cyber Physical Systems are understood to be the fusion of emerging sensor and computing technologies with building control systems, fire fighting equipment, and other safety measures. Cyber Physical Systems are based on the following fundamental concept: [25].

1. Collect data;
2. Process data; and
3. Deliver data.

This three-fold structure is a core principle for CPS, and it holds universal merit despite its stark simplicity. This framework supports the (1) collecting and aggregation of large quantities of data from a range of sources; (2) the processing, analysis, and predictions using that data; and (3) dissemination of the useable results for targeted decision-making. Cyber Physical Systems will revolutionize fire protection and emergency response, with significant potential to positively impact infrastructure improvements.

Figure 3.6 illustrates this core principle and three-fold structure [26]. Despite the technical and implementation barriers, this framework will transform the safety infrastructure from the current state of information-limited decision-making to a

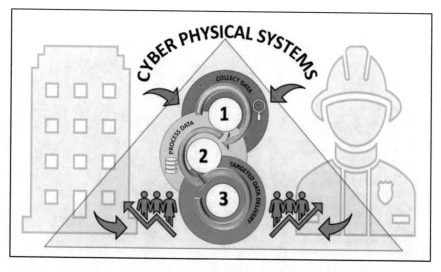

Fig. 3.6 Three realms of cyber physical systems [26]

sensor-rich environment with ubiquitous data collection, analysis, and communication, ultimately leading to data-driven and scientifically informed decision-making

The quest to gather and collect data is becoming less and less of a formidable task in today's sensor-rich world. The rapid escalation of data collection is riding the wave of proliferation that we see today in the "internet of things" (IoT) or as it is sometimes called, the "internet of everything." Sensors and data collection devices of all kinds have become both more powerful and more affordable, allowing them to be embedded everywhere. For fire protection and emergency response applications, data is captured through four primary pathways: [27].

- Community-based;
- Occupant;
- Building, and.
- Emergency responder.

From a data management perspective, this involves a confluence of different data-sources that might be needed during an emergency. An example is a fireground incident commander accessing, processing and utilizing data that is: (1) informing the weather wind activity (community-based), (2) occupant cell-phone location transmissions (occupant), (3) position of building elevators (building), and (4) location of fire fighters using remote locator technology (emergency responder). The integration of this data into targeted deliverables remains a noteworthy challenge.

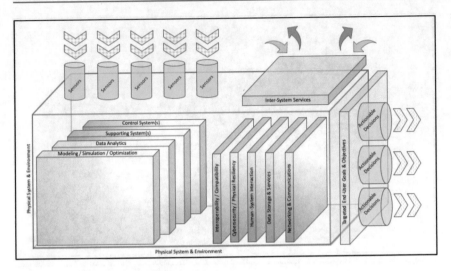

Fig. 3.7 Example CPS components [29]

3.3.4   Architecture and Design of Cyber Physical Systems

Cyber Physical Systems can be distinguished from stand-alone devices and embedded systems based on their inherent use of collective networks of interacting physical inputs and outputs. This is closely related to sensor networks and robotics, though with integral computational intelligence mechanisms [28].

There multiple concepts, characteristics and sub-components found in a Cyber Physical System [29]. Figure 3.7 provides a case study sample of the combination of various designs and architectures that can be seen with a typical CPS. The various elements in Fig. 3.7 are described in further detail in Table 3.2.

Today, we have the growing ability to significantly enhance the fire resiliency of the modern built infrastructure by exploiting new opportunities in technology development. We are entering a new age where the digital world meets the physical world, with technological advances that at times are dramatic leaps forward. We are only beginning to understand the potential value-added of game-changing concepts like internet of things (IoT), artificial intelligence, machine learning, and so on.

As these advances proliferate, through applied research we will be challenged to effectively manage their evolution. With every step forward also comes the possibilities of finding new ways to step backwards. New threats will emerge, and some, like cybersecurity, represent not simply new singular hazards but whole new frontiers that will require enormous resources to address. It is a new era, but nevertheless, an exciting time for civilization.

Table 3.2 Details of CPS component case study [29]

Fundamental characteristics

• Physical System and Environment: The engineered physical system that operates inherently with the CPS in the physical environment

• Sensors: Devices that acquire data from the physical environment and transmit it to storage, measurement and/or control devices

• Targeted End-User Goals and Objectives (with Actional Decisions): Targeted decision-making actions based on end-user goals and objectives that are specific, measurable, action-oriented, realistic, and timely

Architectural layers

• Control System(s): A key CPS sub-system or distributed sub-systems that acquires data from sensors and perform local processing and manage actions to produce a prescribed state of the physical system in the physical environment

• Supporting System(s): CPS sub-systems that support the control system(s)

• Data Analytics: The process of coordinating and processing data for intelligent decision making

• Modelling/Simulation/Optimization: Developing and maintaining dynamic performance-based computational models that utilize the data analytics in support of targeted decision-making

Cross-cutting functions

• Interoperability/Compatibility: The capability of multiple networks or their components to readily interoperate without user intervention, or as a minimum to functionally co-exist (i.e., to be compatible if not interoperable)

• Cybersecurity/Physical Resiliency: The processes and protective measures that provide cybersecurity (e.g., virus protection) and physical safety (e.g., electrical surge) of the CPS

• Human-System Interaction: The interfaces that facilitate interaction between humans with the CPS

• Data Storage and Services: The capability and resources to store and manage data

• Networking and Communications: A means to efficiently, effectively and securely transport data and information across the system architecture

• Inter-system services: Facilitates interactions with other CPS control systems

3.4   Systems Integration

The natural evolution and proliferation of Cyber Physical Systems is based on the advantageous capabilities of new emerging technologies to address real-world problems. Cyber Physical Systems are providing empowerment to provide realistic solutions for realistic problems and challenges.

3.4.1 All-Hazards Approach

Today's safety infrastructure represented by fire protection and emergency responder professionals is evolving into an all-hazards focus. This is based on societal expectations that safety professionals will address safety issues well beyond simply unwanted fire, with emergency responders expanding their response protocols.

The spectrum of hazards that are being addressed by responders is already expansive and ever widening. These fire protection and emergency event applications provide a wide spectrum of problems and challenges that require attention and resolution. All of this ultimately serves as a rich backdrop for the integration with Cyber Physical Systems.

3.4.2 Unified Functionality

In its simplest terms, system integration is the combining of sub-systems to deliver unified functionality in support of real-world applications. For a fire-resistant infrastructure, this is fundamentally the confluence of the sub-systems that ultimately address and solve real-world problems, in this case for needs of the fire protection and emergency response communities.

Systems integration has been an approach that has been successfully applied in other arenas, functioning on the strength of combined and unified systems [30]. These applications often are found addressing expansive societal problems, which have a natural gravitational pull of systems integration [31].

Facilitating divergent sub-systems is often not straight-forward and presents unique challenges with data exchange protocols and other factors. Data exchange barriers include divergent data processing approaches, unwillingness to share data, intellectual property questions, data confidentiality, cost of integration, and so on [32]. Various approaches have evolved with their own methodologies used for integration, such as: vertical integration (that integrates based on distinct silos, bins or buckets); horizontal integration that involves a dedicated communication or networking sub-system; and other approaches that enable common data handling formats [33].

Real-world examples are useful to provide a backdrop of the possibilities for systems integration for fire-resistant infrastructures. An example are the multiple software packages that might be utilized by a typical municipal fire department. The duties and tasks for such an organization are relatively extensive, and how should all respective systems and sub-systems best co-exist?

Data development and records management for the fire service is addressed in a standardized manner in NFPA 950, Standard for Data Development and Exchange for the Fire Service [34]. In particular, Chap. 8 of NFPA 850 outlines key data related issues important for a fire service organization. This is tabulated in Table 3.3. This summary symbolizes the relatively extensive software needs of a modern fire department. It is not meant to be all-inclusive but instead represent the most common software needs and activities.

Table 3.3 Examples of fire department software [34]

Software category	Specific application
A Record collection and record management	• Emergency incident documentation • Fire investigation documentation • Emergency incident record quality assurance
B Personnel scheduling, safety, and administration	• Scheduling and rostering • Health, wellness, and medical records • Occupational exposure and injury tracking • Training, certification and qualification • Human resources and hiring • Emergency service billing
C Equipment and facilities	• Maintenance • Equipment inventory • Apparatus and equipment inspection and testing • Hydrant and water supply management
D Incident command and situational awareness	• Incident command software • Responder notification/alerting • Pre-incident planning
E Analytical and decision	• Geographic information systems (GIS) mapping • Business intelligence • Analytical
F Risk reduction	• Inspection records • Mobile inspection • Plans review workflow management • Community risk reduction management • Mobile integrated healthcare/community paramedicine • Public education/outreach • Permit and license management
G Imaging	• Vehicle-mounted imaging • Body/PPE-mounted imaging • Deployable device-mounted imaging • Augmented reality/virtual reality
H Sensor	• Apparatus-mounted sensor • Body-worn sensor • Equipment-mounted sensor
I Community risk awareness	• Hazardous materials emergency awareness • Inspection, testing, and maintenance notification • Hazardous materials pipeline location • Railroad carrier inventory • Community risk assessment • Citizen notification, AED location and emergency alerting
J Digital alert warning systems for emergency response vehicles	• Responder-to-vehicle citizen motorist alerts • Responder-to- responder apparatus alerts
K Building information modeling	

3.4.3   Time Critical Events

Systems integration for emergency events has the particular challenge of time-criticality. Pre-event and post-event activities have the luxury of relaxed time frames to perform and deliver, but not so for the during-event time frame. Secure and robust networks are needed to process and handle data that are typically generated under different oversight.

Near real-time data protocols must be reliable and secure to deliver functionality during an emergency. Protocols involving well-defined priorities are needed to assure that the proper information is delivered to the appropriate emergency responder, providing them with what they need to make targeted decisions, and not flooding them with superfluous information that blinds them during critical time-sensitive periods of an emergency [35].

3.4.4   Communication Pathways

For systems integration to be effective, it must have robust connections and pathways that can fully handle the transmission and exchange of data at all times. The ability to transmit large amounts of data in short time frames is often taken for granted, since networks are typically well-sized to process data exchange under normal or anticipated peak conditions.

But are those peak conditions realistic for all anticipated emergency scenarios? Will fire protection measures and emergency responders have the communication pathways they need to be effective?

A vivid example can be addressed by asking the question of what happens when, for example, a major airliner ditches in the river of a crowded metropolitan center on a clear, sunny day? Such was the case in January 2009 when US Airways Flight 1549 ditched in the Hudson River in New York City, witnessed by many [36].

Understandably the public communication systems were immediately over-loaded. But of course, the emergency response network still needs to fully function, and their radio communications and other communication pathways need to be resilient and maintain full functionality in the face of any and all events.

Key steps have been taken to address these critical pathways. In the USA one of the most prominent is the FirstNet initiative, formally known as the First Responder Network Authority under the National Telecommunications and Information Administration (NTIA) with the U.S. Department of Commerce. FirstNet is an independent authority established by the United States Congress in 2012 with a mission to develop, build, and operate the nationwide broadband network that equips first responders to save lives and protect U.S. communities [37].

FirstNet uses a significant portion of the radio spectrum, dedicated to emergency responders, which is valuable real estate in the cyber world. A helpful analogy to assist with this visualization is that it is similar to a dedicated 12 lane highway carved through a busy urban traffic center. During non-emergency conditions it is leased out for normal everyday use in a public/private partnership with a telecommunication partner, which via a public auction is AT&T.

If and when an emergency occurs, all other users of this dedicated radio spectrum move aside and off the proverbial highway, with dedicated priority given to emergency responders. FirstNet provides a nationwide, interoperable, broadband network for the entire emergency response community. The FirstNet story is an example of far-reaching efforts well beyond smaller independent, innovative activities, with sweeping positive implications to the overall built infrastructure [38].

Today the worlds of fire protection and emergency response are becoming more integrated with the physical world. Our body of knowledge is rapidly growing, and we are finding novel ways to implement multiple new approaches that leverage the latest technology. We are now in a new era where we are collecting data globally, processing the information centrally, and distributing the results locally.

3.5 Case Study Scenarios

The consideration of a fire-resistant infrastructure implies the consideration of an emergency of some kind, either pre-event, during-event or post-event. The type of emergency is dictated by the word "fire," and this operative word leads to consideration of numerous associated actions, activities, and applications.

This is partially addressed by the three traditional emergency response organizations (i.e., Fire, EMS, and Law Enforcement) who today have evolved into "all-hazards" responders. In our modern world we know that if things "go bad" and help is needed, you can pick up a phone and call, and help is on the way. Just like the immune system of any biological creature, our first responders stand at the ready and immediately respond to address the threat or hazard.

3.5.1 Case Study Emergency Event Scenarios

The worlds of fire protection and emergency response, and the world of high-tech computer and data scientists, are quite different. They have different focus points and unique appreciation for the genuine needs of end-user applications and how to effectively and efficiently address these applications. The new era of Cyber Physical Systems brings us to a crossroads that are blending these different worlds.

As we address a fire-resistant infrastructure for today's world, and the emergencies associated with it, a fair question to ask is: "what emergencies are we talking about?" and "what do emergency responders address each day, month, year?" For example, if our single-family residential home in suburbia has a kitchen fire, we call the fire department and they address this unwanted fire. But what else do fire fighters do?

For many these are fair questions, and it is helpful to reflect upon the types of emergency events we are trying to address. Conveniently, this was previously addressed by the "Research Roadmap for Smart Fire Fighting," which provided a set

Table 3.4 Case study emergency event scenarios [40]

	Emergency scenario	Scenario historical basis
1	*Wildland Urban Interface Fire* (with retirement community evacuation)	Waldo Canyon Fire, Jun/2012 in CO (w/2 civilian fatalities and 346 buildings destroyed) and Yarnell Hill Fire, Jun/2013 in AZ (w/19 FF LODDs and 129 buildings destroyed)
2	*Residential Structure Fire* (wind driven fire)	Marsh Overlook Structure Fire, Apr/2007 in Prince William County VA (w/1 FF LODD) and Houston TX Residential Fire, Apr/2009 (w/2 FF LODDs) and Pittsburgh PA House Fire, Feb/1995 (w/3 FF LODD)
3	*Hi-Rise Apartment Fire* (wind driven fire)	Vandelia Ave 10-Story Apartment Fire, Dec/1998 in NYC (w/3 FF LODDs)
4	*Vehicle Crash* (ICEV and EV with entrapment)	Per NFPA statistics of U.S., with 17 vehicle fire per hour and 287,000 vehicle fires per year, throughout U.S.
5	*Train Derailment* (with fire and toxic hazmat event)	Lac-Mégantic Train Derailment, Jun/2012 in Quebec (w/47 civilian fatalities and 30 buildings destroyed
6	*Hi-Challenge Warehouse* (with extreme height-area)	Food Product Warehouse, Dec/2007 in Hemingway SC (w/2-day fire and warehouse destroyed)
7	*Night Club Code Compliance* (with enforcement)	Happy Land Social Club Fire, Mar/1990 in NYC (w/87 civilian fatalities), and Station Nightclub Fire, Feb/2003 in West Warwick RI (w/100 civilian fatalities)
8	*Tornado* (with post-event action)	Joplin MO Tornado, May/2011 (w/158 civilian fatalities and ~ $2.8 billion loss) and Moore OK Tornado, May/2013 (w/25 civilian fatalities and ~ $2.0 billion loss)
9	*Terrorist Bombing* (large scale EMS event)	Boston Marathon Bombing, Apr/2013 in Boston (w/3 civilian fatalities, 260+ injuries)
10	*Elevator Rescue* (metro city power failure)	Sub-station electrical fire causing widespread city center power failure in 2012, Boston MA (w/hundreds of elevator rescue calls)

EV electric vehicle, *FF LODD* fire fighter line of duty death, *ICEV* internal combustion engine vehicle

of ten key case study scenarios to provoke dialogue and provide the understanding and appreciation on the types of emergencies that should, as a minimum, be considered [39].

These case study scenarios are illustrated Table 3.4, with the intent of enabling the vision of Cyber Physical Systems to support the needs of fire protection and emergency responder activities [40]. The case study scenarios are based on actual historical events with similar basic characteristics, to facilitate and stimulate consideration of novel Cyber Physical System applications to solve real-world problems.

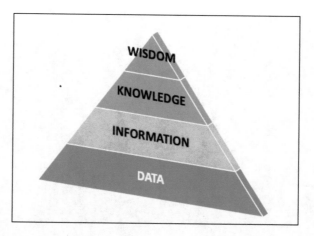

Fig. 3.8 The DIKW hierarchy [42]

These are oriented around low-probability high-severity events, with the intent of maximizing consideration of when "things go bad" and resources become fully stretched. Thankfully, these are relatively rare events, but they exemplify the criticality of this overall exercise. Here, proper execution and delivery literally equate to the difference between life and death.

3.5.2 The DIKW Hierarchy

A convenient conceptual tool to explore and learn from case study scenarios is the DIKW Hierarchy (a.k.a., DIKW as Data-Information-Knowledge-Wisdom). This conceptual tool has been a guidepost for decades in the computer science field and has become referred to in a number of ways such as the "Knowledge Hierarchy," the "Information Hierarchy," or the "Knowledge Pyramid." [41] This provides a useful framework to demonstrate the harvesting of data for targeted decision-making.

This basic elements of the DIKW Hierarchy are illustrated Fig. 3.8, and it is shown in its most elemental form without the variations and finer details that have evolved through the years [42]. This provides a structure for the harvesting of data as a raw ingredient, that is given meaning as information, that is then provided context as knowledge, that finally in turn delivers wisdom for targeted actions.

Thus, we see a process of collecting and processing data, cognitively transforming it into information, further transitioning into knowledge that allows judgment, and finally evolving into wisdom that supports shared understanding and actionable decision-making. Applications of the DIKW Hierarchy have been widely used to apply data in ways that result in deeply informed actionable strategies that move beyond traditional tribal-oriented decision-making [43].

Applying the DIKW Hierarchy for events and related activities that are the overall focus of a fire-resistant infrastructure is useful. Table 3.5 provides ten

Table 3.5 Examples of the DIKW hierarchy for fire service applications

	Scenarios	Data Raw elements	Information Providing meaning	Knowledge Providing context	Wisdom Applied actions
1	Traditional computer science example	Traffic light	The color "RED" is RGB 255.1.1	The east traffic light at first and D streets has changed to 'red'	The traffic light before me has turned "red", indicating stop	Apply the brakes and stop the vehicle
2	Structural fire	Line fire fighter hose and nozzle	Inadequate hose line performance	Water is not reaching the seat of the fire	Need more pressure	Send signal for more pressure from pump operator
3	Structural fire	Structural fire rapid intervention (mayday, lost FF needs help)	Mayday alert signal with GPS (global positioning system) location	Fire fighter needs assistance	Immediate intervention is needed based on fire condition and fire fighter location	RIT (rapid intervention team) deployed on side B near second hallway door
4	Structural fire	Incident command of structure fire	Quadcopter visual on side C which has limited accessibility	Fire appears on multiple floors	Fire is spreading beyond resources on the fire ground	Call for additional resources
5	Vehicle incident	Vehicle extrication technical rescue	Telematics transmission of critical vehicle data involved in crash	Vehicle is an electric vehicle with a fully charged large format battery	Extrication will also involve potential shock hazard	Use specific work-around to avoid battery and associated cabling
6	Large scale evacuation	Wildland urban Interface (WUI) fire evacuation	Weather wind directional data	Wind shifts sharply coming from the east	Fire will change direction, creating possible entrapment	Evacuate immediately to the south
7	Exposure concerns	Fireground contamination and field Decon'	Truck fire involves hazardous substances	Vehicle placards indicate corrosive materials	Fire fighters are at risk of exposures to contamination	Engage in immediate field decon' after exposure
8	Exposure concerns	Long term health and wellness exposure tracker	Building permits indicate basement fire had transformer with PCBs	This fire included PCB chemical contaminants in smoke	Fire fighters experienced direct exposure to PCBs	Track this specific field exposure for future health and wellness
9	Pre- and post-event	Prevention and enforcement	Noise complaint data on college campus	Noise complaint data is unusually high in certain residential areas	Noise complaints match incidence of fire and higher fire risk	Pre-emptively conduct fire inspections
10	Pre- and post-event	Fire investigation	Door is damaged after a fire	Specific burn pattern on the door on both sides	Doorway was in a certain position in the fire primary flow path	Door was opened prior to event, corroborating other "digital dust" from other sources

examples, with the first being the classical computer science example of distilling the raw data of a red traffic signal from data to information to knowledge to wisdom, ultimately supporting the action to apply the brakes and stop the vehicle. The assortment of topics addressed are intended to exemplify this process of capturing and processing data into meaningful actions, which is the heartbeat of Cyber Physical Systems.

3.5.3 Innovative Applications

Examples of specific technology development to support a fire-resistant infrastructure continue to evolve all around us. Often, these are simply applying known or proven technological advances that are blossoming in other arenas and adapting them for specific targeted uses in fire protection and emergency response.

A common example is the use of small unmanned aerial vehicles (UAVs), a.k.a., drones. They are finding wide use in the emergency response arena, with perhaps the most common use to provide visual data to support fireground incident command [44].

However, with UAVs researchers and others in the professional community are actively considering a multitude of other novel applications. Some of which are manifesting as prototypes or already finding their way into full use, such as:

- HazMat Event: Deployment of environmental sensors on tethered drones around a hazardous materials event, powered indefinitely and with continual streaming data.
- High-Rise Building Fire: Deployment of an external hose line for a high-rise structural fire, such as those involving combustible exterior wall assemblies.
- Medical: EMS defibrillators flown to remote medical events, with voice communications to guide civilian use.
- Structural Fire Fighting: Indoor swarming to rapidly chart indoor structures involved in a fire (e.g., warehouse or commercial/mercantile/industrial property).
- Technical Rescue: Rope deployment during swift water technical rescue or ice rescue.
- Wildland: Wilderness grid-pattern search over large areas in support of remote rescue.

Other novel applications are being widely explored. Most commonly are the efforts to improve situational awareness, which is an obvious need during an emergency for emergency responders and civilians alike. On a fire ground, the situation and conditions are continually changing and highly dynamic.

3.5.4 Next Generation Cyber Fire Fighter

During a fire the safe operation of fire fighters and rescue of victims is a challenge. As an example of efforts to address these challenges, one on-going research project at the University of New Mexico in collaboration with the Fire Protection Research Foundation and supported by the National Science Foundation is addressing these issues by exploring novel use cases for sensors to improve the safety of firefighters on the fireground [45]. The effort seeks to make fundamental technical and algorithmic advances using artificial intelligence and machine learning to provide direct benefits to fire fighters and incident commanders during a fire. This includes a focus on the following five key topic areas: [46].

1. *Examples of Fire Department Software.* Establish a practical personal area network (PAN) using a PPE sensor network and a local area network (LAN) involving a fireground local area data communication system, with a mesh structure based in Wi-Fi communications but extended to other communication methods.
2. *Fireground Sound Discrimination.* Capture and identify critical fireground sounds (e.g., PASS device or Mayday) and discriminate and filter these sounds from other fireground noise using algorithms supported by machine learning to implement specific automatic fireground actions.
3. *Fire Fighter Exhaustion Prediction.* Capture, identify, and process speech features through a central computer to determine the level of stress and exhaustion of a fire fighter. Combined with respiration estimation procedures, this data can be used for actionable measures, such as estimation of remaining quantities of available SCBA air, or as a supplement to other physiological indicators.
4. *Navigational Image Search Techniques.* Adapt algorithms that utilize imaging techniques and are supported by machine learning for fire fighter locator navigation, that ultimately benefit key fireground activities dependent on locator technology such as search and rescue or use by RIT (rapid intervention teams).
5. *Thermal Imaging Human/Object/Event Recognition.* Algorithms identify specific target entities using thermal imaging such as a fire victim or downed fire fighter. With support from machine learning, this can be transformed into knowledge-based actions for fire fighters.

3.5.5 Las Vegas Active Shooter Case Study

Large scale events provide important lessons learned for future preparations. These events test the limits of available resources and identify and magnify the weaknesses and knowledge gaps. Similarly, they provide clear hindsight opportunities of the potential benefits of Cyber Physical Systems.

A useful and powerful case study scenario that exemplifies a wide range of potential positive attributes of Cyber Physical Systems, is the Las Vegas Shooting at the Mandalay Bay hotel in October 2017. This mass casualty event resulted in 58 fatalities and 413 injuries. While not a fire, this was a classic "all-hazards" event that includes deep involvement of all branches of the emergency response spectrum and other parts of the safety infrastructure [47].

A key aspect of this case study is the flow of data among all those involved as the event was unfolding. This included the critical task used by emergency responders to "size-up" the event, which is a term used by responders as they define the threat and collectively gather as much data on the event as possible to inform their decision-making. During an emergency, the time frames for action become extremely condensed. It is imperative to have access to the fundamental details of the event, including all dynamically changing variables, to guide targeted decisions that have positive impact.

The complexities of gathering, processing, and delivering the data with near real-time exchange of changing data sets is a significant challenge from a data science perspective. Key response agencies (fire, EMS, and law enforcement) each have starkly different roles and objectives that sometimes overlap but regardless must always work in unison. Law enforcement mitigates the threat and processes the crime scene, while the fire service and EMS implement victim rescue and triage. In the understandable chaos of the event, i.e., under live gunfire and immediately after with the threat unclear, performing the assigned job performance tasks was a monumental effort [48].

At the Las Vegas active shooter event, knowing as early as possible the number and location of the shooter or shooters was obviously critical. Law enforcement worked as rapidly as possible to secure the setting, but what additional valuable data from the built infrastructure could have helped them?

For example, building systems information could have confirmed broken high-rise windows and their location though anomalies such as the impact on building air handling systems. Victim medical data compiled by hospitals provided important information on the nature of injuries that was valuable to on-site responders, but access to this data struggled with data processing barriers and non-technical issues like confidentiality protocols. Addressing these barriers requires understanding and defining the changing variables and using algorithms to aid the decision process.

It is a tall order to obtain, process, and deliver data for targeted decision-making to separate and disparate emergency responders and agencies. This is a challenge that Cyber Physical Systems are well suited to address. This knowledge exchange must be done rapidly, reliably, and securely. When decisions make the difference between life and death, it is imperative to get the necessary information to the right emergency responder, the way they need it, when they need it, and nothing more, i.e., without paralyzing them with information overload.

Past events provide important lessons as we explore new opportunities for integrating modern technologies. The case studies provided herein exemplify the complex fabric of dynamically changing details involved with emergencies. Establishing a more fire-resistant infrastructure is a challenge, but it is one that Cyber Physical Systems and related technologies will help us achieve.

3.6   Cybersecurity: The New Frontier

In the face of emerging technologies with great promise, we find the present safety infrastructure. It is well-established with a deep history, but it has been built on painful lessons of the past. As we implement new technology, we most certainly do not want to "re-learn" these painful historical lessons.

3.6.1   Need for Resiliency

The detailed and extensive requirements in the model codes and standards and other regulatory documents provide assurance that fire protection measures of the built infrastructure are reliable, resilient, and effective. Man-made and natural disasters have continually tested the strength and durability of these measures. When the worst calamity hits, such as, for example, an earthquake, the fire protection measures cannot get knocked out and must continue to function. Passive and active fire protection systems are hardened to assure they will always function as intended under any scenario of duress.

But how well are we guarding the many back doors into the cyber kingdoms of fire protection systems? What steps are being taken to prevent cyber hackers from incapacitating systems remotely? Can false alarms be sent at random times simply to tax resources? Can multiple systems be activated to serve as a distraction to allow another nefarious act to proceed unchecked? Can a terrorist disable all safety and security systems in an active shooter event? Are we taking the same defense-in-depth approaches with cybersecurity as with other aspects of fire protection systems to guarantee operability?

3.6.2   The Interconnectedness Risk

Today, built-in fire protection measures are everywhere, including traditional systems like fire alarm, sprinklers, special suppression, and so on. These systems are becoming more and more interconnected and integrated in the era of Cyber Physical Systems. But this interconnectedness also raises a monumental concern that fire protection systems are exposed as the soft underbelly of the cyber safety infrastructure. The issues of cybersecurity represent an entire new frontier for the safety infrastructure [49].

Cybersecurity breaches are not uncommon. In one major case in November 2013, sophisticated computer hackers found their way into the computer systems of a major retailer and made off with information on more than 40 million credit cards and other critical personal data of customers. The resulting financial losses for the retailer soared into the tens of millions of dollars, while the lost time and dollars for consumers affected by the breach was significant [50].

It is noteworthy that the hackers did not directly attack the retailer's financial or personal data systems. Instead, they broke into the networks of a Pennsylvania based HVAC contractor with direct links to the retailer's system in order to remotely monitor and control the retailer's local building systems. Over a period of weeks, they uploaded trojan software into the computer systems of the retailer in multiple steps and ultimately became implanted on every cash register throughout their retail empire, directing skimming credit card information. This specific case was initially breached through a local HVAC contractor, and the concern for similar potential upheaval likewise exists with all safety infrastructure measures.

The interconnectedness of today's world is growing at an impressive rate. Fire protection systems are increasingly exposed to the public-facing internet, based on their design to work with the Internet of Things, building control systems, and other networks. Cyber vulnerabilities are a real threat, and attacks on fire protection measures have the potential to have significant consequences. This is a new domain requiring dedicated attention [51].

3.6.3 Other Threats

Organizations directly involved with manual intervention techniques are equally vulnerable, like fire departments and other emergency response organizations. In addition to destructive or malicious hacking and cases of espionage on their extensive software systems, some have been held for ransom or subjected to extortion. Many emergency response agencies have already increased their focus on cyber threats on their systems, with law enforcement, the fire service, and EMS all recognizing that they are direct targets [52].

The Internet of Things and the interconnectedness of fire protection systems are bright lights on tomorrow's horizon and hold enormous promise. But just as we do for other perceived threats and hazards, we must prepare for the worst and make sure our built-in fire protection measures and emergency responder services are never intentionally compromised.

3.7 Future Directions

The world is rapidly evolving with positive technological advances, fueling the emergence of Cyber Physical Systems. These are yielding potentially great advantages for the safety infrastructure, and shine with promising advances for built-in fire protection measures and manual intervention techniques by emergency responders and others.

3.7.1 The Lexicon Gap

As we march forward, one interesting twist is the lack of a universal lexicon due to the relatively rapid evolution of the digital era. The data and computer science worlds have not yet achieved a mature and universally recognized glossary of key terms, which continue to rapidly evolve in recent years. Searches in the literature for oft-used terms reveals multiple variations, much of which are generally credible, though each lacking the inherent gravitational pull that supports their universal recognition. Examples of such terms are: Application Programming Interface (API); Artificial Intelligence (AI); Augmented Intelligence; Big Data; Data Mining; Data Science; Deep Learning; and Machine Learning [53].

Fortunately, when using these terms we generally understand and accept the intended meaning, despite multiple slight and not-so-slight variations. But going forward, the safety infrastructure will need to settle on universally recognized terminology, at least for its specific applications. This is a knowledge gap for us all.

3.7.2 Smart Fire Fighting Research Priorities

Other knowledge gaps will become more apparent as technology continues to advance, shining a light on attractive new opportunities (e.g., artificial intelligence supported algorithms) as well as revealing new and unexpected threats (e.g., cybersecurity). Filling these knowledge gaps will require research and clear research agendas. One effort to address the pathway forward was the "Research Roadmap for Smart Fire Fighting." Conveniently we can use the summary observations from this effort as a preliminary guidepost [54].

Figure 3.9 illustrates the Roadmap's research priorities [54]. These are based on key elements consolidated from multiple sources

The recommendations address four primary categories: Standardization; Developmental Gaps; Broad Conceptual Gaps; and Solution Approaches. Standardization is recognized as a cross-cutting gap affecting all other areas. Gaps that are developmental or broadly conceptual are candidates for traditional top-down research efforts. Solution approaches are different in that they alternatively coordinate from a bottom-up crowd-sourcing perspective rather than traditional top-down research. They have the distinct advantage of identifying and promoting novel and impactful technological advances that already exist elsewhere or are emerging for other applications.

3.7.3 Non-Technical Barriers

An important footnote as we press forward, reaping the benefits of Cyber Physical Systems and technology, is that this is much more than only advancing science

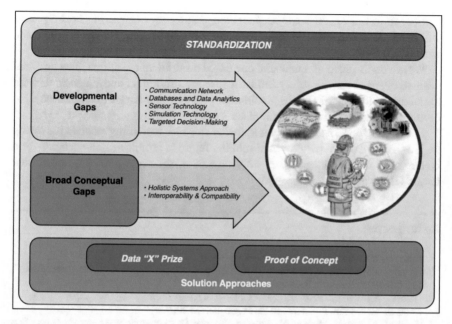

Fig. 3.9 Research priorities for smart fire fighting [54]

and technology. Arguably the most formidable obstacles to full implementation are rooted in issues that are legal, political, ethical, and so on.

In the parallel universe reflecting that "data is the new oil," we can understand that the construction of an oil pipeline from Canada to Mexico is not overly challenging from an engineering standpoint, but instead struggles with intense political and legal hurdles. Similarly the barriers that confront data exchange are often raising supremely important questions on issues such as: confidentially (e.g., health data); legality (e.g., intellectual property rights of who "owns" the data); labor (e.g., displaced workforce due to advances); marketplace (e.g., companies put out of business), and so on.

3.7.4 Final Thoughts

Civilization fully recognizes that emergencies are a real part of our world. The safety infrastructure is essential to a genuinely resilient world, displaying durability and the ability to bounce-back after taking a hit. Almost all data becomes important during an emergency. This is one dimension that sets the safety infrastructure apart from most other applications.

Enabling Cyber Physical Systems and advancing technology and will involve experts from many disciplines. As technology advances, as networks get faster, as sensor are more prevalent, as computers become more powerful, and other barriers to advancement wane, applications that harvest Cyber Physical Systems will continue to expand. The question is not if it will happen, but rather when, and what it will look like.

We are on the cusp of a new era, one based on the data revolution. The potential positive changes for the safety infrastructure are profound., and we look forward to maximize the use of data as a commodity in new and hard-to-imagine ways, to ultimately make our world a better place.

Integrating modern technologies to realize fire-resistant infrastructures is a reality we live today, and the path forward into the future is promising. The benefits for a better world are within grasp. A world absent of the scourge of fire and related hazards is a worthy and noble goal.

References

1. Evarts B, "Fire Loss in the United States During 2018", NFPA, October 2019.
2. Zhuang J, Payyappalli V, Behrendt A, Lukasiewicz K, "Total Cost of Fire in the United States", Fire Protection Research Foundation, October 2017.
3. Grant C, "Connection Risk", Research Column in NFPA Journal, May/June 2019.
4. Grant C, "Smart Fire Fighting", SFPE Magazine, Q3 2019.
5. Grant C, Hamins A, Bryner N, Jones A, Koepke G, "Research Roadmap for Smart Fire Fighting", NIST Special Technical Publication 1191, Page 9, May 2015.
6. Bryner N, Grant C, Hamins A, Jones A, "Realizing the Vision of Smart Fire Fighting", IEEE Potentials Magazine, Volume: 34, Issue 1, January/February 2015.
7. Grant C, "Key Data, Crucial Seconds", Research Column in NFPA Journal, May/June2018.
8. NFPA 550, Guide for Fire Safety Concepts Tree, National Fire Protection Association, Quincy MA, 2017 edition, page 550-8.
9. Grant C, Hamins A, Bryner N, Jones A, Koepke G, "Research Roadmap for Smart Fire Fighting", NIST Special Technical Publication 1191, May 2015, pages 129 & 139]
10. Janssens M. L., "Basics of Fire Containment", Section 2, Chapter 5, Fire Protection Handbook, Volume I, 20th Edition, 2008, page 2-60
11. Watts J. M., "Systems Approach to Fire-Safe Building Design", Section 1, Chapter 9, Fire Protection Handbook, Volume I, 20th Edition, 2008, page 1-159
12. "Core Competencies", Recommended Minimum Technical Competencies for the Practice of Fire Protection Engineering, SFPE: Society of Fire Protection Engineering, Bethesda MD, website: https://www.sfpe.org/page/CompetenciesforFPE?&hhsearchterms=%22core+and+competencies%22, cited: 19/Feb/2020.
13. Subcommittee on Professional Competency and Credentialing, "Recommended Minimum Technical Core Competencies for the Practice of Fire Protection Engineering", SFPE: Society of Fire Protection Engineering, Bethesda MD, 19/December/2018, Table 1, page 9, website: https://www.sfpe.org/page/CompetenciesforFPE, cited: 20/Feb/2020.
14. Cote, A.E., Grant, C.C., "Codes and Standards for the Built Environment", Fire Protection Handbook, 20th edition, National Fire Protection Association, Quincy MA, Sect. 1, Chap. 3, 2008, pg. 1-60
15. "The Bill of Rights", National Archives and Records Administration, Website: www.archives.gov/exhibits/charters/bill_of_rights_transcript.htm, cited 19/Mar/2020.
16. "Regulations Governing the Development of NFPA Standards", Rules and Regulations, Section 3.3.6.1, National Fire Protection Association, Quincy MA, 2020 edition, pg. 15, Website: https://www.nfpa.org/-/media/Files/Codes-and-standards/Regulations-directory-and-forms/RegsGovDevStds_2020.ashx, cited 12/Mar/2020.

17. Grant C, "Reaching the U.S. Fire Service with Hydrogen Safety Information: A Roadmap", Fire Protection Research Foundation, September 2009, Figure 1-25, Part 1, Page 53.
18. Grant C, "Reaching the U.S. Fire Service with Hydrogen Safety Information: A Roadmap", Fire Protection Research Foundation, September 2009, Figure 1-9, Part 1, Page 26.
19. Evarts B., Stein G., "Fire Department Profile 2018", National Fire Protection Association, Quincy MA, February 2020, website: https://www.nfpa.org/-/media/Files/News-and-Research/Fire-statistics-and-reports/Emergency-responders/osfdprofile.pdf, cited: 11/Mar/2020.
20. Grant C, "Reaching the U.S. Fire Service with Hydrogen Safety Information: A Roadmap", Fire Protection Research Foundation, September 2009, Figure 1-21, Part 1, Page 38.
21. Lee E., "Cyber Physical Systems: Design Challenges", Center for Hybrid and Embedded Software, University of California Berkeley, January 2008, website: http://citeseerx.ist.psu.edu/viewdoc/download?doi=10.1.1.156.9348&rep=rep1&type=pdf, cited: 6/Mar/2020.
22. "Cyber Physical Systems", NIST National Institute of Standards and Technology, website: https://www.nist.gov/el/cyber-physical-systems, cited: 6/Mar/2020.
23. Grant C, Hamins A, Bryner N, Jones A, Koepke G, "Research Roadmap for Smart Fire Fighting", NIST Special Technical Publication 1191, May 2015, page 7.
24. Humby, C. "Data is the New Oil!" DunnHumby, ANA Senior Marketer's Summit, November 2006, website: https://ana.blogs.com/maestros/2006/11/data_is_the_new.html, cited:3/Feb/2020.
25. Grant C, Hamins A, Bryner N, Jones A, Koepke G, "Research Roadmap for Smart Fire Fighting", NIST Special Technical Publication 1191, May 2015, page 9.
26. Grant C, Hamins A, Bryner N, Jones A, Koepke G, "Research Roadmap for Smart Fire Fighting", NIST Special Technical Publication 1191, May 2015, Figure 1.4.2, page 9.
27. Jones A., Subrahmanian E., Hamins A., Grant C., Human's Critical Role in Smart Systems: A Smart Firefighting Example", Physical-Cyber-Social Computing, May/June 2015.
28. "CPS Framework", CPS Public Working Group, Release 1.0, NIST National Institute of Standards and Technology, website: https://pages.nist.gov/cpspwg/, cited: 6/Mar/2020.
29. "Ptolemy Project", EECS: Electrical Engineering and Computer Sciences", University of California Berkeley, website: https://ptolemy.berkeley.edu/projects/cps/, cited: 29/Feb/2020.
30. Hobday M., Davies A., Prencipe A., "Systems Integration: A Core Capability of the Modern Corporation", Industrial and Corporate Change, Volume 14, Issue 6, December 2005, Pages 1109–1143, website: https://doi.org/10.1093/icc/dth080, cited: 20/February/2020.
31. Liu J., Mooney H., Hull V., Davis S., Gaskell J., Hertel T., Lubchenco J., Seto K., Gleick P., Kremen C., Li S., "Systems Integration for Global Sustainability", Science, Volume 347, Issue 6255, February 2015, website: https://science.sciencemag.org/content/347/6225/1258832, cited: 18/Feb/2020.
32. Gulledge T., "B2B E-Marketplaces and Small- and Medium-Sized Enterprises", Computers in Industry, Volume 49, Issue 1, September 2002, pages 47-58.
33. Gold-Bernstein B., Ruh W., "Enterprise Integration: The Essential Guide to Integration Solutions", Addison Wesley Longman Publishing, Redwood City CA, July 2004.
34. NFPA 950, Standard for Data Development and Exchange for the Fire Service, National Fire Protection Association, Quincy MA, 2020 edition, Chapter 8.
35. Benson K., Bouloukakis G., Grant C., Issarny V., Mehrota S., Moscholios I., Venkatasubramanian N., "FireDex: A Prioritized IoT Data Exchange Middleware for Emergency Response", presented at: Middleware '18, 10-14/December/2018 at Rennes France, Association for Computing Machinery, 2008.
36. "Aircraft Accident Report, Loss of Thrust in Both Engines After Encountering a Flock of Birds and Subsequent Ditching on the Hudson River, US Airways Flight 1549, Airbus A320-214, N106US, Weehawken, New Jersey, January 15, 2009", NTSB: National Transportation Safety Board, Washington DC, NTSB/AAR-10/03, 4/May/2010, website: https://www.ntsb.gov/investigations/AccidentReports/Reports/AAR1003.pdf, cited: 5/Jan/2020.
37. "First Responder Network Authority", website: https://www.firstnet.gov/about, cited: 5/Jan/2020.

38. Gallagher J. C., "The First Responder Network (FirstNet) and Next-Generation Communications for Public Safety: Issues for Congress", Congressional Research Service, 27/Apr/2018, website: https://fas.org/sgp/crs/homesec/R45179.pdf, cited: 7/Jan/2020.
39. Bryner N, Grant C, Hamins A, Jones A, "Realizing the Vision of Smart Fire Fighting", IEEE Potentials Magazine, Volume: 34, Issue 1, Page 39, January/February 2015.
40. Grant C, Hamins A, Bryner N, Jones A, Koepke G, "Research Roadmap for Smart Fire Fighting", NIST Special Technical Publication 1191, Table 1.4, page 13, May 2015.
41. Rowley, J., "The Wisdom Hierarchy: Representations of the DIKW Hierarchy", Journal of Information Science, April 2007, website: https://journals.sagepub.com/doi/abs/10.1177/0165551506070706, cited: 28/Feb/2019.
42. Ackoff, R., "From Data to Wisdom.", Journal of Applied Systems Analysis, Volume 16, 1989, page 3-9.
43. Conger S., Probst J., "Knowledge Management in ITSM: Applying the DIKW Model", from Engineering and Management of IT-Based Service Systems. Intelligent Systems Reference Library, Volume 55. Springer Berlin, October 2013.
44. NFPA 2400, Standard for Small Unmanned Aircraft Systems (sUAS) Used for Public Safety Operations, National Fire Protection Association, Quincy MA, 2019 edition.
45. Jordan R., Martinez-Ramon M., "Next Generation Smart and Connected Fire Fighter System", University of New Mexico Technical Notes, Fire Protection Research Foundation, Quincy, MA, December 2021, pages 1–2.
46. Grant C., "Smart Fire Fighting", Fire Protection Engineering, SFPE: Society of Fire Protection Engineers, Bethesda MD, Issue #83, Q3, page 26, 2019.
47. Lombardo J., "LVMPD Criminal Investigative Report of the 1 October Mass Casualty Shooting", Las Vegas Metropolitan Police Department, Event Number 171001-3519, 3/Aug/2018, website: https://www.lvmpd.com/en-us/Documents/1-October-FIT-Criminal-Investigative-Report-FINAL_080318.pdf, cited: 15/Jan/2020.
48. Grant C, "Key Data, Crucial Seconds", Research Column in NFPA Journal, May/June2018.
49. Grant C, "Connection Risk", Research Column in NFPA Journal, May/June 2019.
50. Krebs B., "Target Hackers Broke in Via HVAC Company", Krebs on Security, website: https://krebsonsecurity.com/2014/02/target-hackers-broke-in-via-hvac-company/, cited: 29/Jan/2020.
51. Cichonski J., Marron J., Hastings N., Ajmo J., Rufus R., "Security for IOT Sensor Networks", Building Management Systems Case Study, NIST: National Institute of Standards and Technology, Gaithersburg MD, February 2019.
52. "Protecting Against Cyberattacks: A Guide for Public Safety Leaders", Joint Publication of IAFC, FireRescue1 and American Military University, February 2019, website: https://www.iafc.org/topics-and-tools/resources/resource/protecting-against-cyberattacks, cited: 20/January/2020.
53. Grant C., "Smart Fire Fighting", Fire Protection Engineering, SFPE: Society of Fire Protection Engineers, Bethesda MD, Issue #83, Q3, page 26, 2019.
54. Grant C, Hamins A, Bryner N, Jones A, Koepke G, "Research Roadmap for Smart Fire Fighting", NIST Special Technical Publication 1191, Figure 14.1, Page 210, May 2015.

Intelligent Science Empowers: Building Fire Protection Technology Development

4

Fang Li

4.1 A Look into Fire Prevention in Buildings

Building's fire prevention strategy tailored with the creation of urban buildings design has developed rapidly while promoting the more advanced modern cities. After nearly a hundred years of experiences in fire protection, the concept of building's fire protection of the so-called active fire protection design and passive fire protection design has gradually been reformed. With the development of high-speed computer technology and the improvement of modern science and technology, a new concept of the so-called performance-based fire protection design could be first traced down as early as 1970s when the goal-oriented approach to building fire safety was developed by the U.S. General Services Administration. The design concept uses the computational fluid dynamic (CFD) software to obtain the movement data and the trace of fire and smoke, temperature distribution, gas concentration's distribution cloud map, and other parameters [1]. The emergence of this new technology enlarged the traditional prescriptive-based fire protection design approach with the alternative design method if code complaint approach does not fit the design best. The fire protection design of complex buildings can use this technology to perform the optimized design or justification / verification analysis quantitatively and obtaining more sound results and conclusions. However, the focus is still on how to fight and control the fire while preventing the fire as the second. With the new Performance Based Design (PBD) method, the fire risk could be mapped and quantified in order to have a better weight in analysis toward a better fire and life safety tailored design. This could be even more achieved with the new tools including the information technology, Internet of Things, cloud computing, Big Data, and artificial intelligence. The PBD approach has been used in other

F. Li (✉)
FPE, Worcester Polytechnic Institute, Worcester, MA, USA
e-mail: fli4@wpi.edu

© Springer Nature Switzerland AG 2022
M. Z. Naser, G. Corbett (eds.), *Handbook of Cognitive and Autonomous Systems for Fire Resilient Infrastructures*, https://doi.org/10.1007/978-3-030-98685-8_4

engineering disciplines for longer time and approved to be a scientific based sound approach, with the new empowered information cutting-edge technology captioned above, fire protection design can take the advantages without leaving behind. "Smart Building" is a new thing that is being nurtured in the current era. It is a building that provides safety, efficiency, comfort, and consumes less energy for building, people, and environment by using computer technology and Internet technology. Under the ideal condition, smart buildings are supposed to be safe and the level of fire safety can meet people's expectations based on the reliable risk management tools. The implementation includes how to recognize the building features to identify the risk levels and areas, sending signals alerting the building's brain for further diagnosis and action. The smart building concept is getting more and more attention with penetration and coordination of various engineering disciplines.

4.2   Build Intelligent Fire Detection and Response System

In traditional fire prevention concepts, possible fire scenarios are assumed. If a fire does occur, a reliable and effective fire defense system is provided to activate to achieve the purpose of extinguishing or controlling fire. This type of fire protection design concept works well and has been adopted in modern fire protection design for dozens of years. However, because fire is still a major threat to urban development and constraints on architectural design, is it possible to prevent fire before the occurrence of fire? For example, to perceive fire hazards in buildings in the early stages of the fire. Traditional fire detection refers to the detection of fires that have occurred after a fire, and in this case, "perceive" refers to the perception of risk factors that may cause or scale to cause a fire before afire occurs. If we can achieve this with the help of today's science and technology, then we can provide a corresponding response system that will eliminate the fire hazard before it develops into a fire. This technology may reduce the incidence of fire, which is also used to be called intelligent fire detection with very early identification of fire risk, this has made the traditional building fire prevention system full of angles. In this chapter, the reader will be guided to analyze and understand the current intelligent fire protection technology and the future trend of the development, and in the intention of using the less to get the more and invite all the stake holders to support such research and applications.

4.2.1   Fire System Based on Internet of Things

The so-called Internet of Things is based on the current and well developed all combined system, that is, the process of converting various physical data/information such as detected water pressure, flow, signal sound, gas concentration, lighting illuminance, etc. into the digital data through the sensors and transmitting the data output to the network for analysis At present, the technology is also used by major fire equipment manufacturers to monitor the fire control system equipment component such as water pressure sensors data, flow sensors data, control cabinet signal transmission devices, and sound level, etc. In US the major fire alarm manufacturers are working on using the internet to inspect and test devices. NFPA 72 is being updated to allow some remote monitoring. On the suppression side there have been a move to monitor pressure, flow, electrical power to fire pumps follow system performance. Remote inspection and testing is growing with AHJ's. Big Data is not a trend yet in US due to the fact lacking the code standard support as well as the testing approval. Also most owners do not want to share their system status. The use of the internet based cutting edge technology applied to the fire alarm and detection system design will not happen until the devices are UL listed for that use and NFPA 72 allows that use.

Building's fire protection system based on Internet of Things is the underlying data acquisition hardware of the entire intelligent fire pyramid, its function is described as converting some parameters related to the operation of a fire-fighting system into electronic/digital data signals, just a dialogue between the hardware and its media for example, fire water pressure. If the value of the system fire water pressure is collected separately, it will be meaningless for the system for example, the water pressure difference between a multi-story building and a high-rise building is very significant. It is necessary to use the water pressure to determine that the fire water supply system is incomplete, but the reliability of the system judged from the pressure also requires some data analysis, which is assisted by the evaluation of the pre-set up criteria rules entered in advance. For example, building height will be required to assist judgment, for a 50 m high building, the water pressure of the first floor of the building should be about 0.5 MPa, if it is lower than this value, an alarm will be activated. Therefore, many software manufacturers have developed a variety of applications for fire-fighting equipment to analyze the collected data.

Because of the complexity of the fire protection system, the collection of hardware data alone is not enough to completely evaluate the quality of the whole fire-fighting systems. For example, the fire water supply system which does not activate during the normal condition, the same is true of the smoke exhaust system, therefore collecting and analyzing the data from the system key component will be very essential to verify the system's status as normal or abnormal. This could not be worked out without the smart system.

The current fire system based on Internet of Things has been able to collect some hardware data, but there is still some hardware data cannot be collected.

4.2.2 Fire Risk Monitoring System Based on Big Data Analyses

Despite the fact that there is very limited data collected by the fire system based on the Internet of Things, at present, many software vendors are able to further extract other more valuable information based on the available Big Data statistical analysis For example, through the pressure change to analyze the opening and closing status of the system valve in the fire water system as well as through the failure frequency smoke detectors to analyze the system cycle time during its operation.

The key of Big Data is to use the vast amount of data accumulated in the system to diagnosis the characteristic and the mechanism aiming to predict the outcome based on the assumption. Compared with the traditional way to analyze fire hazard which relies mostly on experience, data-based analysis has distinctive advantages for predicting the fire risk and hazards in a long run.

4.2.2.1 Big Data Sources

The advantage of big data analysis is that it combines all the relevant data to establishing the database for proceeding with the analysis and recommendations.

There are several things you can do about how to build a fire-fighting database:

1. Building information model BIM basically contains architectural components, interior design materials.
2. Electrical-mechanical system contains the essential data of the mechanical and electrical system in the building.
3. Power supply system, data for fire alarm and camera, occupant load number at specified locations such as the main entrances, etc.
4. Daily Management Operations Data, including fire and non-fire system (examples of Sect. 4.1.2.1)

For the traditional fire monitoring technology, the alarm activation threshold is set up by detecting the temperature and smoke concentration generated. The criteria/parameters for determining the fire are not analyzed as a whole. Therefore, this mechanics applied is not sophisticated enough to be a good solution usually causing the false alarm or does not send the warning signal. This is the situation when the electrical fire hazard detection system equipped with the big data analysis function collecting and uploading the real time data from the end of the sensor device and communicating and transmitting data through the sensor network for analysis and determining the potential hazard origin.

Through the Big Data analysis, if there is a static deviation degree at the corresponding sensing data (combination of multiple sensors) of an electrical line over a period, this is a static safety hazard indicator. Static deviation softens can distinguish the differences between the types of hazards common to electrical lines and the causes of hazards. Static security risk only need to remind users to check and adjust in advance, it does not need real-time warning and sending alarm. Static safety hazards reflect the safety of electrical lines and other output which belong to the category of electrical line risk assessment. Therefore, the large data-based electrical fire hazard management system, compared with the traditional electrical fire monitoring detector, increased the electrical line risk assessment accuracy as defined as smarter and boarder range.

Dynamic safety hazards analysis process can determine the reason for the abnormal deviation of the sensor data against the abnormality of the sensor data that occurs quickly in a short period of time (not a single sensor data threshold change, but a variety of sensor data calculated and analyzed through a mathematical model), such as socket short circuit, overload operation. Once the dynamic deviation occurs, timely warning and alarm are required. Therefore, the system of the management of electrical fire hazards based on Big Data is the intelligent diagnose of the data model rather than the simple criteria threshold evaluation in the mechanism of early warning the electrical fire.

4.2.2.2 Big Data Fire Risk Forecasting Process

Compared with the traditional risk analysis, which relies on the people's experience, the Big Data-driven analysis has significant advantages in digging and grasping the key factors to be able to locating and predicting the fire related risk via the more advanced tools, software beyond isolated discipline. The output is more user friendly and well managed. Even currently there is some resistance from the owner of sharing the data depends on the sensitivity level of the data and possible impact.

In US, the application of big data still majorly only for the big fire alarm manufactures doing the remote testing and inspection task, particularly since the pedantic, NFPA 72 has started the process considering of updating the code to reflect the situation allowing some automatic system devices to be monitored remotely such as pressure, flow, and the power status etc. The process is underway with a slow pace, the bottleneck is the owner/developers willingness to share the data. The driven force now is mainly from the fire service side who is only part of the stakeholders.

Using machine learning algorithms via the correlation analysis based on the historic data, the intelligent tracking and prediction model is established at the conception process as described as the following:

1. The huge amount of data from different sources, such as historical building fires, potential fire sources, and defensive measures.
2. According to the specific applications using the parallel computing architectural information to analyze and process historical massive data, through the data machine learning production prediction model.
3. By inputting relevant dynamic data to the prediction model, the fire prediction results of each unit building are produced.

4.2.3 Fire Inspection Based on Indoor A Geographic Information System (GIS Technology)

In additional to taking the operation data of building and equipment, the data of fire safety management plays an essential role for serving the system goal. The reliability of the building's system performance is considered the top priority for the fire system's success defending the fire. The data collection during the process can help the end user to understand the risk factors and weak areas in order to take the proactive approach. Fire inspections are confirmations of the status of facilities and equipment, which also include environmental status and information about the status of some facilities and equipment that are temporarily unable to be collected by IoT technology.

Traditional fire protection system's ToE (Testing and Inspection) of a facility uses the paper records for filing and management, which requires the skillful and efficient aministration's capability

The Big Data approach even with the very complicated building system could be far more friendly and easily to be used. With the breakthrough of the indoor GIS technology, we can utilize the cutting-edge technology to take the inspection and management of fire-fighting facilities task indoor positioning system (IPS) is a system used to locate objects and people inside buildings where GNSS signal is usually unavailable. There is a number of techniques for indoor positioning and most of them falls into one from two categories: Infrastructure-based or infrastructure-less Infrastructure-based systems assume that building is known in advance and special measurements/installation can be performed beforehand. Because of this reason exploiting infrastructure IPS during Search and Rescue (SAR) operations is limited and can be connected with substantial cost of implementation [2]. Infrastructure-less solution, on the other hand, operates without prior knowledge of building. Most promising solutions are based on dead reckoning using foot-mounted inertial sensor or SLAM systems. Since core of the latter is based on mostly on optical system usage of SLAM in smoke (without any visibility) can be limited (moreover there is no evidence in scientific literature that examined this technique SAR scenario). Dead Reckoning using inertial-based system has the most widest area of application but it introduces significant error over time that need to be corrected by the means of different systems of procedure. Nevertheless IPS for firefighters needs to be integrated with SAR procedures and designed with special consideration of specific needs of smoke divers, situational awareness of commander, command chain, and communication.

Supporting firefighters during search and rescue operations has been actively studied in the Main School of Fire Service in Warsaw. Experimental Risk Management System has been developed in which sensory data (including local isolation data was studied). A number of specific problems have been already investigated relating to sensory data (acquisition, monitoring, correction, in integration, analysis). So far significant progress has been made and the university has developed some technical solutions which can now be further tested, improved, and integrated with available BIM solutions.

System ICRA is a wider concept that tests a number of new ideas. It has been developed with sensors allowing for data acquisition independent of the commander and firefighters.

The localization system for fire brigade for determining and reporting the position of firefighters in real time in the building during the search and rescue operations. For this purpose, the system consisted of number of radio beacons (or anchors) installed in the building (with a density of approx. 1–2 beacon for a room, coinciding with the density distribution of fire detectors). Changes in the propagation of the signal strength induced mostly by the walls is called fingerprinting. A large number of measurements need to be taken at known positions that can be later processed using machine learning techniques. The system consists of two elements:

The data is processed in real time of data acquisition unit integrated with clothes and transmitted to the module ICRA by radio. From there, the firefighters are getting into the system and are by the commander. Position is determined by geographical coordinates. This system operates only in the building. If operate in the outdoor open area is used the GPS receiver.

Heading correction is not based on the Zero-Velocity phase assumption which introduces significant error. The experiments conducted on ground-truth data shows that the proposed approach outperforms state-of-the-art solution by reducing systematic and modeling errors and also provides better heading estimation.

An ultra-wide band (UWB) radio positioning system is extensively studied subject in the area of IPS and can provide high accuracy precision even with millimeter precision. UWB technology can be used in accordance with inertial-based system to enhance precision that degrade over time. The reason is that IMU based systems are characterized by a propagating error, while radio-based systems are either expensive or inaccurate or require a large infrastructure.

The application of the building's fire protection system inspection based on the indoor GIS technology can collect the inspection information, identify the abnormal issues, and summarize the relevant findings reporting back to the system brain. The so-called smart fire-protection system's inspection can be divided into inspection device for the use of inspector and management platform running on the server. Hand-held inspection device is used to collect and update the data of the fire protection system, by scanning the system's component electronic label and barcode, the unit information such as the brand of manufacture, production date, location, and inspection history, etc. will be available by the viewer.

In the traditional firefighting and rescue operation, the success of the fire-fighting relies on the quick and accurate information the commander can obtain from various sources. The commander not only needs to understand the basic situation of the fire scene from the beginning. The practical operational plan depends on the circumstances changes is one of the critical essential part for the fire fighting success.

The building with the intelligent fire awareness and response system combined with the communication technology has the capability to recognize the digital fire rescue plan, which also recognize the transformation from two-dimensional to three-dimensional, from macro to micro, from one-sided information to model visualization and three-dimensional, and recognize the visualization of fire control data, scientific decision analysis and intelligent firefighting and rescue.

The structure of the digital plan platform for firefighting and rescue consists of a physical support layer, a data layer, an application support layer, and an application layer, which is connected up and down and supported by each other. The physical support layer builds the cloud computing platform by building a distributed server cluster, which provides the computing and storage power of dynamic, easy-to-scale, and virtualized. The data layer collects all kinds of basic data to provide data support for the application support layer, and the application support layer provides the interface of data query and display for fire fighters. The user layer is divided into roles and permissions for the purpose of command decision-making and desktop deduction simulation.

1. 3D analysis: through BIM 3D analysis function, information such as above-ground, underground, facade, internal structure, and so on can be visualized. Visual access to the fire control center, evacuation facilities, fire water supply system, fire-fighting system, floor plan plays a key role in the firefighting and rescue action.
2. Force deployment: by setting the disaster level, the best evacuation and rescue routes, fire vehicle stops are deployed, combat areas and units are divided, equipment and equipment are allocated, specific rescue plans are established, emergency linkage unit location is clearly defined to ensure that personnel and equipment configuration is accurate and synchronous.
3. 3D navigation: it can be divided into outdoor and indoor navigation. Outdoor navigation combined with road traffic information can analyze and find out the best rescue routes, and accurately and rapidly search for the nearest water source. Using the BIM model and based on positioning technology, indoor navigation can navigate large spaces and large buildings, accurately and quickly reach the rescue site and the concerned area to fight the fire.
4. Desktop projection: 3D simulation and desktop projection can be carried out at the disaster site, including command exercises, force mobilization, force deployment, on-site rescue, etc., and the platform can automatically give scores and qualitative descriptions according to the situation of deduction, to carry out statistical and evaluation later.

In addition, the digital fire prevention and rescue plan platform can be used to integrate into the smart city platform to play a greater role. The intelligent traffic of the intelligent cities can detect traffic information on city roads. Through the control of traffic lights, it implant the precise planning and scheduling of fire truck rescue routes, as well as the priority and optimal path of fire truck roads to arrive the site as quick as possible 1.5 AI, robot and automatic fire equipment.

With artificial intelligence, the perception and responsiveness of intelligent fire awareness/alert and response systems could be optimized. Modern developed robots can help human beings to deal with challenging situations such as nuclear facility and other dangerous areas under fire for firefighting.

According to the reference and literature among all the countries in the world, fire intelligent robot is divided into three major categories of fire detection, firefighting, and rescue.

4.2.3.1 Detective Robots

Fire detection robot can be divided into two categories as potential risk and hazard detection robot and fire source detection robot. According to the fire department statistics, most of the fires are caused by ignorable fire and ignition source, if could be identified in advance and propose the appropriate fit in measurement. Therefore the fire incidents could be prevented in a large scale, such as excessively thin fuses, broken circuits, cigarette ends, etc. is very important to be one part of the successful tools.

Now with the combined drone technology and tilt photography technology, the first hand spatial view could be collected from multiple angles of vertical and tilting through the drone aerial photography obtaining comprehensive details and present users with a real world. The use of the current lysing drones equipped with tilt photography cameras for terrain mapping with automatic modeling systems can bring revolutionary efficiency improvement to the field of mapping.

The following illustration is a 3D geographic information model built through this technology.

4.2.3.2 Fire-Fighting Robots

Fire-fighting robot as it named for fire-fighting purpose is a special fire-fighting equipment as the alternative or even the main role for fighting the fire on the scene instead of the fire fighters. It fits the hazardous fire incident involving nuclear facility, chemical toxic plants, and fire rescue effort for inaccessible areas/spaces, etc. The task of the fire-fighting robot is relatively straightforward, as shown in the figure below.

1. Sensing technology: whether the robot can successfully check the fire source in each room, accurately find the source of the fire and extinguish, the key depends on the sensor. Therefore, the design of the system selects infrared ranging sensor.
2. Walking technology: The robot should be able to avoid obstacles during walking, finds the source of the fire and extinguish at the earliest. Therefore, the main program control adopts the barrier/obstacle avoidance algorithm, which determines the obstacle by reading multiple PSD sensor signals, thus controlling the robot to walk along the wall without collision.
3. Other technologies: The fire-fighting robot's own device has embedded specific structural component to facilitate the installation of various types of equipment to cope with various types of fire.

4.2.3.3 Rescue Robots

Fire rescue robot refers to a robot that can do the fire-fighting task for the fire fighters.

Nowadays, the existing fire rescue robots are mainly used to rescue fires in high-rise buildings. With the rapid development of urban construction in China and elsewhere, we see more and more skyscrapers have been and will be built. This has brought significant challenges for the traditional fire-fighting tactics. The fire-fighting robot has significant advantages in terms of accessible, tolerance, and endurance, etc. even with less advantages of flexibility, communications, and coordination, etc.

Rescue robots' body needs to have an adequate level of fire resistance, fire prevention features to keep the integrity of its body. It is equipped with far-infrared flame sensors, the effective detection with distance of visible and infrared light up to 2 m. After entering the room, the rescue robot automatically moves toward the darker light to avoid direct contact with the flames. The robot equipped with the well insulated clothing to preventing the temperature rise due to the fire. The rescue robot is equipped with an LED spotlight set. After the trapped person is found, the light is turned on according to the feedback signal, and the infrared ranging sensor and far-infrared flame sensor continue to operate for the direction of illumination, leading people to leave the fire scene through the safe path.

4.3   Combined with MEP System

4.3.1   Temperature and Humidity Warning and Control

Normally the smoke detection system acts as the first line to sense the fire followed by the thermal detectors due to the temperature changes. Pre-fire ambient temperature and humidity awareness can help intelligent sensory systems with machine learning and Big Data analysis. By comparing the data of ambient temperature, humidity using the same criteria during the fire mode, the correlation between ambient temperature and humidity is found, and an early warning message is delivered when environmental conditions may trigger a fire.

4.3.2   Intelligent Electrical Fire Warning System

The elctrical related fire accounts for a large percentage fire incidents worldwide. One key feature of the building's intelligent smart system is to incorporate the detection of the specific electrical fire parameter into the diagnosis process, and the intelligent power monitoring system terminal adopts the existing electrical parameters monitoring technology to recognize the monitoring of the three-phase current value, residual current value and temperature of the electrical line, so the monitoring of the status of electrical lines, transmitting of electrical line status parameters in real time, and issuing of alert and warnings when electrical parameters appear abnormally.

The intelligent power monitoring system hardware consists of the following components:

1. A current transformer that collects the current value of a three-phase current.
2. The remaining current transformer for collecting the remaining current value.
3. Temperature sensors for collecting the temperature data of electrical cabling wiring and the environmental surrounding.
4. Converter which converts the captured analog current into digital current output to the CPU.
5. CPU is the core of the intelligent power terminal, which is adopted to control the detection device of the three-phase current, sample leakage current, and other data as well.
6. As conducting data processing, alarm output, communication with the cloud server, and LCD display.

Intelligent power monitoring systems have the following characteristics:

4.3.2.1 Cloud Service Platform

A cloud server platform is built by mobile and Internet networks. For the electrical parameters value collected by the monitoring system, through the monitoring device host for signal conversion and logical conversion, each parameter can be uploaded to the cloud service provider through mobile network technology, to establish the electrical monitoring cloud service platform. The cloud service platform can upload real-time data that reflects the electrical status to the monitoring display panel and mobile phone. By viewing the monitoring display on the production site, the running status of the enterprise's electrical circuits can be viewed. A user can also remotely check the status of the facility's electrical circuit and its loop through the mobile phone terminal with real-time info.

4.3.2.2 Big Data Analysis

Command line usage, through the big data analysis, electrical safety risk and hazards are automatically identified and established the notification system. The system determines the maximum current of the line and the temperature limit according to the manufacture's technical sheet The system can also analyze the monitored facility's alarm history and the daily operation status to identify the problems related to the electrical system including excessive load, electrical line aging, and ground fault issues, etc.

4.3.2.3 Early Warning of Electrical Accidents

The system compares the electrical data collected in real time with the pre-set alarm values of the intelligent power terminal, and immediately issues an alarm if the value exceeds the threshold. Accident alarm is divided into three levels: One is the on-site monitoring equipment alarm, the second is the monitoring display alarm, and the third is mobile phone SMS alarm.

4.3.3 HVAC Control

4.3.3.1 Pre-accident Prevention

Gas sites used in buildings, such as kitchens, need to maintain adequate ventilation to reduce the concentration of flammable gases that leak gradually during the daily operation. In the warehouse where flammable materials are stored, appropriate ventilation measures are also required to ensure the safety of the environment according to the characteristics of the goods stored.

4.3.3.2 Post-incident Response

For the combination of fire and ventilation air conditioning systems, the most common example is the combination of smoke exhaust systems and ventilation air conditioning systems. HVAC systems are installed throughout the building, especially in underground garages. Therefore not only because of the insufficient

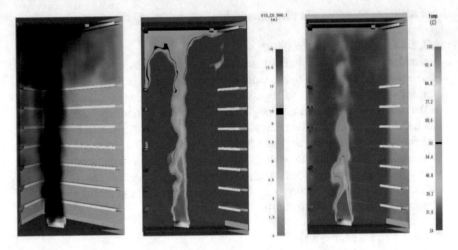

Fig. 4.1 FDS smoke view output (source: FDS user's manual)

space to propose the smoke exhaust pipes separately, but also because the cost concerns for such approach. It is often the case to switch the working state of the system by setting fire-proof smoke exhaust valves at different temperatures. In addition, in some ultra-high and large space, fire exhaust system can take use of air conditioning air supply system to organize air flow, improve the efficiency of smoke control.

Kitchens and places that use flammable gases will activate the ventilation system in the event of a flammable gas leakage or the concentration exceeding the safety level, and it is necessary to cut off the supply valve to mitigate the risk (Fig. 4.1).

4.3.4 Elevator Assisted Evacuation

4.3.4.1 Elevator Assisted Evacuation

For the sky scraper fire and other emergencies even for normal population, evacuation through stairs could be a long unexpected process due to the complicated fire scenarios, individual's physical fitness as well as the group characteristics. Therefore different countries, regions have different local code requirements regarding the approach of either defend in the places or entire building evacuation etc. In China and some other Asian countries, the concept of the area of refuge for super high rise buildings are mandatory for the building over 100 meters which require the occupant to travel to the AOR floors for further instruction during the emergency evacuation process.

Lessons learned from some super high rise buildings worldwide's studies, the fast, efficient and well organized evacuation is the key for the egress success. US Nist's fire investigation report of September 11 recommended that to design the supplemental egress means including the dedicated stairs for fire fighters or using the assisted elevator for egress.

As an intelligence awareness and response system for building, supplemental evacuation elevator is a response technical measure for the safe evacuation considering of the overall building features. The elevator to be used as means of egress shall be designed with special features including the water proof and reliable power supply, two way communication, material fire resistance rating and smoke ventilation etc. The full evacuation strategy using the elevator as part of the egress process fits well with the area of refuge floor (AOR) as the occupants uploading and downloading areas.

4.4   Combined with Security System

4.4.1   Camera Assisted Fire Detection and Confirmation

Video surveillance plays an important role in the security system, and because of its intuitive, convenient, and rich information content, it is widely used in buildings. Due to the large number of cameras installed in the building, the use of image technology and fire detection together can greatly improve the accuracy and detection range of the fire. The Fire detection system combining the surveillance technology as becomes more and more popular for the occupany area such as the Museum, exhibition etc.

This technology can provide the following functions in an intelligent fire awareness and response system.

4.4.2   Fire Smoke Detection Equipment

This technique is to detect a specified area through a computer. The visual characteristics of smoke are visible, relatively dispersed in volume, with a strong sense of edge, and the ability to move with air circulation. Therefore, we can edit this characteristic of the smoke and organize the operation of the smoke into a signal that can be alerted when the camera finds the relevant signal formed in the monitoring network area.

4.4.3   Image Flame Monitoring System

It can monitor the flame depending on the shape. Its monitoring system consists of an infrared ICCD (infrared charge-coupled device) camera and a color CCD camera, which can capture images and record infrared light during monitoring. In the process of identifying the flame, the system can make a judgment according to the continuous images, and perform video extraction of the images in the video. Video extraction can detect the color, motion form, growth mode, temperature distribution,

Fig. 4.2 Fire detection system based on the image processing technology (source: website)

and strobe characteristics of the flame, which allows an organic combination of fire detection and image monitoring. For each form of detection, the detection is carried out by two different cameras, and the fire spot can be positioned in the probe. While the other camera can show the actual situation of the fire spot through the video and take appropriate fire-fighting strategy to extinguish the fire (Fig. 4.2).

4.4.4 The Combination of Fire Detection and Video Systems

In the building, the combination of fire detector and video system can form a linkage camera, which can not only carry out fire detection, but also compatible with monitoring devices for common use. The monitor can also record while displaying an alarm image and provide security systems with monitoring services in the building.

4.4.5 Fire Monitoring Confirmation

For fire monitoring, it is generally necessary for the fire control room personnel to go to the site to check and confirm. However, the integration of security and fire protection systems solves the difficult problem of traditional fire alarm confirmation, which greatly improves the efficiency of building configuration equipment operation.

4.4.6 Access Management and Personnel Positioning

The current computer technology, communication technology, and electronic information technology provide reliable technical support for the effective combination of the video surveillance system, access control system, and video intercom system in the fire protection system and security protection system, which can enhance the perception and response ability of building fire awareness and response system. The fire protection system and security system combined to form a subsystem, and it is included in the BAS together with the equipment operation management and monitoring subsystem, so that the building automation system is further perfected. If a fire occurs in a place and trigger the alarms, the fire alarm system will link the video monitoring system and transfer the images of the relevant video monitoring points at the fire site to observe the fire situation and the movement of people in real time, and determine the authenticity of the fire alarm. The linkage of access control system and visual intercom system plays a key role in the evacuation of personnel and the positioning of trapped persons. After the fire alarm, the access control system will open all the access locks to make the system lock open and ensure the evacuation of personnel. The most important thing is that in the event of an accident, the building automation system can share information from several subsystems, providing the heavy mold information to rescue workers in the first place, effectively reduces the rescue time. This "on-site and property-type" building automation system is a prerequisite for rescue and property management, and a favorable tool for obtaining reliable data resources in the process of accident handling.

Today's IoT smart buildings include two types of connected technologies: information technology (IT) and operational technology (OT). The backbone for all systems, IT protects sensitive corporate data, connects critical IT infrastructures, and is normally managed by IT experts. OT includes the connection of smart devices and other critical building functions, which rely on the IT, and provides information through connected devices [3]. Fire alarm control units, intrusion detection systems, mass notification systems, and access control systems typically reside on the OT side and normally managed by facilities operations. Both systems have vulnerabilities, which commonly include equipment tampering as well as inside and outside threats. Firewalls and other cyber protection processes and devices can help mitigate the potential for a widespread attack and protect the individual components of the IT or OT systems. The discussion here today introduced the current and future applications for the smart system, also shared the view of the system vulnerability.

4.5   Building DNA Mapping

4.5.1   Foundation for All Smart Building Applications

Just like any human being, every building has its own specific DNA. Even though two buildings with the same exterior design features, same building heights and floors, they still have two different and separate DNA (Figs. 4.3, 4.4, and 4.5).

How to convert physical buildings into readable datasets is the key to smart applications. Before we develop digital applications, such as smart space management, facility management, and energy saving solutions, we have to digitize the building at first.

Fig. 4.3 Building DNA function powered by SpaceBox platform. Source: SpaceBox technical whitepaper

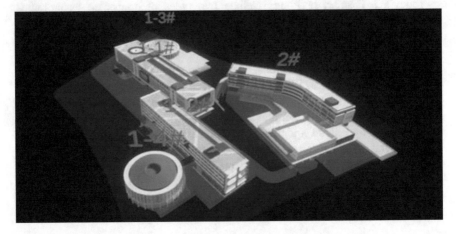

Fig. 4.4 Data-driven 3D mapping generator. Source: SpaceBox technical whitepaper

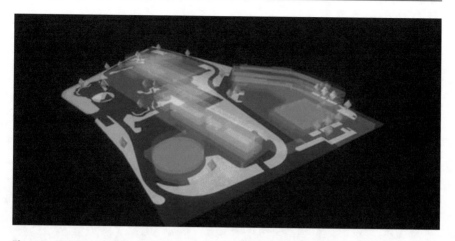

Fig. 4.5 Buildings and featured Landscapes with unique DNA in a Campus. Source: SpaceBox technical whitepaper

What is building digitization? Is it a BIM model or something else? The answer is it is not only a physical model but also building operation logic. The philosophy of such design concept of such tailored functionalities is the sprit of each building. But can people read the inner core value of the building (sprit) beyond the building's facade and interior spaces? Of course. But the building could be read if the shop drawings are provided as well as all the necessary documents and sheets as the basic info. Even with this info, the CPU and algorithms are not capable to identify and realize the building's code as described as its DNA.

There is an essential reason for us to explore a methodology of digital conversion from drawings and documents to readable datasets, and of course, that is not enough. For data visualization, we need mapping. Data-driven mapping enhances us to combine the dynamic data from IOT and IBMS together with physical models.

Mapping is the most efficient way to describe a building with geometry and logic. CAD drawings and BIM models are parts of mapping. Now for smart building applications, lots of teams have been working on integration of physical model and sensor data with a 3D map.

The fundamental functions of DNA should include at least the following contents.

- Bill of material of all assets including spaces and MEP devices.
- Relations for spaces such as how the circulation spaces connect rooms together with specific functionalities.
- Logic descriptions for Routes of MEP systems based on services for different spaces.
- Location of all equipment and the target destination of each MEP system.

There are big differences between data-driven mapping and physical model-driven mapping.

4.5.2 Data-Driven Mapping

Map is rendered by Data, including the geometry static data and dynamic data from sensors.
Features:

- Very light weight
- Easy to be edited
- Cost effective maintenance.

4.5.3 Model-Driven Mapping

Building geometric data from BIM physical models Data from sensors will combine the physical model.
Features:

- Heavy models rely on lightweight software.
- Everything relies on BIM model and is very hard to be edited
- expensive with limited channels to smart building applications.

The SpaceBox online Building DNA data-driven mapping platform is a tool to help project owners and maintenance teams to achieve the digitized goal of buildings with or without BIM models. The DNA datasets could be built up online step by step, that means you needn't initialize all DNA map at one time. You can start from the basic steps and move forward to the deeper operation logic world later as needed.

DNA maps will be rendered instantly from DNA data sets so the team can view physical models without BIM physical models. All the data sets could be edited back and forth and then connected to different smart building applications through standard APIs.

We will start from spaces through extraction data from drawings or models, and then go for MEP systems with families of products. All will be stored in a structured database and will provide services in one package with APIs.

As we mentioned, buildings are like human beings that are assembled with different elements and work together in specific logic. Each building has spaces with mechanical products that provide services in different systems. Once we realize the basic operation theory of Building DNA elements and have powerful tools to digitize them, we will find a bright path towards all smart applications.

Step 1
Spaces extraction

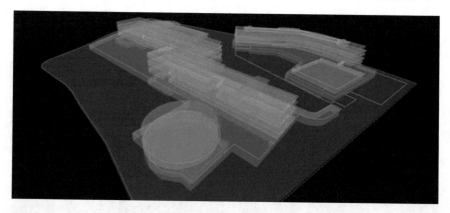

Fig. 4.6 Spaces profile in campus level. Source: SpaceBox technical whitepaper

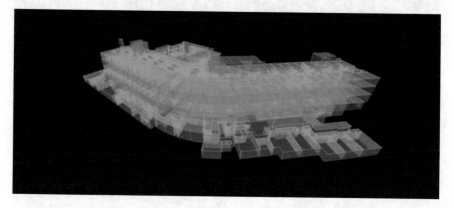

Fig. 4.7 Spaces profile in building level. Source: SpaceBox technical whitepaper

Space data can be extracted from drawings and BIM models or even with online tools, and then save all spaces data in the Building DNA database. As we mentioned, you can view the spaces' physical models as long as you successfully extract data of spaces and save them into the DNA database (Figs. 4.6 and 4.7).

Step 2

MEP Assets extraction

Assets data can be extracted from drawings and BIM models or even with online tools, and then save all assets data in the Building DNA database. Assets can be categorized with special tools and to be defined into families with KPI (Fig. 4.8).

Step 3

Tagging system

All assets (Spaces and MEP equipment) should be tagged with different systems. All the systems could be self-defined by the end user through the understanding of drawings and documents. Every single system will have specific KPI and will be impacted by building components

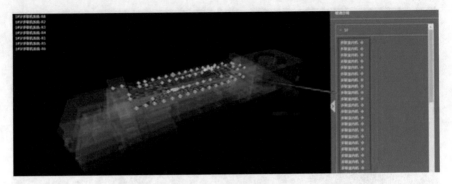

Fig. 4.8 BOM maps for MEP systems. Source: SpaceBox technical whitepaper

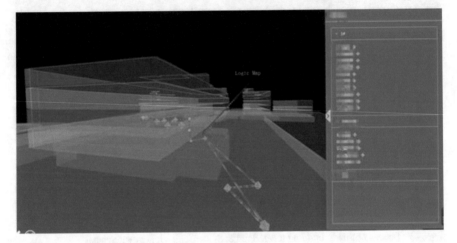

Fig. 4.9 Routes maps for MEP systems. Source: SpaceBox technical whitepaper

Step 4

Logic mapping

All assets will work together with logics from design side and maintenance side. So developing the logic connection between different assets is most essential to Building DNA map. SpaceBox platform has tools to edit logic between assets and try to shape the logic relationship database as easy as possible (Fig. 4.9).

Step 5

Data visualization

Building DNA map includes two essential contents, one is an asset map and one is a logic map. With the SpaceBox rendering tool, users can get these two into one 3D map rendered through the visual engine instantly from the database developed in the process. Furthermore, physical BIM models or any other kinds of 3D models can be input as a canvas of the real Building DNA model. The most important issue is the DNA map can be edited any time whenever the user is authorized (Fig. 4.10).

Fig. 4.10 DNA map. Step 6 Source: SpaceBox technical whitepaper

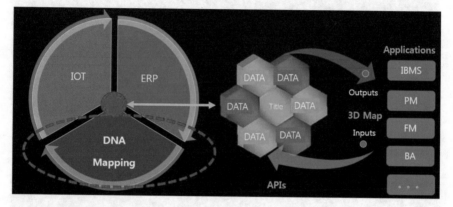

Fig. 4.11 Building DNA mapping delivers 3D interactive maps with APIs. Source: SpaceBox technical whitepaper

Building DNA map is an integrated Dataset developed through an interactive platform and can be rendered instantly to provide services for different smart building applications. Numerous applications such as IBMS, IOT, property management, energy saving, and emergency systems such as intelligence firefighting can get the services from Building DNa through standard APIs.

4.5.4 What Building DNA Map Delivers

Building DNA map delivers DNA data through APIs together with 3D map as following,

1. Spaces list with level of details and categorized by function use.
2. MEP Bill of materials.
3. Logic relationships between building assets.
4. 3D maps describe locations of all assets and operation routes.

All the presentations will be data driven so that other applications can easily call specific datasets or iframe lightweight 3D DNA maps from the platform (Figs. 4.11 and 4.12).

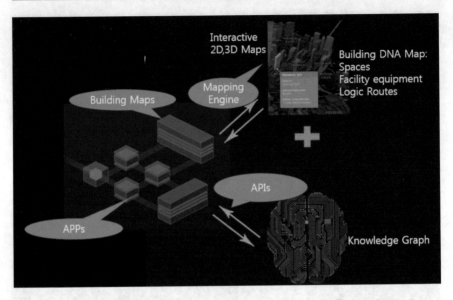

Fig. 4.12 All Smart Building applications can easily collaborate with Building DNA. Source: SpaceBox technical whitepaper

Fig. 4.13 Water tank and route for fire hydrant in basement. Source: SpaceBox technical whitepaper

4.5.5 Fire-Fighting Instance

With Building DNA map, we can figure out equipment maps for fire hydrant systems in a building.

The 2D Map and 3D Map will list the location of hydrants and the relationship between the valve and equipment (Figs. 4.13, 4.14, 4.15, 4.16, and 4.17).

IBM and IoT sensors can combine the live Data with Building DNA map with unique ID, then Fire Hydrant sys live data such as pressure can be read directly on the 3D map.

For such application applying to the building's maintenance and management operation, users can propose an inspection schedule through the Map with the ERP tool and the inspection report can be easily combined into the platform with a unique asset ID.

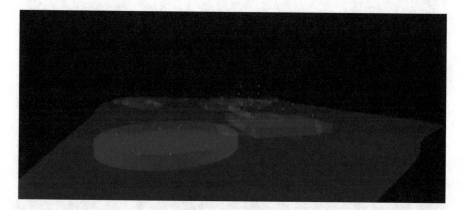

Fig. 4.14 All assets map in one fire hydrant sys in master plan view. Source: SpaceBox technical whitepaper

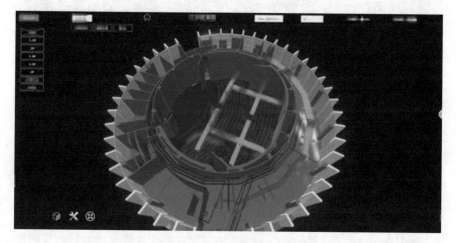

Fig. 4.15 Building BIM physical model. Source: SpaceBox technical whitepaper

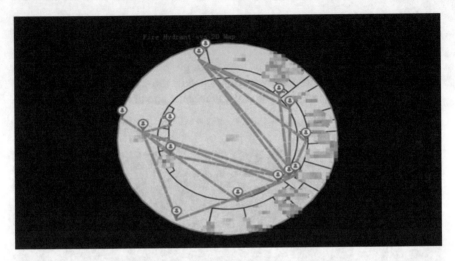

Fig. 4.16 Fire hydrant assets 2D map in a building. Source: SpaceBox technical whitepaper

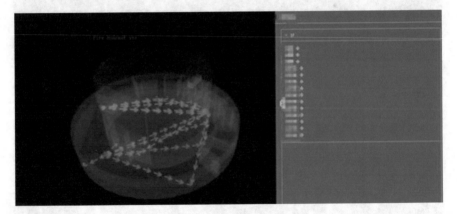

Fig. 4.17 Fire hydrant assets 3D map in a building. Source: SpaceBox technical whitepaper

References

1. Hurley, Morgan J., and Eric R. Rosenbaum. Performance-Based Fire Safety Design, Taylor & Francis Group, 2015. Performance based fire protection
2. Visual Localization system for fire brigade using BIM technology Final Report funded by the SFPE educational $ Scientific Foundation Chief Donald J Burns Memorial Research Grant, Feb 28, 2018
3. Cyberseurity for Fire Protection System by Lou Chavez FPE-Q42020

Building Codes and the Fire Regulatory Context of Smart and Autonomous Infrastructure

5

Kevin J. LaMalva and Ricardo A. Medina

5.1 Introduction

The concept of Smart and Autonomous Infrastructure (SAI) and the legacy of building/fire codes remain in philosophical conflict. Specifically, prescriptive building/fire code provisions are structured in a layered or defense-in-depth fashion rather than a holistic and synergistic manner. Consequently, many designers have been conditioned to evaluate systems in isolation with little thought of their synergistic potential. Also, these provisions are primarily based upon past experience or precedent and collectively represent an immense barrier to entry by new technologies. However, developments in performance-based design of buildings and infrastructure constitute viable outlets for holistic and emerging technologies such as SAI. Existing performance-based design methodologies provide owners and stakeholders with quantitative information to make risk-informed decisions for facilities exposed to multiple hazards (e.g., fire following earthquake) and design features based on predefined performance goals (e.g., life safety, continued functionality). Conceptually, such methodologies include components that can be readily adapted to permit the incorporation of SAI concepts. Ultimately, the implementation of SAI into building design could result in designs with enhanced reliability to (1) safeguard the public health, safety, and general welfare of buildings occupants, fire fighters, and emergency responders, as well as (2) mitigate damage to property.

In the vast majority of facilities, owners and stakeholders must ultimately deal with three competing factors when planning, designing, constructing, operating,

K. J. LaMalva (✉)
Holmes Fire, Lancaster, MA, USA
e-mail: kjlamalva@gmail.com

R. A. Medina
Simpson Gumpertz & Heger Inc., Waltham, MA, USA

© Springer Nature Switzerland AG 2022
M. Z. Naser, G. Corbett (eds.), *Handbook of Cognitive and Autonomous Systems for Fire Resilient Infrastructures*, https://doi.org/10.1007/978-3-030-98685-8_5

maintaining, and when required, recommissioning their facilities: *Time* (investment period), *Function* (efficient use of investment), and *Risk* (hazard potentials and consequences of failure). Buildings are designed with life safety as a baseline performance target, whether explicit (e.g., smoke control design) or implicit (e.g., structural fire resistance design). For essential facilities, continued functionality is another performance objective that must considered. Any application of SAI into design practice will provide quantitative, near real-time data to aid in the development of designs, operations and monitoring plans, and building performance that result in an acceptable balance between time, function, and risk during the expected design life of the facility. However, inevitable inertia would exist for designers to employ such technologies unless specifically required by the codes. Instead, it is more likely that codes will eventually provide an option for SAI integration with only specialist engineers engaging in the near term for certain high-profile and/or demanding projects. Nonetheless, if the societal benefits become apparent to the public, a wider adoption of SAI within the building design paradigm could certainly take hold in the longer term. This would be especially true if related industries (e.g., automotive industry) are able to prove certain envisioned concepts (e.g., active structural integrity monitoring) through at-hand use by the public at large (e.g., real-time display within an automobile's dashboard).

This chapter will discuss some of the most relevant opportunities and challenges associated with the incorporation of SAI into building/fire codes while concurrently accounting for time, function, and risk. On the hazard side, the focus will be on fire. This chapter will address a general approach for integration of SAI into performance-based approaches to fire safety engineering, considering structural fire protection, building evacuation, damage assessment, and building recommissioning.

5.2 Active Fire Protection

Active fire protection differs from passive fire protection (discussed in Sect. 5.3) in that it requires mechanical/electrical or manual action to be effective. The following subsections describe common active fire protection systems and their potential for integration with SAI.

5.2.1 Fire Suppression Systems

Fire suppression systems involve the use of an agent that is dispersed through space to control and perhaps extinguish a fire. By far the most common type of fire suppression system is water-based suppression (referred to as fire sprinkler systems). Water-based systems rely on the energy absorption ability of water droplets as they evaporate upon heating in order to work effectively. Other types of fire suppression systems include those that distribute the following agents: dry chemical, foam, clean agent, or water mist. Water mist systems differ from common water-based systems in that they depend on the mist to interrupt the chemical

Fig. 5.1 Fire sprinkler head [2]

reaction of the fire itself as opposed to its energy absorption capability, which would be much smaller compared to water droplets.

Fire sprinkler systems are designed in accordance with the NFPA 13 [1] standard as referenced by most building codes. This standard prescribes the number of sprinkler heads (Fig. 5.1) in the hydraulically-most-remote area of the system that should be assumed to be releasing water during a fire, usually between about 4 and 8 heads. Based upon basic hydraulic calculations to be executed by the designer, this prescribed method results in economically sized distribution piping and water delivery demands that are experientially and empirically shown to be effective at controlling a single ordinary fire in a single location. Fires that originate due to extraordinary events (e.g., explosion) or in multiple simultaneous locations are not contemplated, nor is the explicit assessment of energy absorption by the water droplets. In this respect, if SAI were able to automatically detect and identify the number of sprinkler heads that have actuated during a fire event (e.g., with the aid of infrared cameras), this information could be relayed to first responders. This information would inform arriving personnel on the likelihood of the fire being controlled or otherwise uncontrolled. As long as such diagnostics do not interfere with the fire sprinkler system's mechanical operation, such an approach would not necessarily constitute a building code issue but may require explicit approval by the building authority. Also, many engineers have experience designing either system type, but would need to adapt their mindset on how they can interact with each other.

Fire sprinkler systems are usually required to be outfitted with alarm and monitoring devices connected to a fire alarm panel (Fig. 5.2). For instance, water paddles are installed in main piping intermittently to trigger the fire alarm system upon water flow (referred to as water flow switches). Also, water shut off valves are equipped with tamper switches to indicate if the system may be inadvertently shut off. Such equipment and their connectivity to a central panel could conceivably be integrated with SAI with simply electrical connectivity. However, the building

Fig. 5.2 Fire alarm control panel [2]

code requires that such equipment be listed by Underwriters Laboratories, so such an integration may pose approval issues. Hence, any SAI would likely be subject to comprehensive vetting by approving agencies and may require modifications to NFPA 13 or the governing NFPA standard for the given delivery agent (e.g., NFPA 2001 for clean agent systems [3]).

Performance-based approaches to fire sprinkler design may allow for more direct integration of SAI; however, such approaches are very rare for fire sprinkler systems currently. Notably, accurately characterizing the energy absorption of water droplets based upon a given sprinkler spray distribution for a given realistic fire is a very challenging endeavor, and the current state-of-the-art in computer modeling (e.g., fire dynamics simulator) is not particularly effective at this type of prediction. Hence, fire sprinkler design requirements remain grounded in empirical testing and prescriptive guidance at the moment. Nonetheless, SAI could be harmonized with such a paradigm if it can be demonstrated that its interface with the system is completely non-intrusive. However, computational capabilities would remain on the critical path in these respects, not necessarily the inertia of design methods.

5.2.2 Fire Detection/Alarm Systems

Fire detection/alarm systems involve the sensing of fire and/or smoke as well as alerting building occupants and off-site authorities (e.g., central station which connects to the local fire department). The most common types of detection systems are smoke and heat detection systems. Smoke detectors provide earlier detection compared to heat detectors but may not be appropriate for use in certain areas such as kitchens or bathrooms which may result in frequent nuisance alarms. Heat detection is particularly well suited to roadway tunnels and other specific applications. Other types of detection systems include flame infrared detection and video monitoring.

Fire detection/alarm systems are designed in accordance with the NFPA 72 [4] standard per the building codes. This standard prescribes device spacings/coverage, electrical components, and other specific system attributes. Along with fire sprinkler system monitoring devices (described in Sect. 5.2.1), detection/alarm devices require connection to a fire alarm panel (Fig. 5.2), which could conceivably be integrated with SAI with simply electrical connectivity. Again, the building code requires that such equipment be listed by Underwriters Laboratories, so such an integration may pose approval issues. Hence, any SAI would likely be subject to comprehensive vetting by approving agencies and may require modifications to NFPA 72. For this application, designers may not need to adapt their skills, for the SAI integration with the fire alarm signaling could be subcontracted to a SAI specialist and would not necessarily have any bearing on the nominal performance of the fire alarm system.

Similar to fire sprinkler systems, performance-based approaches for fire detection/alarm design are rare. Notably, explicit assessment of device sensing is difficult to achieve, especially considering many of the proprietary elements of such devices. Nonetheless, SAI that is demonstrated to be non-intrusive to such systems certainly have potential for ready implementation. Though designers would need to begin to engage SAI specialists (at least at first) for such endeavors from project conception.

5.3   Passive Fire Protection

Unlike active fire protection, passive fire protection is effectively immune from system malfunctions and/or loss of electrical power. If properly installed at the outset and maintained free of damage/deterioration, passive fire protection systems remain ready to perform. Also, such systems reinforce the defense-in-depth philosophy of building codes. Passive fire protection can be classified into two primary types: compartmentalization and structural fire protection as discussed in the following subsections. Historically, passive fire protection has remained one of the least technologically advanced aspects of building fire safety, as century-old technologies still form much of the basis (e.g., standard furnace testing). Hence, overcoming the inertia of tradition would be heightened for this aspect of building design.

5.3.1 Compartmentalization

Compartmentalization involves the strategic placement of vertical (walls) and horizontal (floors) barriers to impede fire spread across a floor plan or from floor to floor, respectively. Fire barriers are qualified based on standard fire testing which exposes mock-up specimens to an intense heating exposure under controlled conditions. Upon exceedance of a test failure criterion (e.g., temperature rise on the back- or top-side of the barrier beyond a certain threshold), the given test is concluded, and the assigned level of fire resistance is based upon the test clock time (e.g., 1-hr fire resistance). The UL Fire Resistance Directory [5] serves as a catalog of qualified fire barrier listings.

The location and required fire resistance rating of such barriers is prescribed in building codes. For instance, 2-h fire resistance rated walls may be required to surround a mechanical room in order to contain the spread of a fire originating in such a space. Another instance, 1-h fire resistance rated floors may be prescribed based upon the building's height and/or area exceeding thresholds for which horizontal fire separation becomes required.

Unlike most active fire protection systems, compartmentalization barriers do not nominally include any monitoring means. Hence, direct integration of SAI with such systems is not apparent. However, innovative methods could possibly be integrated without compromising associated building code requirements. For example, a small computer chip could be embedded within each unique fire resistance rated assembly (e.g., 1-h fire barrier and each respective 1-h firestopping assembly within the barrier as shown in Fig. 5.3), which links the given assembly to the specified UL listing. Aside from assisting with inspections and/or assessments generally, such technology could be interfaced with SAI to determine the extent to which a given building's compartmentalization is compromised given feedback from damage diagnostics which may be included in the SAI scope nominally. Given their non-intrusive nature,

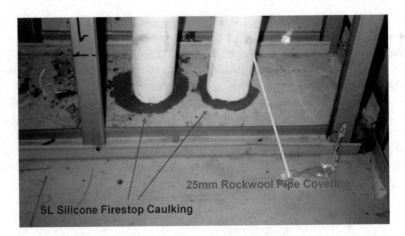

Fig. 5.3 Fire barrier with firestop assemblies [2]

Fig. 5.4 Fire door [2]

such embedment of computer chips would likely not violate UL listings nor present any building code compliance issues. Also, such non-intrusive means could prevent design method inertia delays of adoption.

In order to maintain building functions, fire barriers inevitably will have openings for occupants to pass through to access spaces of a given floor plan. Building codes recognize this and allow for openings within fire barriers to a certain extent, and such openings usually require a fire-rated opening protective. The most common example of this would be a fire-rated door perhaps with fire-rated glazing (e.g., for a viewing panel as shown in Fig. 5.4).

Since the effectiveness of fire barrier opening protectives depends on them being in the closed position during a fire event, building codes usually require these to be self-closing or on magnetic hold-open devices that demagnetize upon fire alarm actuation. Granted, an obstruction as simple as an innocently placed door wedge could compromise this safety aspect of a fire barrier. This aspect of compartmentalization perhaps has the greatest potential for SAI integration. For instance, SAI could be used to monitor and map the open/closed status of opening protectives in real-time during a fire event. This information could be relayed to first responders to assist in their identification of areas that are safe for them to stage their attack of the fire. Similar to fire alarm systems, these aspects could be subcontracted to a SAI specialist and would not necessarily have any bearing on the nominal performance of the fire barrier systems.

Fig. 5.5 Uncontrolled building fire [2]

5.3.2 Structural Fire Protection

The frequency of uncontrolled fire in an engineered building is low (e.g., CCTV Tower as shown in Fig. 5.5), but the consequences can be severe. When structures are heated by fire, they experience thermal effects that are not contemplated by conventional structural engineering design. Under these conditions, the functionality of building safety systems assumes the structural system will remain stable. However, prescriptive building codes do not require explicit assessments of structural fire safety. Rather, structures are usually required to be insulated (or "fireproofed") to reduce their heating, with the intent of mitigating the risk of structural collapse due to thermal loading effects. This is unlike consideration of hurricanes and earthquakes by building codes, in which structures are explicitly engineered to withstand these natural phenomena.

As described in Sect. 5.2, fire sprinkler systems significantly reduce the probability of uncontrolled fire in buildings. However, we must appreciate the fact that these systems are generally not effective against large fires, like one due to an explosion. Fire sprinkler systems are only designed for a limited number of contiguous sprinklers (e.g., five sprinkler heads) to activate and spray water onto a single small fire. Moreover, the effectiveness of a system drops precipitously when more than ten sprinkler heads are activated, since this would begin to tax the water delivery [6]. Fire sprinkler systems can also be rendered ineffective under the

Fig. 5.6 Fire sprinkler
system valves [2]

following possible cases: water supply valves (Fig. 5.6) are accidently left shut (e.g., after maintenance), the system was not properly maintained (e.g., pipe blockage), damaged mechanical components, and other cases. In these rare instances, structural fire protection is relied upon and represents a critical backup safety system.

During an uncontrolled fire event, a safe evacuation of building occupants is expected as well as protection of those that take shelter in designated building refuge areas. Accordingly, it must be appreciated that the total evacuation time increases as the height of the building increases since occupants must traverse stairways from upper floors to reach the ground level. However, building codes do not necessarily require more structural insulation as the height of the building increases (e.g., from 30 stories to 200 stories). Building codes limit the distance that occupants must travel to reach exit stairways, but they do not regulate the total evacuation time. Also, the presence of designated refuge areas is more common in tall buildings, which are expected to remain safe for a period of time during and after a fire event. These performance expectations are predicated on the assumption that the structural systems supporting evacuation routes and refuge areas in buildings remain stable. Beyond occupant life safety, it is generally expected that buildings possess a certain level of resiliency when exposed to an uncontrolled fire. However, building codes do not require confirmation of this, but rather they prescribe certain levels of insulation to mitigate the risk (Fig. 5.7).

Since prescriptive structural fire protection does not consider structural system performance nor explicit performance objectives, potential interfaces with SAI would be limited. However, emerging performance-based regulations may pave the way for harmonious integration with SAI. Notably, ASCE/SEI 7 [7] is the parent standard for structural engineering for the International Building Code (IBC) [8],

Fig. 5.7 Traditional structural insulation ("fireproofing") [2]

and now contains guidance that addresses structural fire protection design alternatives. ASCE/SEI 7 Section 1.3.7 stipulates that structural fire protection design shall be conducted in accordance with the prescriptive requirements of the applicable building code (without extrapolation) or in accordance with the performance-based design requirements of ASCE/SEI 7 Appendix E (per the discretion of the building authority). Also, the new ASCE/SEI Manual of Practice No. 138 [9] serves as a companion guideline to ASCE/SEI 7 Appendix E. Most importantly, ASCE/SEI 7 Appendix E requires explicit consideration of structural system response under fire conditions per required performance objectives. Guidance provided by materials organizations (e.g., AISC 360 Appendix 4 [10]) serve to complement ASCE/SEI 7 provisions; however, these resources should not be relied upon exclusively for performance-based structural fire design. [10] AISC 360: Specification for Structural Steel Buildings, American Institute of Steel Construction, Chicago, IL, 2016.

Structural diagnostics implemented as part of SAI for a given building could be conceivably tailored and implemented based upon damage limits derived from a performance-based structural fire design. For instance, such a design approach may depend upon the floor slab mesh remaining below a certain tensile strain limit in order to uphold stable tensile membrane action during fire exposure. If SAI were implemented to monitor mesh strains, such real-time information could be compared with performance-based design acceptance metrics (perhaps in an automated fashion) to inform first responders if a building is safe to enter with a reasonably degree of engineering certainty. Also, such measures could beharnessed during site reconnaissance operations after the fire event, which would dramatically increase the speed and efficiency of such operations. Currently, performance-based structural fire design remains in its infancy in the U.S., so this discipline would require greater traction in the industry before SAI integration could be employed widely. Nonetheless, there is nothing that would prevent specialist designers to integrate both concurrently, especially if the SAI aspects are non-intrusive.

5.4   Means of Egress

Means of egress systems are meant to provide occupants safe passage from the given building or to refuge areas within the given building. To this extent, active- and passive-type system components are used. For instance, illuminated exit signage (Fig. 5.8) on standby power is an active means to direct occupants toward exits. Also, fire-rated exit stairway enclosures are passive means of maintaining exits to be free of fire and smoke. Various aspects of means of egress systems and their potential for interface with SAI are discussed in the following subsections.

5.4.1   Notification and Directives

Aside from fire detection/alarm systems as discussed in Sect. 5.2.2, certain buildings may be equipped with an emergency voice and communication (EVAC) system to enhance exiting efficiency. Typically, such systems are reserved for higher risk buildings such as high rises. EVAC systems comprise the placement of audio speakers to achieve a certain level of audibility throughout occupied spaces and may involve adaptation of existing equipment. These speakers are used to transmit prerecorded or live messages to building occupants conveying evacuation instructions. Such instructions are particularly critical for high-rise buildings, which often rely on staged evacuation since full evacuation of the building may not be practical.

Fig. 5.8 Illuminated exit signage [2]

EVAC systems have great potential to interface with SAI due to their flexibility in terms of technology as well as code requirements. Aside from audibility distribution and equipment requirements, EVAC systems offer general flexibility in terms of messaging and monitoring. For instance, SAI could actively adjust, or augment, EVAC instructions based upon other related SAI stimuli (e.g., compromised fire barriers in near vicinity to occupants) in a real-time basis. Since the SAI integration would impact the potential effectiveness of EVAC systems, the designer-of-record would need to be versed in SAI and the building authority would need to concur with the SAI aspects (notably any augmented messaging).

5.4.2 Exit Passage

In order to facilitate safe exit passage of occupants through building spaces toward exits or refuge areas during a fire event, building codes prescribe spatial limitations, such as follows:

- Maximum walking distance from a remote portion of a given floor to a point at which two exit paths become available (common path of egress distance).
- Maximum walking distance from a remote portion of a given floor to the entrance of an exit (e.g., exit stairway as shown in Fig. 5.9) (exit travel distance).
- Minimum width of exit corridors.
- Minimum width of walkways/aisles through fixed seating areas.
- Minimum door clear width.
- Minimum ceiling height.
- Other specific requirements (e.g., maximum door swing into a corridor).

Common path of travel requirements are meant to provide at least one option for exiting in the case that fire has compromised or is blocking the other option. Exit travel path requirements are meant to prevent excessive evacuation times to exits. Other spatial requirements are meant to provide safe and efficient exiting paths for occupants during a fire event.

If the exit passage intent of a given building is properly documented, SAI could potentially inform real-time conditions of which in order to inform building occupants and/or first responders. Similar to what is described in Sect. 5.3.1, SAI could be harnessed to determine the extent to which a given exit path is compromised given feedback from damage diagnostics, which may be included in the SAI scope. Given that such an interface would not alter the means of egress system, such measures would unlikely present any building code compliance issues (and product listings would not be directly relevant). Hence, a SAI specialist could be subcontracted to "append" a layer of safety to the exiting designs instead of altering them. In this case, the code-prescribed minimum level of safety/compliance is maintained. However, advocacy for a higher level of safety may be difficult for project stakeholders beyond the building authorities to accept given the likely

Fig. 5.9 Exit stairway [2]

sizable added costs. Nonetheless, such measures are more likely to be integrated into high risk structures such as high-rise buildings.

In addition to spatial aspects of exit passage, exit signage systems are used to direct occupants along safe designated paths during a fire event. Building codes typically prescribe the form of such signage including its illumination, dimensions, and that it statically reads "EXIT." Also, building codes prescribe the distribution of such signage including maximum viewing distances and directionality. As mentioned, exit signage is a static system that uses longstanding technologies. Yet if such signage were able to harness more modern technology, such as deliberate and real-time messaging (e.g., similar to those used in airports as shown in Fig. 5.10), SAI would have a significant potential for interfacing. For instance, SAI diagnostics of fire barrier and exit path conditions (as discussed in Sect. 5.3.1 and herein, respectively) could influence the message of exit signage, which could read "TURN AROUND" if the path ahead is deemed unsafe. However, unlike EVAC systems, such changes to traditional exit signage systems would certainly require changes to building codes which could be a significant challenge given the longstanding use of traditional signage. However, justifications for added costs may be achievable if such signage could be used for building functions during service, such as general announcements or advertising.

Fig. 5.10 Digital signage [2]

5.4.3   Exit Discharge

Once occupants have reached a safe exit, it is important they are able to continue their travel toward a safe location (unless they are within a refuge area). For instance, it would not be appropriate for an exit stairway to discharge occupants into an underground tunnel or a gated yard in most cases. Accordingly, building codes usually require that each exit of a given building connects to public right of way (e.g., a public roadway as shown in Fig. 5.11). Also, building codes limit the following aspects of these paths, but certainly less stringently than exit paths within the building:

- Walking path levelness and continuity.
- Obstructions.
- Proximity to the building.
- Other specific aspects.

Since building code requirements for exit discharge are significantly more generalized and flexible as compared to those for exit paths within a building, there would be less barriers to entry for SAI interfacing. For instance, real-time signage messaging (Fig. 5.10) that interfaces with SAI building diagnostics would pose little or no code compliance issues. Also, traffic signage could perhaps be integrated to assist law enforcement personal during emergency conditions (e.g., automatic roadway shutdowns to allow for exterior gathering of evacuated occupants). For this application, SAI specialists (and perhaps transportation engineers) could lend their skills and experience to the exiting discharge design per concurrence of the building authority.

Fig. 5.11 Building connectivity to a public right of way [2]

5.5 Damage Assessment

Building damage assessment is not regulated. In fact, nationally adopted code requirements or standards for engineering assessment of damage after a fire event are non-existent. Standards for other hazards, e.g., earthquakes, windstorms, and floods, are currently available. For instance, the Applied Technology Council (ATC) originally developed in 1989 a standardized set of procedures for evaluating damage to buildings after an earthquake to determine if they are safe to be occupied, i.e., ATC-20 [10]. The ATC also developed a standardized set of procedures for the safety evaluation of buildings after windstorms and floods to determine whether damaged or potentially damaged buildings are safe for use, or if entry should be restricted or prohibited, i.e., ATC-45 [11] Both ATC-20 and ATC-45 represent the standard of care in these areas and are generally used in post-disaster training classes. Although the abovementioned procedures for the post-earthquake, windstorm, and flood safety evaluation of buildings have been implemented for decades in the United States, they have not been translated into regulatory requirements. One of the most significant barriers for regulation is the fact that such evaluations are highly subjective and based on engineering judgment.

In the context of fire engineering, the intensity and duration of a fire can cause damage to structural components, non-structural components, and contents. Such damage could compromise the structural integrity, stability, and function of a building. Post-fire building structural damage assessment is usually performed by a forensic engineer based on visual inspection, in some cases with the aid of testing. On the contrary, damage assessment of non-structural components and contents is typically based on visual inspection alone. Knowledge of the characteristics of a fire, the state of fire suppression systems, and quantitative measures of structural response would be of immense help in this process. Thus, SAI presents the engineering community with an opportunity to inform, via sensing and diagnostics procedures described in preceding sections of this chapter, more reliable and quantitative post-fire building assessments. Procedures to perform these assessments could also be incorporated into the development of post-fire building evaluation standards.

The implementation of damage detection and quantitative characterization of the response of buildings to fires can be incorporated into structural health monitoring (SHM) practices that have been implemented in buildings and bridges throughout the world. SHM involves the quantification of changes to the material properties, boundary conditions, and configuration of structures over time using sensor data, data processing, modeling approaches, and statistical analysis. In practice, SHM is complemented with maintenance programs, and in some cases on-demand visual inspections, to assess damage potential and prognosis during the design life of the building. SHM can also be used to provide information necessary to evaluate the structural integrity of structural systems after an extreme event such as an earthquake. SAI can provide a mechanism to readily incorporate short-term fire damage assessment into SHM approaches currently available for buildings. In this process, some of the main challenges relate to (1) reliable data collection in the presence of elevated temperature conditions, and (2) the availability of algorithms able to provide fire damage diagnosis and short-term damage prognosis given the initial state of the system obtained from long-term SHM data.

This section will focus on providing a brief discussion on the two most relevant components of building damage assessment: site reconnaissance and data collection and synthesis. Although the concepts presented herein can be applied to different hazards, the discussion will focus on fire hazards.

5.5.1 Site Reconnaissance

In concept, site reconnaissance has a threefold objective: (1) to collect information on building damage, (2) to provide an initial assessment of the stability and safety of the building, and (3) to recommend short-term and/or long-term protective measures, if needed. This process is generally driven by visual surveys that include documentation based on notes, photographs, videos, communications with facility operators, and at times, limited testing. Site reconnaissance is the first step in a post-event investigation (e.g., Fig. 5.12) and can be used to provide initial recommendations on whether a building, or some of its portions, is deemed safe for re-occupancy or continued entry for additional site investigations.

Fig. 5.12 Post-earthquake investigation [12]

The process of site reconnaissance would be greatly facilitated by the availability of quantitative information on the performance of fire suppression systems, structural components, as well as the state of the means of egress in the building. This in area in which SAI, via sensing, would be positioned to provide field inspectors with more detailed and up-to-date information on the overall state of the building, which would make site reconnaissance a safer, more rapid, effective, and reliable process.

5.5.2   Data Collection and Synthesis

The information collected during site reconnaissance can be synthesized in conjunction with data on the original design, construction, history of the building (i.e., maintenance, use, modifications), and SHM data to quantify the extent of post-fire damage in a building. In addition, this information can be used to develop a plan for more extensive data collection when preliminary evaluation suggests that additional data is required.

SAI has the potential to deliver data on the integrity of fire barriers and fire suppression systems, temperature history, and structural members strains to serve as complement to visual inspection and testing to inform a more objective assessment of building damage. SAI will provide inspectors, engineers, and analysts with quantitative information to conduct more detailed building damage assessment. SAI is not meant to replace good engineering judgment; however, SAI can aid

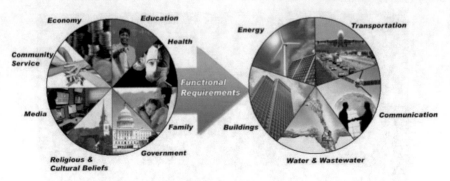

Fig. 5.13 Functional requirements of a community's buildings and infrastructure defined by the social functions of a community [13]

engineers make more informed assessments of building safety by reducing the inherent uncertainty associated with the subjective evaluation of the condition of a structure.

Another important contribution of data collected using SAI is the availability of information on actual building condition and response to permit a building assessment that does not focus exclusively on structural integrity but also on building functionality and loss of service. Quantification of building functionality and loss of service, especially for essential facilities, is critical to achieve performance goals related to recovery. Recovery is a fundamental component of building resilience in the presence of uncertain and disruptive events. This is a component of resilience where SAI promises to make significant contributions given that recovery is a process with duration that should not exceed a predefined tolerable amount of downtime and requires reliable quantification of recovery goals as well as target levels of post-event performance as a function of time. Pre- and post-event functional requirements for a building can be established by building owners or in the case of essential facilities or facilities critical to the resilience of a community, by society as schematically illustrated in Fig. 5.13.

5.6   Building Recommissioning

Building owners and stakeholders must deal with three competing factors when planning, designing, constructing, operating, and maintaining, and when required, recommissioning their facilities: time (investment period), function (efficient use of investment), and risk (hazard potentials and consequences of failure). Accounting for these factors requires a clear identification of performance targets suitable for the facility of interest during both commissioning and recommissioning, e.g., life safety, structural stability/integrity, continued functionality. SAI can provide quantitative, near real-time data to aid in the development of designs, operations and monitoring

Building Commission Process

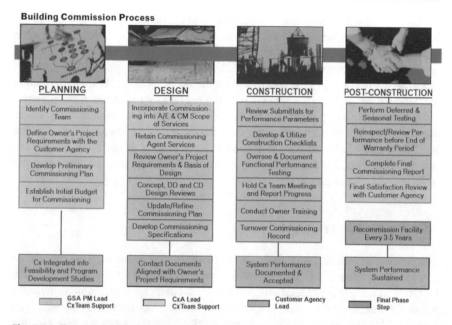

Fig. 5.14   Example building commissioning process [14]

plans, and building performance that result in an acceptable balance between time, function, and risk during the expected design life of the facility.

In general, building recommissioning occurs when a building that has been commissioned requires another commissioning process that is the result of a change in ownership, use, energy-efficiency goals, or adverse consequences from a disruptive event such as a fire. Figure 5.14 illustrates the overall building commissioning process, including recommissioning as part of the post-construction phase. The goal of recommissioning is to ensure that the previously commissioned building can be operated safely as intended and that operation and maintenance staff are properly trained and equipped to operate systems and equipment. The conventional recommissioning process includes inspecting, restarting or upgrading (when applicable), and testing the appropriate operation of fire protection systems, which involves interactions with the Fire Marshall; other plumbing systems, HVAC systems, lighting systems, electrical systems, mechanical systems, communications systems, alarm systems, control systems, and other relevant systems such as elevators. For buildings that have experienced structural damage due to a disruptive event, recommissioning also incorporates the implementation of pertinent repairs and alterations to ensure structural stability, structural integrity, and acceptable performance of building systems.

Standards and guidelines are available for building commissioning, including NFPA 3 [15] and ASHRAE/IES 202 [16]. Additional guidelines and best practices for building recommissioning are described in documents such as those published by the Building Commissioning Association [17]. Standards and guidelines for building recommissioning of facilities with SAI technologies would either need to be developed or quality assurance for these technologies be incorporated into existing building commissioning ones. The adoption and implementation of SAI technologies will require the continued evaluation of these technologies to inform the need for upgrades. This process will impact building recommissioning, especially when technology improvements are expected to develop continuously, which may translate into multiple upgrades over the design life of the building. Current initiatives and efforts to encourage the commissioning of building automation systems (BAS) constitute a viable path for recommissioning practices that explicitly include SAI technologies.

5.6.1 Condition Analyses

The process of building recommissioning involves evaluating whether the performance of a building and its systems is commensurate with the owner's and/or society's needs and expectations. Post-fire recommissioning includes building condition assessment of fire suppression systems, fire barriers, and structural systems. This assessment should be informed by on-site testing of systems and their functionality (when feasible) as well as data collected and synthesized following a post-event site reconnaissance. The main objective is to evaluate and identify building components and systems that experienced damage during a disruptive event to determine whether they need to the repaired, replaced, or simply upgraded. An important component of post-event condition analyses is the evaluation of the condition, functionality, and interaction between building systems, other buildings (when applicable), and utilities and infrastructure lifelines systems, e.g., those that supply water and power to the building, which may or may not have been compromised during the event. A comprehensive post-event building condition analysis cannot be performed independently of the functionality of infrastructure lifelines systems given their strong interdependence.

Ideally, enhanced condition analyses would include quantitative information on fire intensity and duration, structural member strains, the integrity of fire barriers and the performance of fire suppression systems. These analyses should be based on reliable data on structural integrity and functionality, first principles, and sound engineering judgment to guide the development of design and recommendations for repairs and alterations. The successful implementation of SAI would provide the necessary data to assist in this process and facilitate a more reliable quantification of expected recovery times and costs associated with these efforts.

5.6.2   Repairs and Alterations

Repairs and alterations to structural systems that would result from the condition analyses described in Sect. 5.6.1 must be consistent with procedures to estimate demands and capacities as specified in relevant model codes, structural codes, and standards. The sensing capabilities of SAI have the potential to play an important role in the quantification of demands and capacities of building systems and structural components. This would require incorporation of SAI technologies as part of acceptable analysis and design strategies in relevant codes and standards. In the case of building codes, at this time, this effort seems to be a long-term objective given the inertia historically associated with changes to code requirements. However, the successful development of performance-based design requirements of ASCE/SEI 7 Appendix E provides a blueprint, and perhaps a suitable avenue, for potential incorporation, albeit incremental, of technologies that would foster repair and alteration strategies that take advantage of and account for SAI to improve fire safety.

Other building systems may also require repairs, alterations, or replacement as informed by condition analysis. This presents an opportunity for upgrades that incorporate SAI, e.g., advanced sensing technologies, to improve efficiency, monitoring, and reliability. As stated in Sect. 5.6, building recommissioning practices should be modified to be able to account for the presence of SAI technologies in a consistent manner.

5.7   Retrospective

This chapter provides a brief summary of relevant opportunities and challenges related to the incorporation of SAI into building/fire codes. The discussion focuses on the identification of elements of building codes and the fire regulatory context where SAI can enhance the performance and resilience of fire suppression systems, means of egress, and increase life safety. This chapter also addresses two areas that can accommodate SAI and would greatly benefit from it: building damage assessment and building recommissioning. It is recognized herein that in the vast majority of facilities, owners and stakeholders must ultimately deal with three competing factors when designing, constructing, operating, and maintaining, and when required, recommissioning their facilities: *Time* (investment period), *Function* (efficient use of investment), and *Risk* (hazard potentials and consequences of failure). Buildings are designed with life safety as a baseline performance target. For essential facilities, continued functionality is another explicit performance objective that must considered. Any application of SAI into design practice will provide quantitative, near real-time data to aid in the development of designs, operations and monitoring plans, and building performance that result in an acceptable balance between time, function, and risk during the expected design life of the facility.

Even if explicitly permitted by code, designers would need to extend their skills (or acclimate to subcontracting with specialists) in order for SAI to truly take hold in the building industry.

It is the intent of the authors to cover in this chapter important components of the abovementioned topics, albeit not comprehensively, with the goal of providing basic suggestions and foster new ideas to increase fire safety via SAI. The successful design, construction, and performance of SAI promise significant advancements in fire safety, but the road to seamless integration with entrenched methodologies may be less certain. At a minimum, this evolution will require time and a concerted effort from researchers, engineers, architects, regulators and building officials, manufacturers, insurance companies, building owners, and all relevant stakeholders to continue to develop these technologies, test them, and ultimately incorporate existing and future SAI components into a fire regulatory framework conducive to implementation.

References

1. NFPA 13: Standard for the Installation of Sprinkler Systems, National Fire Protection Association, Quincy, MA, USA, 2019
2. Image from Wikipedia (open source), 2020
3. NFPA 2001: Standard on Clean Agent Fire Extinguishing Systems, National Fire Protection Association, Quincy, MA, USA, 2018
4. NFPA 72: National Fire Alarm and Signaling Code, National Fire Protection Association, Quincy, MA, USA, 2019
5. UL Fire Resistance Directory, Underwriters Laboratories, Northbrook, IL, 2020
6. Hall, J., "U.S. Experience with Sprinklers and Other Automatic Fire Extinguishing Equipment," National Fire Protection Association, Quincy, Massachusetts, 2010
7. ASCE/SEI 7: Minimum Design Loads and Associated Criteria for Buildings and Other Structures, American Society of Civil Engineers: Structural Engineering Institute, Reston, VA, 2016
8. International Building Code (IBC), International Code Council (ICC), Washington, DC, 2018
9. LaMalva, K.J., ASCE/SEI Manual of Practice No. 138: Structural Fire Engineering, American Society of Civil Engineers: Structural Engineering Institute, Reston, VA, 2018
10. ATC 20, Report: Procedures for Post Earthquake Safety Evaluation of Buildings, ATC 20, prepared by the Applied Technology Council, Redwood City, California, 1995.
11. ATC 45, Procedures for the Safety Evaluation of Buildings after Windstorms and Floods, ATC 45, prepared by the Applied Technology Council, Redwood City, California, 2004.
12. NIST GCR 14-917-29, A Framework to Update the Plan to Coordinate NEHRP Post-Earthquake Investigations, prepared by the NEHRP Consultants Joint Venture, National Institute of Standards and Technology, Gaithersburg, MD, 2013.
13. NIST SP 1190, Community Resilience Planning Guide for Buildings and Infrastructure Systems, National Institute of Standards and Technology, Volume I, Gaithersburg, MD, 2016.
14. GSA, The Building Commissioning Guide, Public Building Service, Office of the Chief Architect, U.S. General Services Administration, 2005.
15. NFPA 3: Standard for Commissioning of Fire Protection and Life Safety Systems, National Fire Protection Association, Quincy, MA, USA, 2018.
16. ASHRAE/IES 202: Commissioning Process for Buildings and Systems, American Society of Heating, Refrigerating, and Air-Conditioning Engineers, Atlanta, GA, USA, 2018.
17. http://www.bcxa.org/wp-content/pdf/BCA-Best-Practices-Commissioning-Existing-Construction.pdf

Perspectives of Using Artificial Intelligence in Building Fire Safety

6

Xinyan Huang, Xiqiang Wu, and Asif Usmani

6.1 Introduction

Urban infrastructure has become a defining characteristic of modern human societies. The density, scale, condition, and complexity of the urban infrastructure are reliable measures of socio-economic development on a national and global scale. Driven by globalization, the fast urbanization, especially in developing countries, and the vast growth of wildland–urban interface (WUI), disastrous fires in such dense and complex urban environments (Fig. 6.1) pose extreme challenges to fire safety design, firefighting, emergency response, and recovery. For example, the 2017 Grenfell Tower fire in London resulted in more than 70 fatalities, and the 2018 California Camp Fire caused at least 85 fatalities and destroyed more than 18,000 structures. Today, the cost of fire on society is estimated at 1–2% of global annual GDP [1, 2] and correlates well with per capita GDP and human development index; therefore, the cost is likely higher for industrialized countries and regions.

High-density urban areas rely on complex interconnected infrastructure (e.g., high-rise buildings tunnels and underground spaces), while the high-density populations stretch the capacity of infrastructure resilience in the event of an urban disaster requiring extremely efficient emergency management. These infrastructures are therefore highly vulnerable to major disruptive fire events (e.g., facade fires and the increasing global WUI fires) and derived hazards (e.g., the large-scale urban fire after earthquakes) and require innovative solutions to be resilient.

Over the past decade, smart techniques of artificial intelligence (AI), Internet of things (IoT), cloud and edge computing, sensor and communication networks are gradually adopted in the building and construction area [3, 4], as introduced in other chapters. So far, most AI methods and applications in fire engineering focused on

X. Huang (✉) · X. Wu · A. Usmani
The Hong Kong Polytechnic University, Kowloon, Hong Kong
e-mail: xy.huang@polyu.edu.hk

© Springer Nature Switzerland AG 2022
M. Z. Naser, G. Corbett (eds.), *Handbook of Cognitive and Autonomous Systems for Fire Resilient Infrastructures*, https://doi.org/10.1007/978-3-030-98685-8_6

Fig. 6.1 (**a**) London Grenfell Tower fire, 2017 and (**b**) California Camp Fire, 2018 (source: Wikipedia)

fire detection, fire risk assessment [5–7], and structure analysis [8–10]. Only limited studies adopted AI to forecast the fire behaviours in compartments [11–17] and tunnels [18–20]. For example, Hodges et al. [11] used a transpose convolutional neural network (TCNN) and simulating results conducted by FDS to predict the temperature distribution inside compartment rooms. Arjan et al. [13] adopted a logistic regression model to predict the occurrence of flashover in a compartment using the information of fuel thickness, burning intensity and duration. Wang et al. [17] applied the machine learning method to predict the flashover of a compartment fire with a simulation database. Lee et al. [21] compared the performance of various models and demonstrated the feasibility of their faster regional CNNs in detecting fire, meanwhile reducing false positives. More recently, Wu and co-workers [18–20] explored the methodology of using deep learning AI methods to detect and predict critical tunnel fire events.

Today, most of the fire safety budget for a city is spent on the firefighting actions (i.e., withstand) and post-fire investigation and reconstruction (i.e., recovery), especially in the wildland–urban interface (WUI) [1, 2]. In contrast, only limited attention is paid to the pre-fire preparation, and it is a common problem in every country, as illustrated in Fig. 6.2. Aided with AI technology, the fire safety related budget could be reduced and used more cost-effectively. The infrastructures could be designed with AI in a smarter way to reduce the fire risk. The AI can also help monitor the fire development and guide the fire evacuation, so that firefighters will be more prepared for the fast-changing fire, extinguish the fire faster, rescue more people, and minimize the casualty. Eventually, the fire casualties and damages can be reduced, so that the cost in reconstruction will be minimized. As Murphy's law says, "*whatever might happen will happen.*" The best thing we can do is to be better prepared, and AI can make our preparation more wisely.

This chapter will review the latest understanding of AI applications in fire safety engineering by comparing with the application of computational fluid dynamics (CFD). The latest developments of fire databases and algorithms for reconstructing fire scenarios are also reviewed. We also propose three new concepts for applying

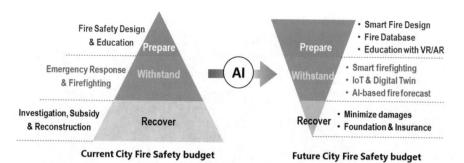

Fig. 6.2 The vision of the AI's role in shaping the fire safety of the future city

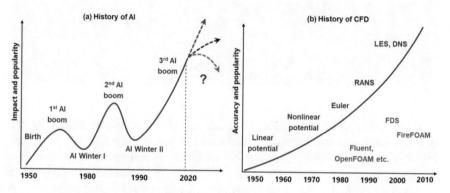

Fig. 6.3 The history of (**a**) AI research, and (**b**) CFD and application in fire engineering

AI in building fire safety, (1) the AI-based fire engineering design to improve the structure fire safety, (2) the building fire *Digital Twin* to monitoring the fire risk and fire development, and (3) the Super Real-time Forecast (*SuRF*) of the fire evolution.

6.2 Foundations of AI-Based Fire Engineering

6.2.1 AI vs. CFD in Fire Engineering

The concept of artificial intelligence (AI) was arguably first proposed in 1956 that aims to use computer algorithm to handle specific tasks, like language understanding, data storage, and pattern matching [22]. However, the progress in the AI algorithms and its application are not always promising and booming (Fig. 6.3a), which is different from conventional research areas, such as CFD. Because of the enormous gap between the ambitious expectation of AI's application and the limited computational resources, there were two "AI winters" in the 1970s and the late 1980s, when most of the research fund on AI was withdrawn.

Table 6.1 The comparison between CFD and AI application in fire safety engineering

	Facts of CFD applications in fire	Predictions of AI applications in fire
Principle	Solving partial differential equations	Pattern matching with the database
Process	Geometry, Boundary & Initial Conditions → CFD → Temporal & spatial of flow, smoke, temperature	Building & Fire → AI + Database → Temporal & spatial of flow, smoke, temperature
Process time	Hours to days	Seconds to minutes
Development	Mature and commercialized	Early stage
Users	Led by fire experts (not by CFD experts)	Led by fire experts (not by AI experts)
Fire safety design	Widely used in the building fire safety performance-based design (PBD)	Smart AI-based building fire safety design will be the future trend
Fire forecast	Rarely used to predict real fire events (costly and questionable)	Enable fire prediction and forecast based on big fire database and digital twin
Firefighting	Difficult to support the firefighting (slow computation and not real-time)	Play a key role in the smart firefighting (data-driven and real-time forecast)

In the early 2000s, the third booming of AI research was driven by the development of powerful computer hardware and the new algorithms. The emerging deep learning models, including convolutional neural networks (CNNs) and recurrent neural networks (RNNs) [14, 23], imitate the human brain and have been successfully applied to many problems in academia and industry. Compared with conventional AI models, such as machine learning models, the greatest strength of deep learning models is automatic feature extraction, although this often puts forward a higher requirement on the volume and variety of the data for updating the model through training. Although AI has been widely used in every aspect of scientific and engineering research, the question of "Is Another AI Winter Coming?" still remains. Currently, the use of AI in fire safety is still new and far from mature, which requires more multidisciplinary research. Nevertheless, it is also a great opportunity for AI to reshape traditional fire safety research and firefighting strategies and promote the concept of resilient city and smart emergency response.

To understand the role of AI in building fire safety and fire research, we can refer to the history of adopting CFD by the fire community over the last few decades and make a comparison (see Table 6.1). As a conventional engineering discipline, fire engineering over the past century used empirical formulas and correlations for design and education. These empirical laws are mathematically derived from experimental data and supported by analytical and scale modelling formed by fire researchers [24]. Since the late 1990s, the emerging CFD tools and software, e.g., the Fire Dynamics Simulator (FDS) developed by NIST and FireFOAM developed by FM Global, revolute the entire fire community. Today, CFD tools are commonly used by fire engineers in fire engineering analysis (FEA) or performance-based design (PBD) for innovative architectural design, such as super high-rise buildings, large atriums, and long tunnels. The CFD fire modelling is a part of the education programme for fire engineers, a strong support for fire investigations [25], and a valuable tool for fire researchers.

It is worth noting that the major users of CFD fire engineering tools are fire safety scientists and engineers, rather than the mathematics and computer scientists who developed the differential equation solvers of the governing equations or the CFD software developers. Although there are many problems with regard to the accuracy, quality, and reliability of modelling results, a user with a good knowledge of fire dynamics and fire engineering principles can use CFD-based fire models to reveal fire phenomena and support fire safety design. Therefore, it is expected that AI will also become a powerful tool for future fire engineering application, fire research, and even firefighting. Moreover, fire engineers and researchers will be the primary user of these new AI tools, rather than the AI and computer scientists. Nevertheless, developing these AI tools for the fire community needs the multi-lateral collaboration of computing and IT communities, frontline firefighters, regulators, and policymakers. For fire engineers and researchers, it is also encouraged to have good knowledge about the AI algorithm, instead of treating AI entirely as a "black box." Knowing the principle of AI methods helps users understand the reliability and limitations of AI's output.

Moreover, the future application of AI will be much broader than conventional CFD tools. One of the significant limitations of CFD fire modelling is the long-lasting and costly computational process, especially for large and complex buildings. Therefore, CFD fire modelling can only support the PBD design, fire investigation and research, which are less sensitive to time. Even for these areas, only limited numbers of CFD fire simulations are conducted, as limited by the computational power and cost. Most of fire simulations are done case by case repeatedly without creating a database, so that the knowledge gained from massive fire simulations cannot be shared by the community. For the firefighting activities and emergency response, the contribution from the CFD fire modelling is negligible because it cannot generate real-time information, even if the most powerful computer is adopted. On the other hand, with the assistance of big data, IoT, drone, and advanced communication system, the data-driven AI technology can effectively resolve these issues by (1) forming a large experimental and numerical database that can train a powerful AI, and (2) the trained AI, fed by the data from IoT system connected to fire scenes, can achieve a super real-time fire forecast that can support the decision making of the firefighting operation.

6.2.2   Establishment of Fire Scenario Database

A large and reliable training database is crucial to develop a well-performed AI model. To enable AI model to learn the hidden features of fire phenomena, the quantity and diversity of the database should be sufficiently large and representative [26]. This could be achieved by considering the influences of different parameters and scenarios in fire tests and simulations. The database can be constructed based on previous data from fire tests and numerical simulations. In general, the establishment of the database is a multi-step task, (1) data collection by searching all available literatures and extracting all useful data from these documents, (2) data pre-processing by data quality check, outliers and noises removal, and filtering, and (3) data mining by extracting valuable information.

Because of the complex nature of fire phenomena and data, it is still challenging to properly store, manage, maintain, and analyse a big fire database. Most of the fire test data, especially for commercial products, are not open to the public. Moreover, the current fire test and research community has a poor habit of unwilling to share data, making the establishment of a big fire database even harder. Therefore, it is urgent to establish a global standard and guideline for organizing, presenting, and sharing fire research data as a community effort. There have been some initiatives on collecting experimental and numerical data for the fire modelling validation and verification, such as the IAFSS working group of Measurement and Computation of Fire Phenomena (MaCFP) [27]. New initiatives for establishing fire database for AI applications are needed in the future.

6.2.2.1 Experimental Fire Database

For a typical fire test, a number of parameters are commonly measured, and the data can be classified into two main categories. One category is sensor data, such as the temperature by thermocouples or other heat detectors, the presence of smoke by detectors, CO and CO_2 by gas detectors, and heat flux by a radiometer or plate thermometer. Sensors are often installed at one or many locations for long-term measurement. In other words, sensor data are often time sequences of point or line measurements, which have both spatial and temporal dimensions. The other category is the visual data, such as the video from closed-circuit television (CCTV) and infrared cameras that are commonly installed inside the building. The visual data also include airborne cameras carried by airplanes and drones and the processed satellite images [28], particularly for large-scale fires in urban, wildland, WUI [29], and informal settlements [30].

Currently, the database can only be established with all available sensor data that can be accessed online, such as journal publications and technical reports. Nevertheless, because most of the existing fire test data initially are not produced and presented under the principle of sharing, it is extremely challenging to extract all the necessary information and data from the accomplished and ongoing tests, as well as to make a fair comparison among existing data. Figure 6.4 illustrates the process to construct a database for tunnel fire based on the published articles and reports [19]. Through a thorough literature review, the literature can first be categorized by key parameters in tunnel fire, such as the fire size, smoke layer thickness, critical velocity, back-layer length, and plug-holing. Afterwards, all related experimental data were extracted from these documents. To enforce a convenient search in the big database, all raw data and images should be well named, organized and indexed with a detailed description. Since the fire process is generally complex and influenced by multiple factors, it is essential to provide adequate test information before the data process.

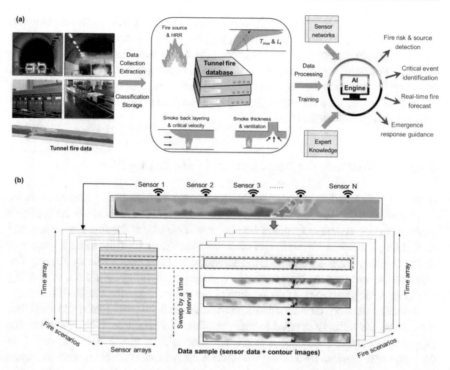

Fig. 6.4 (**a**) Processes of constructing an experimental and numerical database for smart tunnel firefighting [19], and (**b**) the constructing a numerical database for tunnel fire [20]

6.2.2.2 Numerical Fire Database

It would be ideal if adequate sensor data can be obtained from full-scale fire experiments or real fire incidents. However, the existing tunnel fire data from the literature are not large enough or sufficiently well organized to support deep learning. Alternatively, a database formed by massive CFD-based numerical simulation could enable AI-based fire forecasting. Particularly, the open-source programme of FDS [31] has been extensively verified and validated and used as a standard tool for performance-based fire safety design and fire research. Moreover, the CFD modelling results can provide much more detailed temperature information than any experiment, which are ideal for the verification of the established AI method for the fire forecast.

For numerical simulations, the visual data refer to the contours and videos generated from computational results. These time-sequence data (e.g., images and videos) can directly show the real-time scene and scale of the fire, evacuation process, distributions of smoke and temperature, and firefighting activities. Compared to sensor data in fire tests, the 2-D or 3-D numerical data are several orders of magnitude larger, which can show more details about the fire process. In contrast, the video data in fire tests are quite rare and mostly unpublic. Even if there are fire test videos, they are less uniform because of various angles of the shot, and their large size and complexity prohibit further data processing and scientific analysis. Figure 6.4b shows a typical construction process of a numerical database for tunnel fire [20], where the point sensor data matrix and 2-D temperature contour data are coupled. The point sensor data are used because in the fire test and real application, the point temperature data by thermocouples are most common. Such a coupling enables a database that combines limited experimental data and more abundant and detailed numerical simulations.

6.2.3   AI Methods for Detecting and Forecasting Fire

In a real building fire event, the hot, toxic, and fast-spreading smoke is more dangerous than the flames. The rapid spread of heat and smoke flows in the confined space not only endangers human life but also challenges the fire evacuation and firefighting strategies. A quick and accurate identification of the location and size of the original fire source is of great scientific and practical value in guiding fire suppression and rescue. Nevertheless, in a real firefighting operation, it is a big challenge to acquire the size and location of fire points inside the smoke-filled space.

Recently, data-driven AI methods, especially artificial neural networks (ANNs), have been increasingly applied in fire detection. For example, Choi and Choi [6] proposed a system including a flame detector with a visual light camera. Fire detection was handled by solving a nonlinear classification problem with a support vector machine (SVM) classifier, but it has a high false-alarm rate. To solve this problem, deep convolutional neural networks (CNNs) have also been applied [32, 33]. Without detailed engineering feature extraction, CNNs were claimed to automatically increase the fire detection accuracy via deep learning. Deep CNNs also showed their capability in detecting building fire with sensor data [34–37] and large-scale wildland fire with geoinformatics data [38].

Recurrent neural networks (RNNs) are specially designed for the prediction or classification of temporal data. As illustrated in Fig. 6.5a, the loop structure can be unfolded into a chain of cells. It should be noted that the RNN cells are the same cell at various states rather than different cells. The order-dependent data are put into the structure and reserved, and then the information hidden inside the sequence of data can be interpreted as the output of the structure. Though it has been used in many areas, such as speech recognition [39], the major limitation of RNN is the difficulty in treating long pieces of sequence [40].

To handle this problem, Hochreiter [41] proposed a new type of RNN, i.e., LSTM, by introducing a specific internal structure. LSTM cell is comprised of three gates. When data are imported into the cell, the forget gate, input gate, and output gate together decide whether to block or pass it on, how to treat it, and how to output it, respectively. The information inside the cell will be updated by forgetting and inputting gates. Therefore, after importing a piece of data sequentially, useful information can be remembered and translated for the designed output. LSTM network has been demonstrated as a powerful network to effectively solve complex prediction problems [42]. Wu et al. [18] built up a regression two-layer LSTM

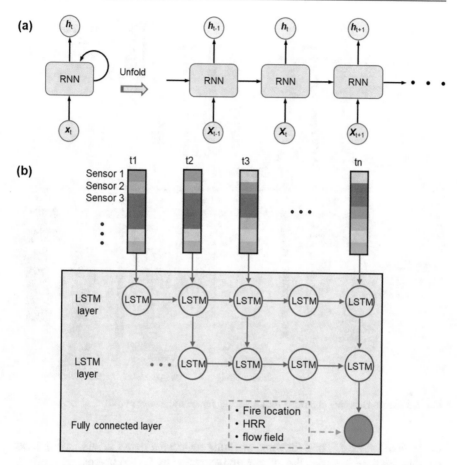

Fig. 6.5 (**a**) Data processing in typical RNN cell, and (**b**) LSTM method for fire detection

network model for identifying fire scenarios in tunnels (see Fig. 6.5b). With a numerical database generated in Fig. 6.4b, the LSTM unit in the input layer receives a vector of temperatures measured by sensors sequentially. The received data is then treated and passed on to the next state. Then, a three-element vector (fire location, HRR, and ventilation speed) can be predicted with an accuracy of 90%. To verify the accuracy of AI-based fire detection, a sensitivity analysis should be carried out to optimize the database configuration and spatial-temporal arrangement of sensors.

Apart from monitoring the current fire scenario, a timely forecast of future fire development is also significant to alleviate the potential casualties, in view of the rapid evolvement of the extreme temperature induced by fire. Based on the LSTM model, we proposed a deep learning model combing an LSTM model and a TCNN model, and the model is composed of LSTM, dense, and TCNN layers, as shown in Fig. 6.6 [20]. The input layer read a data of 32 (No. of the sensor) × 30 (s) (length

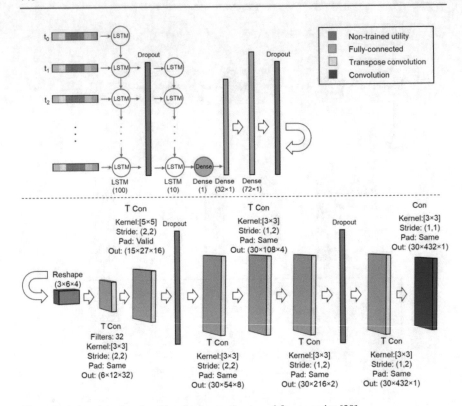

Fig. 6.6 A deep learning algorithm for forecasting tunnel fire scenarios [20]

of the data sample). The following two fully connected dense layers having more neurons are used to enrich the information extracted by LSTM layers.

After up-sampling through for six TCNN layers and one CNN layer with various padding and stride parameters, a greyscale image illustrating its height and width can be generated. The generated image can then be compared with the actual one, and their difference will be minimized during the training process. After establishment, the model is trained with the numerical database constructed, as shown in Fig. 6.4. Then, a real-time forecasting of the spatial-temporal temperature distribution inside the tunnel in the next 1–10 min can be achieved within 1 s [20].

6.3   Applications of AI in Building Fire Safety

This section will provide a vision and show the potential applications of big fire database and AI algorithm to improve building fire safety from the fire engineering in the design stage to the fire safety management in the daily operation, and to the final smart firefighting once a fire occurs.

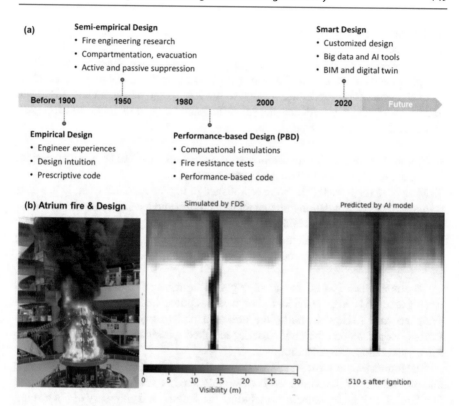

Fig. 6.7 (**a**) The evolution of building fire safety code and design, and (**b**) smoke motion inside an atrium by both CFD simulation and AI prediction [49]

6.3.1   AI-Based Fire Engineering Design

The performance-based design (PBD) is an objective-oriented design approach that goes beyond building prescriptive code requirements [43–48]. PBD for fire safety was emerging from the 1970s to the early 1990s, driven by the innovative architectural design and construction of new materials [45, 46] incentivizes new fire research and engineering tools, such as the CFD and evacuation models (see Fig. 6.7a). Today, the PBD has been widely adopted in fire engineering design globally by consulting company and accepted by the authority having jurisdiction (AHJ).

Ideally, performance-based fire safety design brings design flexibility to the building by introducing acceptable methodologies, analysis tool, and performance criteria. However, the actual PBD in many projects is adopted primarily to reduce the construction and installation cost rather than improving fire safety. Some common technical issues are involved in the fire engineering PBD, for example, (1) a very low design fire, typically lower than 5 MW, (2) smoke management system not verified by actual experiments, and (3) over-use the concept of ASET and RSET without experiments verification and consideration for firefighters. Thus, it is questionable if PBD can really improve the building fire safety, which can be only verified when the fire occurs.

Moreover, for a typical fire engineering PBD project, the engineering-consulting company needs to submit the developed fire strategy and documentation (including CFD and evacuation modelling results) to the AHJ. Both designers and AHJ need to have an extensive experience and scientific knowledge to evaluate the PBD [50], and there are three significant issues.

1. Enormous time and human resources are needed for each PBD case in the design, documenting, review, and approval processes.
2. Many criteria used in PBD were established in limited amounts of early fire tests and numerical simulations, so that their reliability and feasibility for the new built environment are questionable.
3. Fire engineers and designers develop "tricks," e.g., tunning fire modelling parameters and hide poor-performance cases, to bypass the scrutiny of AHJ.

To solve these issues, the emerging development of big data and artificial intelligence (AI) may provide future design approaches and solutions for both designers and AHJ. Potentially, the powerful pattern matching capacity of deep learning could reduce the unnecessarily repeated workload, improve building fire safety, and reduce the construction cost.

The foundation of most AI applications is the pre-establishment of a big database and pre-training of the AI model. For example, the fire smoke motion in the atrium is influenced by multiple factors, including the volume and geometry of the atrium, fire size in terms of heat release rate (HRR), and the ventilation condition of the atrium design (Fig. 6.7b). All these parameters are considered in constructing a fire scenario library representing the potential atrium fire scenarios. The evolvement of the smoke visibility with time inside the atriums are then examined via numerical simulations. The simulated results of all fire scenarios are collected to form a big database for the training of the AI model, which is optimized to achieve the best performance.

Although it takes a long time to form a big database of different fire scenarios and building structures, once accomplished, the trained AI model can be packed into a software tool for direct use by designers and AHJ. Such a process is similar to the development and verification of CFD fire modelling tools, which have been commercialized after many years of developments and updates. The application of AI tools in the PBD also includes three stages:

1. *Data pre-process and database establishment.* Key input factors, such as fire HRR, atrium geometry, and smoke-ventilation designs, are chosen to run thousands of numerical simulations.
2. *AI Prediction.* Given the input information, the fine-tuned AI model can predict the entire temporal and spatial fire process, e.g., the development of smoke motion and visibility profile for the atrium fire. The time for AI prediction takes just a few seconds, like the empirical correlation.

3. *Output and application*. The output of the AI model could calculate ASET, which can be compared with PBD requirement, and determine the possible range of each design parameter, e.g., the maximum fire size under the given ventilation, and provide optimal design options.

Therefore, the laborious and costly trial-and-error process of CFD modelling for every consulting case can be avoided. On the other hand, the AHJ can also apply the proposed AI tool to quickly check the reliability of the conventional PBD without conducting similar CFD fire modelling processes. Ultimately, the design and review process can be reduced from months for conventional PBD to hours and minutes for AI-based PBD.

Such AI-based smart fire engineering design has been demonstrated for the Atrium fire and the prediction of ASET (Fig. 6.7a) [49]. For a given consulting project, instead of repeatedly running selected time-consuming CFD-based numerical simulation cases and guessing the limiting design conditions, AI methods can quickly generate thousands of modelled fire scene images varying with space and time, and then determine if the given fire and building parameters satisfy the fire code. Moreover, AI can determine the optimal multi-dimensional parameter space for different design fires, volume and shape of the building, ASET and REST, safety factor, and fire code.

Currently, the AI-based design software, Intelligent Fire Engineering Tool (IFETool) [51], is under development along with the growing numerical database, and such modelling/design software is expected to be widely used in future smart fire engineering design for different types of buildings. Eventually, the building fire safety can be significantly improved by the AI's powerful optimization capacity; meanwhile, the cost in design, review and approval, construction, and installation can be reduced.

6.3.2 Building Fire Digital Twin

The current "intelligent" building fire protection system connects all fire alarms, smoke detectors, and perhaps the pressure information of the fire sprinkler system, so it can provide an exact location of the fire event. In other words, such a system enables good fire detection in the early stage of the fire and provides an early alarm to residences. However, in terms of fire prevention before the occurrence of fire and the monitoring of fire scene, evacuation and rescue process after the fire events, the existing fire protection system can provide very little information. Therefore, to fundamentally improve the building fire safety system, it is necessary to install all fire safety related sensors and connect them with the building information system and the *Digital Twin* (Fig. 6.8).

The concept of the *Digital Twin* was initiated in the early 2000s and arguably first adopted by NASA in the attempt to improve the physical model simulation of spacecraft in 2010. Today, the digital twin has been applied to aircraft engines, manufacturing processes [52], and large infrastructures like the smart building and

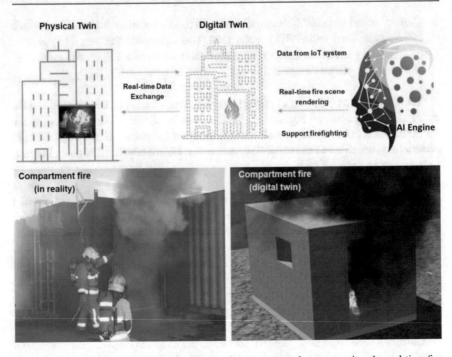

Fig. 6.8 Smart fire digital twin for a building and compartment that can monitor the real-time fire behaviours (Credit: Xiaoning Zhang and Xinyan Huang)

wind turbine, but its overall technology is still far from mature [52]. The *Digital Twin* will be established upon the Building Information Model (BIM) in the design stage and include all sensors and camera information from the IoT system. For most existing buildings without a BIM, a *Digital Twin* could also be constructed by using MicroGIS, Computer Vision, and 3D virtual reality model. However, more advances are needed for applying these technologies together and smoothly. Particularly, different technologies use different software, programming languages, and communication protocols, so that it is extremely challenging to transfer data from real fire scene to its symbolic representative in the digital twin. Once built up, the *Digital Twin* can be regularly updated by the fire risk information, such as the fuel loads in each fire zone and the transport of high fire hazardous materials inside the building.

Once a fire occurs, it is difficult to acquire detailed information from the fire scene, and the heavy smoke from the fire also quickly diminishes the visual information before the arrival of firemen. The key information of fire, such as the locations and heat release rate (or power) of the fire sources and the evacuation progress of residence, plays a critical role in emergency response and firefighting activities. Except for the fire alarm, conventional smoke and heat detectors cannot provide additional information about the fire scene. Nevertheless, the building fire *Digital Twin* will be coupled with a pre-established fire database that is used to train the AI engine. The fire database includes thousands of CFD-based fire simulations, experimental data and empirical correlations of the specific building. With the real-time information, such as data from the temperature sensors and cameras, a pre-trained AI engine can quickly recognize, visualize, and render the fire scene in the *Digital Twin*. A demonstration intelligent digital twin system has been developed for fires in a reduced-scale tunnel with AI and IoT sensor networks to identify the real-time fire size and temperature distribution [53].

Figure 6.8 shows a Fire *Digital Twin* demonstration system for the compartment fire backdraft test facility in the Hong Kong Fire Services Department. Inside the test chamber, 16 thermocouple sensors are installed to monitor the internal temperature distribution. Then, these data are transferred to the cloud and the AI engine. The AI engine is pre-trained with both past experimental data and hundreds of CFD fire simulations. Thus, once the AI is fed with real-time temperature data, its pattern matching capacity can quickly recognize the fire scenario. Such information is further transferred and rendered on the *Digital Twin*, so that the real-time monitoring of the compartment fire can be achieved. More sophisticated systems can be embedded into existing HVAC system and its sensor networks for more complex buildings and infrastructures, so that the real-time fire scene inside the building and behind the heavy smoke can be monitored using an upgraded building facility management system.

6.3.3   Smart Fire Forecast and Fighting

Traditional firefighting does not use live sensor data (or a *Digital Twin*) in support of intervention tactics, because of the limited information collection, inaccurate fire modelling, and slow communication systems. Different approaches have been proposed to support firefighting, such as data-analysis techniques to assist fire modelling [4, 54] and direct deployment of sensors during the event, as reported in a comprehensive state-of-the-review of smart firefighting. The 1st-generation *FireGrid* system [34] applied the analogy of weather forecasting and sensor data assimilation [55]. By feeding the sensor data to the K-Crisp Zone Model [56], the *FireGrid* system demonstrated that super real-time forecasting with up to 10–15 min lead time was feasible, at least on the scale of a full-scale three-room apartment. While the basic aim of forecasting the fire evolution was achieved, the accuracy of the fire forecast is limited by the zone model and only for the small-compartment post-flashover fires. Compared to the simple and analytical zone model, the CFD fire model can provide a more accurate prediction if fed by the same sensor data. Nevertheless, for a more sophisticated CFD-based fire model [27], existing sensors in the building cannot provide adequate inputs to feed the model. Moreover, despite the enormous increase in computational power, CFD simulations can last for hours and days, that is, several orders of magnitude slower than real-time fire development. Therefore, it is not possible to apply the data-driven CFD approach for the real-time fire evolution forecast in complex buildings.

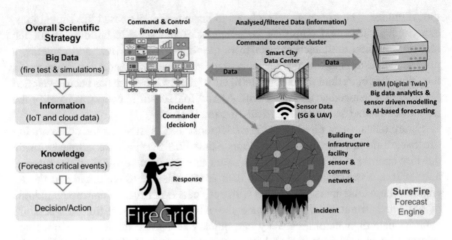

Fig. 6.9 The concept of AI-based SureFire forecast engine and the scope of proposed smart firefighting

On the other hand, once the building fire *Digital Twin* is constructed and in operation, the AI engine (or SureFire Forecast Engine in Fig. 6.9) can go one step further to forecast the fire evolution inside the building. More importantly, it can predict the critical events in real-time and with lead time, such as the flashover, travelling fire, unbearable environmental conditions for residence and firefighters, and the structure collapse. To achieve the super real-time fire forecast with 1–10 min lead time in practice, a few challenges should be resolved.

Firstly, an accurate fire forecast AI algorithm should be developed to predict critical fire events. Forecasting of fire growth and spread is a complex nonlinear multivariate spatiotemporal prediction problem, similar to weather forecasting, including wind [57], temperature [58], and air pollution [59]. The Critical Event Library, based entirely on big data analytics and AI/machine learning techniques, should first be developed, which set the goals of the fire forecast system. During the development of the Digital Twin, a big data-driven modelling framework and associated deep learning AI algorithms shall be developed to render the fire scenes. The AI engine can adopt existing deep learning algorithm, but its structure and user interface should be optimized for the application of fire forecast. With sufficient data filtering, feature extraction, predictive modelling and fire-severity evaluation, the AI engine can achieve *Super Real-time Forecasting* (SuRF) of the future fire scene inside the building at 10–15 min in advance within seconds. More importantly, additional AI model can be coupled with surrogate models (e.g., empirical correlations) and techniques (e.g., genetic algorithms, Kalman filter and Bayesian inferencing), to predict the probability of critical events, such as flashover and structure collapse. Such information will be transfer to the fire command centre to support the decision making of firefighting and rescue.

Secondly, reliable sensor networks for emergency response should be developed and installed to collect data. All sensors (e.g., thermocouples, cameras, smoke detectors) for smart firefighting should be a part of the *Building Automation System* (BAS) [60]. Many existing sensors and network can be used and upgraded for the smart firefighting system. For example, an existing smart building energy saving system will install massive temperature sensors in almost all rooms to control HVAC to ensure both comfort indoor environment and energy saving. These temperature sensor data can also be collected and fed to the smart firefighting system. In case of a fire event, these sensors can automatically switch to the high-frequency collection mode and send data into the AI model for both fire scene rendering and the fire forecast. Similarly, the CCTV camera network that used for the daily security purpose can switch to the fire emergency model to transfer the information of smoke and fire, residence evacuation, and firemen operations back to the cloud data centre as the input of the AI model and improve the accuracy of the building fire *Digital Twin*. More advanced sensors, such as IR camera, laser scanner, Lidar, gas and visibility sensors, can also be installed in the high-value infrastructure, such as the tunnel, warehouse, museum, and airport terminal, to provide more fire data and support the fire forecast.

Thirdly, a stable and self-healing communication network between sensors and digital twin is needed to support firefighting and rescue operation. The wired/wireless sensor IoT system connecting the sensor network and the digital twin can be categorized into infrastructure networks and ad hoc networks. Typically, infrastructure networks are those with fixed major routing nodes intended for long-term use, which should be designed and installed as part of the building services system. The existing BAS infrastructure networks can be fully wired or combining wired backbones and wireless (e.g., 5G cellular [61], WiFi [62], ZigBee, Bluetooth [63]) last-hops to provide balanced *Quality-of-Service* (QoS) and mobility. Because fire can randomly damage existing infrastructure, to communicate sensor data on the fire site, a BAS ad hoc network is a more appropriate choice. Ideally, each firefighter or drone can be equipped with an intelligent dispenser filled with small wireless sensor nodes (breadcrumb MANET [64–66]), which can be dropped in the fire scene to establish a temporal network. Once deployed, the breadcrumb MANET can also accurately localize themselves as well as the firefighter and the trapped people. By combining AI fire forecast with the self-healing sensor and communication network, smart firefighting can be achieved in practice.

6.4   Summary

This chapter reviews the state-of-the-art progress of applying AI technologies in fire safety engineering and smart firefighting. By comparing with the history of applying CFD in fire engineering, the roadmap for AI-based fire engineering is predicted. The methods and guidelines of constructing a large reliable training database utilizing experimental and numerical results are introduced. The AI algorithms and models that have a great potential to detect and forecast fire scenarios are discussed, and the latest research on exploring and developing intelligent firefighting systems are reviewed. Three new concepts are proposed for future intelligent building fire safety applications,

1. AI-based smart fire safety design, once trained by a large fire database, will predict the limiting design conditions within a timescale of seconds, which can minimize the cost of fire protection system and provide sufficient justification for AHJ's examination;
2. The building fire Digital Twin, combine IoT sensor network, fire database, and AI pattern matching engine, can display the real-time rendering of the smoke-covered fire scenes inside the building; and
3. The super real-time fire forecast, driven by the building fire digital twin, critical fire event library, self-healing communication network, and AI forecast engine, can predict the fire development and critical fire events in next 5–10 min, which can support the firefighting and rescue activities and decision making.

Acknowledgements This work is funded by the Hong Kong Research Grants Council Theme-based Research Scheme (T22-505/19-N) and the PolyU Emerging Frontier Area (EFA) Scheme of RISUD (P0013879).

References

1. The Geneva Association Staff, World Fire Statistics, The Geneva Association, 2014.
2. J. HALL, Calculating the total cost of fire in the United States, Fire Journal (Boston, MA). 83 (1989) 69–72.
3. C. Grant, A. Hamins, N. Bryner, A. Jones, G. Koepke, Research Roadmap for Smart Fire Fighting, NIST Special Publication 1191. (2015). https://doi.org/10.6028/NIST.SP.1191.
4. A. Cowlard, W. Jahn, C. Abecassis-Empis, G. Rein, J.L. Torero, Sensor assisted fire fighting, Fire Technology. 46 (2010) 719–741. https://doi.org/10.1007/s10694-008-0069-1.
5. Y. Cao, F. Yang, Q. Tang, X. Lu, An attention enhanced bidirectional LSTM for early forest fire smoke recognition, IEEE Access. 7 (2019) 154732–154742. https://doi.org/10.1109/ACCESS.2019.2946712.
6. J. Choi, J.Y. Choi, An integrated framework for 24-hours fire detection, in: Lecture Notes in Computer Science, 2016: pp. 463–479. https://doi.org/10.1007/978-3-319-48881-3_32.
7. N. Elhami-Khorasani, J.G. Salado Castillo, T. Gernay, A Digitized Fuel Load Surveying Methodology Using Machine Vision, Fire Technology. 57 (2021) 207–232. https://doi.org/10.1007/s10694-020-00989-9.
8. M.Z. Naser, H. Salehi, Machine Learning-Driven Assessment of Fire-Induced Concrete Spalling of Columns, ACI Materials Journal. 117 (2020) 7–16.
9. M.Z. Naser, A. Seitllari, Concrete under fire: an assessment through intelligent pattern recognition, Engineering with Computers. 36 (2020) 1915–1928.
10. M.Z. Naser, Mechanistically Informed Machine Learning and Artificial Intelligence in Fire Engineering and Sciences, Fire Technology. (2021). https://doi.org/10.1007/s10694-020-01069-8.
11. J.L. Hodges, B.Y. Lattimer, K.D. Luxbacher, Compartment fire predictions using transpose convolutional neural networks, Fire Safety Journal. 108 (2019) 102854. https://doi.org/10.1016/j.firesaf.2019.102854.
12. W.C. Tam, E.Y. Fu, R. Peacock, P. Reneke, J. Wang, J. Li, T. Cleary, Generating Synthetic Sensor Data to Facilitate Machine Learning Paradigm for Prediction of Building Fire Hazard, Fire Technology. (2020). https://doi.org/10.1007/s10694-020-01022-9.
13. A. Dexters, R.R. Leisted, R. Van Coile, S. Welch, G. Jomaas, Testing for knowledge: Application of machine learning techniques for prediction of flashover in a 1/5 scale ISO 13784-1 enclosure, Fire and Materials. (2020) 1–12. https://doi.org/10.1002/fam.2876.

14. S. Mahdevari, S.R. Torabi, Prediction of tunnel convergence using Artificial Neural Networks, Tunnelling and Underground Space Technology. 28 (2012) 218–228. https://doi.org/10.1016/j.tust.2011.11.002.

15. E.W.M. Lee, R.K.K. Yuen, S.M. Lo, K.C. Lam, G.H. Yeoh, A novel artificial neural network fire model for prediction of thermal interface location in single compartment fire, Fire Safety Journal. 39 (2004) 67–87. https://doi.org/10.1016/S0379-7112(03)00092-4.

16. R.K.K. Yuen, E.W.M. Lee, S.M. Lo, G.H. Yeoh, Prediction of temperature and velocity profiles in a single compartment fire by an improved neural network analysis, Fire Safety Journal. 41 (2006) 478–485. https://doi.org/10.1016/j.firesaf.2006.03.003.

17. J. Wang, C.W. Tam, Y. Jia, R. Peacock, P. Reneke, E. Yujun, T. Cleary, P-Flash – A machine learning-based model for flashover prediction using recovered temperature data, Fire Safety Journal. 122 (2021) 103341. https://doi.org/10.1016/j.firesaf.2021.103341.

18. X. Wu, Y. Park, A. Li, X. Huang, F. Xiao, A. Usmani, Smart Detection of Fire Source in Tunnel Based on the Numerical Database and Artificial Intelligence, Fire Technology. 57 (2021) 657–682. https://doi.org/10.1007/s10694-020-00985-z.

19. X. Zhang, X. Wu, Y. Park, T. Zhang, X. Huang, F. Xiao, A. Usmani, Perspectives of big experimental database and artificial intelligence in tunnel fire research, Tunnelling and Underground Space Technology. 108 (2021) 103691. https://doi.org/10.1016/j.tust.2020.103691.

20. X. Wu, X. Zhang, X. Huang, F. Xiao, A. Usmani, A real-time forecast of tunnel fire based on numerical database and artificial intelligence, Building Simulation. 15 (2022) 511–524. https://doi.org/10.1007/s12273-021-0775-x.

21. K.B. Lee, H.S. Shin, An Application of a Deep Learning Algorithm for Automatic Detection of Unexpected Accidents under Bad CCTV Monitoring Conditions in Tunnels, in: Proceedings - 2019 International Conference on Deep Learning and Machine Learning in Emerging Applications, Deep-ML 2019, 2019. https://doi.org/10.1109/Deep-ML.2019.00010.

22. S.J. Russell, P. Norvig, Artificial Intelligence: A Modern Approach, Peason Education Limited, Malaysia, 2016.

23. A. Jaafari, E.K. Zenner, M. Panahi, H. Shahabi, Hybrid artificial intelligence models based on a neuro-fuzzy system and metaheuristic optimization algorithms for spatial prediction of wildfire probability, Agricultural and Forest Meteorology. 266–267 (2019) 198–207. https://doi.org/10.1016/j.agrformet.2018.12.015.

24. D. Drysdale, An Introduction to Fire Dynamics, 3rd ed., John Wiley & Sons, Ltd, Chichester, UK, 2011. https://doi.org/10.1002/9781119975465.

25. S. Shyam-Sunder, R.G. Gann, W.L. Grosshandler, H.S. Lew, R.W. Bukowski, F. Sadek, F.W. Gayle, J.L. Gross, T.P. McAllister, J.D. Averill, Federal building and fire safety investigation of the world trade center disaster: final report of the national construction safety team on the collapses of the world trade center towers (NIST NCSTAR 1), (2005).

26. M. Chi, A. Plaza, J.A. Benediktsson, Z. Sun, J. Shen, Y. Zhu, Big Data for Remote Sensing: Challenges and Opportunities, Proceedings of the IEEE. 104 (2016) 2207–2219. https://doi.org/10.1109/JPROC.2016.2598228.

27. A. Brown, M. Bruns, M. Gollner, J. Hewson, G. Maragkos, A. Marshall, R. McDermott, B. Merci, T. Rogaume, S. Stoliarov, J. Torero, A. Trouvé, Y. Wang, E. Weckman, Proceedings of the first workshop organized by the IAFSS Working Group on Measurement and Computation of Fire Phenomena (MaCFP), Fire Safety Journal. 101 (2018) 1–17. https://doi.org/10.1016/j.firesaf.2018.08.009.

28. R.S. Allison, J.M. Johnston, G. Craig, S. Jennings, Airborne optical and thermal remote sensing for wildfire detection and monitoring, Sensors (Switzerland). 16 (2016). https://doi.org/10.3390/s16081310.

29. S.E. Caton, R.S.P. Hakes, D.J. Gorham, A. Zhou, M.J. Gollner, Review of Pathways for Building Fire Spread in the Wildland Urban Interface Part I: Exposure Conditions, Fire Technology. (2016) 1–45. https://doi.org/10.1007/s10694-016-0589-z.

30. A. Cicione, R.S. Walls, C. Engineering, Full-Scale Informal Settlement Dwelling Fire Experiments and Development, Springer US, 2020. https://doi.org/10.1007/s10694-019-00894-w.

31. K. Mcgrattan, R. Mcdermott, Fire Dynamics Simulator User ' s Guide (FDS Version 6.3.0), (2015).
32. K. Muhammad, J. Ahmad, I. Mehmood, S. Rho, S.W. Baik, Convolutional Neural Networks Based Fire Detection in Surveillance Videos, IEEE Access. 6 (2018) 18174–18183. https://doi.org/10.1109/ACCESS.2018.2812835.
33. N.K. Kim, K.M. Jeon, H.K. Kim, Convolutional recurrent neural network-based event detection in tunnels using multiple microphones, Sensors (Switzerland). 19 (2019). https://doi.org/10.3390/s19122695.
34. L. Han, S. Potter, G. Beckett, G. Pringle, S. Welch, S.H. Koo, G. Wickler, A. Usmani, J.L. Torero, A. Tate, FireGrid: An e-infrastructure for next-generation emergency response support, Journal of Parallel and Distributed Computing. 70 (2010) 1128–1141. https://doi.org/10.1016/j.jpdc.2010.06.005.
35. Y. Pei, F. Gan, Research on data fusion system of fire detection based on neural-network, Proceedings of the 2009 Pacific-Asia Conference on Circuits, Communications and System, PACCS 2009. (2009) 665–668. https://doi.org/10.1109/PACCS.2009.134.
36. Y. Yao, J. Yang, C. Huang, W. Zhu, Fire monitoring system based on multi-sensor information fusion, 2010 2nd International Symposium on Information Engineering and Electronic Commerce, IEEC 2010. (2010) 448–450. https://doi.org/10.1109/IEEC.2010.5533209.
37. C.J. Xue, The road tunnel fire detection of multi-parameters based on BP neural network, CAR 2010 - 2010 2nd International Asia Conference on Informatics in Control, Automation and Robotics. 3 (2010) 246–249. https://doi.org/10.1109/CAR.2010.5456677.
38. F. Cui, Deployment and integration of smart sensors with IoT devices detecting fire disasters in huge forest environment, Computer Communications. 150 (2020) 818–827. https://doi.org/10.1016/j.comcom.2019.11.051.
39. D. Lee, M. Lim, H. Park, Y. Kang, J.S. Park, G.J. Jang, J.H. Kim, Long short-term memory recurrent neural network-based acoustic model using connectionist temporal classification on a large-scale training corpus, China Communications. 14 (2017) 23–31. https://doi.org/10.1109/CC.2017.8068761.
40. Y. Bengio, P. Simard, P. Frasconi, Learning Long-Term Dependencies with Gradient Descent is Difficult, IEEE Transactions on Neural Networks. 5 (1994) 157–166. https://doi.org/10.1109/72.279181.
41. S. Hochreiter, Long Short-Term Memory, 1780 (1997) 1735–1780.
42. K. Greff, R.K. Srivastava, J. Koutnik, B.R. Steunebrink, J. Schmidhuber, LSTM: A Search Space Odyssey, IEEE Transactions on Neural Networks and Learning Systems. 28 (2017) 2222–2232. https://doi.org/10.1109/TNNLS.2016.2582924.
43. A.H. Buchanan, Implementation of performance-based fire codes, Fire Safety Journal. 32 (1999) 377–383. https://doi.org/10.1016/S0379-7112(99)00002-8.
44. M. Kohno, T. Okazaki, Performance Based Fire Engineering in Japan, International Journal of High-Rise Buildings. 2 (2013) 23–30. https://doi.org/10.21022/IJHRB.2013.2.1.023.
45. S.C. Tsui, Performance-Based Fire Safety Design in Hong Kong, International Journal on Engineering Performance-Based Fire Codes. 6 (2004) 223–229.
46. V. Beck, Performance-based Fire Engineering Design And Its Application In Australia, Fire Safety Science. 5 (1997) 23–40. https://doi.org/10.3801/iafss.fss.5-23.
47. D.L. Zhao, J. Li, Y. Zhu, L. Zou, The application of a two-dimensional cellular automata random model to the performance-based design of building exit, Building and Environment. (2008). https://doi.org/10.1016/j.buildenv.2007.01.011.
48. Q. Zhang, M. Liu, C. Wu, G. Zhao, A stranded-crowd model (SCM) for performance-based design of stadium egress, Building and Environment. (2007). https://doi.org/10.1016/j.buildenv.2006.06.016.
49. L.C. Su, X. Wu, X. Zhang, X. Huang, Smart performance-based design for building fire safety: Prediction of smoke motion via AI, Journal of Building Engineering. 43 (2021) 102529. https://doi.org/10.1016/j.jobe.2021.102529.
50. C.M. Fleischmann, Is prescription the future of performance-based design?, Fire Safety Science. (2011) 77–94. https://doi.org/10.3801/IAFSS.FSS.10-77.

51. Y. Zeng, X. Zhang, L. Su, X. Wu, X. Huang, Artificial Intelligence Software (IFETool) for Building Fire Safety Design Analysis, Automation in Construction (under review). (2022).
52. F. Tao, H. Zhang, A. Liu, A.Y.C. Nee, Digital Twin in Industry: State-of-the-Art, IEEE Transactions on Industrial Informatics. 15 (2019) 2405–2415. https://doi.org/10.1109/TII.2018.2873186.
53. X. Wu, X. Wu, X. Zhang, Y. Jiang, X. Huang, G.G.Q. Huang, A. Usmani, An intelligent tunnel firefighting system and small-scale demonstration. 120 (2022) 104301.
54. J. Torero, Scaling-Up Fire, Proceedings of the Combustion Institute. 34 (2013) 99–124.
55. T. Palmer, The ECMWF ensemble prediction system: Looking back (more than) 25 years and projecting forward 25 years, Quarterly Journal of the Royal Meteorological Society. (2018).
56. S.H. Koo, J. Fraser-Mitchell, S. Welch, Sensor-steered fire simulation, Fire Safety Journal. 45 (2010) 193–205. https://doi.org/10.1016/j.firesaf.2010.02.003.
57. F. Zhou, B. Young, Web crippling behaviour of cold-formed duplex stainless steel tubular sections at elevated temperatures, Engineering Structures. 57 (2013) 51–62. https://doi.org/10.1016/j.engstruct.2013.09.008.
58. Z. Karevan, J.A.K. Suykens, Spatio-temporal Stacked LSTM for Temperature Prediction in Weather Forecasting, ArXiv Preprint ArXiv:181106341. (2018).
59. A. Russo, A.O. Soares, Hybrid model for urban air pollution forecasting: A stochastic spatio-temporal approach, Mathematical Geosciences. 46 (2014) 75–93.
60. S. Wang, Intelligent buildings and building automation, Routledge, 2009.
61. T.S. Rappaport, Wireless Communications–Principles and Practice, (The Book End), Microwave Journal. 45 (2002) 128–129.
62. H.A. Omar, H. Abboud, N. Cheng, K.R. Malekshan, A.T. Gamage, W. Zhuang, A survey on high efficiency wireless local area networks: Next generation WiFi, IEEE Communications Surveys & Tutorials. 18 (2016) 2315–2344.
63. N.V.R. Kumar, C. Bhuvana, S. Anushya, Comparison of ZigBee and Bluetooth wireless technologies-survey, in: 2017 International Conference on Information Communication and Embedded Systems (ICICES), IEEE, 2017: pp. 1–4.
64. M.R. Souryal, J. Geissbuehler, L.E. Miller, N. Moayeri, Real-time deployment of multihop relays for range extension, in: Proceedings of the 5th International Conference on Mobile Systems, Applications and Services, ACM, 2007: pp. 85–98.
65. H. Liu, J. Li, Z. Xie, S. Lin, K. Whitehouse, J.A. Stankovic, D. Siu, Automatic and robust breadcrumb system deployment for indoor firefighter applications, In: Proceedings of the 8th international conference on Mobile systems, applications, and services., Pp. (2010) 21–34.
66. H. Liu, Z. Xie, J. Li, S. Lin, D.J. Siu, P. Hui, K. Whitehouse, J.A. Stankovic, An automatic, robust, and efficient multi-user breadcrumb system for emergency response applications, IEEE Transactions on Mobile Computing. 13 (2013) 723–736.

Intelligent Firefighting

7

Brian Y. Lattimer and Jonathan L. Hodges

7.1 Introduction

Intelligent firefighting is the inclusion of new technology to enhance firefighter planning and tactics while reducing the exposure of firefighters to hazardous conditions. In intelligent firefighting, electronic based systems are used to provide firefighters with information about the fire that is difficult or dangerous to obtain and use advanced algorithms to assist in interpreting this information. In addition, intelligent firefighting also includes hardware with integrated sensors that can enhance performing firefighting activities (e.g., fire suppression, search-and-rescue, etc.). Overall, intelligent firefighting seeks to make the firefighter more effective and safer.

A significant motivation for intelligent firefighting is to decrease the injuries and deaths of firefighters According to the National Fire Protection Association (NFPA) 2019 statistics [1, 2], there were 60,825 firefighters that were injured in the line of duty and 48 fatalities. The cost of firefighter injuries alone is estimated to be $1.6–5.9 billion annually, primarily to treat firefighter cancer and other occupational diseases including post-traumatic stress injuries. The NFPA statistics indicate firefighter injuries and fatalities have been steadily declining over the last 20 years. The decrease in injuries was in part attributed to better training, new standard practices to assist with safety issues, and improved firefighting equipment [1].

The focus of intelligent firefighting is to reduce the injuries and fatalities of civilians and emergency responders while minimizing the damage to a burning

B. Y. Lattimer (✉)
Mechanical Engineering, Virginia Tech, Blacksburg, VA, USA
e-mail: lattimer@vt.edu

J. L. Hodges
Jensen Hughes, Baltimore, MD, USA
e-mail: jhodges@jensenhughes.com

© Springer Nature Switzerland AG 2022
M. Z. Naser, G. Corbett (eds.), *Handbook of Cognitive and Autonomous Systems for Fire Resilient Infrastructures*, https://doi.org/10.1007/978-3-030-98685-8_7

structure and the community. The majority of injuries and fatalities occur on the fireground, which is where firefighters are attacking the fire, accounting for 39% of injuries and 27% of fatalities. A breakdown on the causes of injury and fatalities on the fireground is shown in Fig. 7.1. The majority of injuries are overexertion (29%), falls (20%), and direct exposures to fire and fire products (20%). For fatalities, exposures to fire products or burns as well as overexertion resulting in cardiac arrest represents 61% of the fatalities. These data indicate that advances in firefighting and new technology to assist in reducing these types of injuries and fatalities could have a significant impact on firefighter safety.

Intelligent firefighting could be included in several aspects of firefighting as outlined in the report on research needs for this area [3] and shown in Fig. 7.2. The initial aspect of firefighting is to determine the state of the fire when the firefighters arrive on the scene (called sizing up) and develop a plan for attacking the fire. After the plan is developed, firefighters are deployed for a variety of activities including suppression, search-and-rescue, and protection of property / other potential hazards. In this stage, firefighter situational awareness is critical to ensure that all activities are being performed safely and most effectively. In addition, efficiently performing firefighting activities is desirable to limit damage, save lives, and eliminate the hazard as quickly as possible. This chapter provides a brief overview of some technologies that have been developed to assist in sizing up/planning, firefighter situational awareness, and firefighter activities to support intelligent firefighting.

7.2   Sizing Up and Planning

The sizing up and planning of a fire has historically required firefighters to walk around and within the structure containing the fire to determine the current state of the fire, potential hazards, and infrastructure that needs to be protected. This is then reported back to incident command to determine a plan for attacking the fire, see Fig. 7.2. This information collection and planning takes time, especially if the structure is large, and puts firefighters in harm's way since the fire hazard is not usually well understood. There have been several types of systems proposed to improve the speed of this information gathering while reducing the firefighter exposure. This includes security cameras, sensors within the building, and cyberphysical systems.

Security cameras have been increasingly used for detecting and monitoring fire conditions [4–6]. Advances in image processing have allowed security camera systems distributed over large areas to assist in understanding the development of fire conditions. This has been proposed for use in transportation systems and other large infrastructure. As a result, these cameras can provide firefighters with real-time updates and information about the development of the fire that occurred before arrival. The challenge with these systems is the access to the video footage, analysis of this real time, and creating a visualization that allows incident command to understand the current scene to assist in planning.

Sensors within a structure can also monitor conditions that are developing during the fire. The most common sensors in fireground operations are in-place smoke

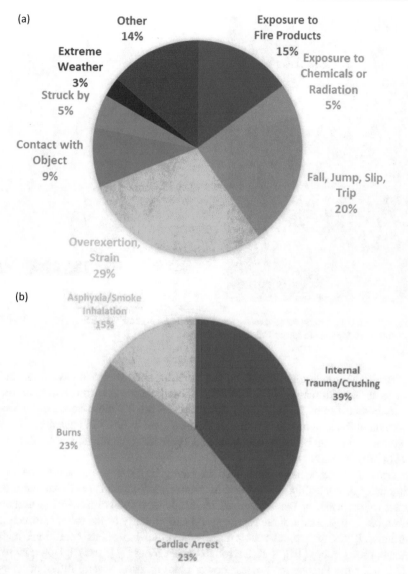

Fig. 7.1 Causes for firefighter (**a**) injury and (**b**) fatalities in the fireground in 2019 with statistics from [1, 2]

detectors and thermal detectors. The time and location of these alarms can be used to understand the progression of the fire in a structure. In addition, the actuation of sprinklers within a building has also been used to assist in assessing the size and growth rate of the fire. By monitoring the water flow rate, the number of sprinklers activated over time can be estimated. Sensors within HVAC systems (oxygen, carbon dioxide, temperature, flow rates, etc.) have also been proposed

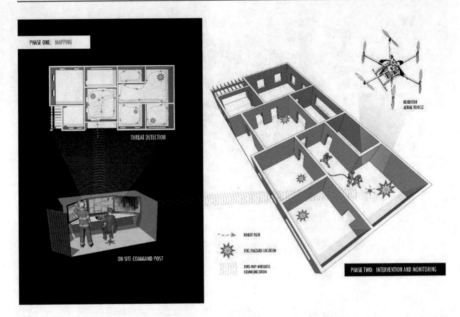

Fig. 7.2 An overview a potential intelligent firefighting workflow where firefighters and robotics systems work together to locate and suppress a fire

to assess the conditions throughout the building and predict the evolution of the fire to assist with planning [7, 8]. One proposed approach is to use measurements in individual rooms and use this information to predict more detailed conditions throughout the structure [7, 9], similar to that shown in Fig. 7.3. Packaging this into a software that can be used and visualized by incident command is a hurdle to this being fully implemented.

Ground vehicles have been used in emergency events for years due to the simplicity of controlling the system by an operator. The most well-known are the track robots used by bomb disposal teams; however, there have been numerous advances in this area to increase the speed and mobility of the robot over complex terrains. Types of ground robots used to explore hazardous conditions include Packbot track robot [10], X-RHex rough terrain robot [11], and BigDog quadroped [12]. After the Fukushima nuclear power plant disaster, several different types of ground vehicles were deployed to determine the conditions inside of the power plant following multiple explosions. Ground vehicle robots had problems with navigating over complex terrains that had stairs, debris, and narrow passages. This resulted in several robots becoming trapped. In addition, the communications with ground robots inside of building was difficult resulting in the use of tether systems to stream back information to an outside operator. Managing the tether so that it did not impede the movement of the robot also limited the effectiveness of these robots. In firefighting applications, the use of ground robots for exploring buildings has not been broadly adopted. This is reportedly due to the robots becoming stuck on

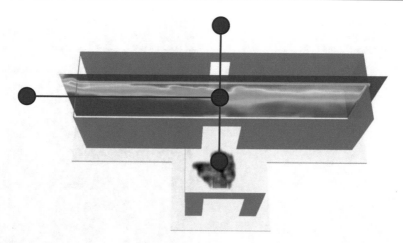

Fig. 7.3 Nodal locations of measured conditions inside a building along with a prediction of the conditions based on a machine learning model

different terrain in the buildings (rugs, stairs, clothes, etc.). While mobility can be a disadvantage of these systems, ground vehicles have an extreme advantage over other cyberphysical systems in the amount of payload that can be pulled/carried, which makes them useful in other types of firefighting activities such as suppression.

The challenges in moving robots through the Fukushima nuclear power plant motivated the development of legged robots, including humanoids (see Fig. 7.4) and quadropeds [12]. Two full-size humanoid robots were developed at Virginia Tech through the Shipboard Autonomous Firefighting Robot (SAFFiR) project which focused on developing a robot to navigate through the tight confines of a ship to support watch activities and perform dangerous duties such as firefighting. The first humanoid developed by Virginia Tech, THOR, was deployed in a proof-of-concept study on a U.S. Navy ship *ex-USS Shadwell* in November, 2014, see Fig. 7.4. The DARPA Robotics Challenge on 5–6 June, 2015 in Pomona, California had 25 teams from around the world competing to complete a series of emergency response tasks using mostly humanoid robots. The first version of this class of robots showcased their flexibility in performing a wider range of tasks and overcoming complex terrains. However, additional work is needed to increase the operational speed and robustness of these systems as well as reduce the complexity of operation and manufacturing expense before they will be widely available for use by firefighters.

Unmanned air vehicles (UAVs) are a cyberphysical system that is being used increasingly in firefighting for several reasons. Due to the general demand for UAVs, the control and imaging on these systems continues to dramatically improve. This allows a firefighter to quickly deploy a UAV and fly it around a structure to rapidly obtain views of the structure which would be difficult/not possible by firefighters on the ground. In addition, UAVs can fly into structures and explore the conditions while limiting exposure of firefighters to potential hazards. Both the speed and the ability to overcome complex obstacles (stairs, uneven terrain, equipment, etc.) make

Fig. 7.4 THOR robot
developed as part of the
SAFFiR robot for the U.S.
Navy

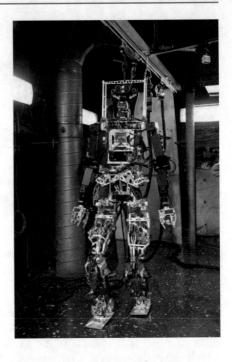

them very effective. A draw back to these systems has historically been the limited payload that they can carry. However, as sensor technologies have continued to miniaturize, there is the ability to put multiple sensor modalities on these systems. This includes thermal imagers, which have been shown to be useful in imaging through dense smoke [13–15], locating fires [16–18], and assisting with locating people [16–18]. UAVs have also been used to create maps and track where they have been within structures [19, 20], which would be useful to assist firefighters in planning how to most efficiently attack the fire. Due to the increased interest in using UAVs in firefighting, NFPA has produced a guide on the use of UAVs by firefighters [21]. Maintaining communication with the UAV when it goes into a building with the operator outside or far from it is still a challenge that needs to be addressed. DARPA initiated a Subterranean (SubT) Challenge [22] that could potentially assist in this, where robots were required to performing mapping of a long tunnel where communications with a command control station would be lost using routine communications.

7.3   Firefighter Situational Awareness

Situational awareness of firefighters is perhaps the most important goal in intelligent firefighting. In this context, situational awareness encompasses both the firefighter's own awareness of their surroundings and health as well as the incident commander's awareness of the health of all firefighters and their location relative to potential hazards and objectives. Many technologies have been developed and deployed to address different aspects of situational awareness.

The earliest of technology to identify changes in firefighter health during an emergency is the Personal Alert Safety System (PASS) device. This device detects the continuous motion of a firefighter and will automatically sound an alarm if no motion is detected or can be manually activated if the firefighter gets lost or trapped. The distress signal can then be used by other firefighters in a search-and-rescue operation to find the lost, trapped, or unconscious firefighter.

Firefighters commonly use self-contained breathing apparatus (SCBA) to prevent them from breathing in toxic gases generated by fires. To protect from heat, firefighters have high temperature suits and helmets. While these technologies protect the firefighter from the fire environment, the weight and internal temperature of the equipment increase the risk of overexertion in an emergency. Enhancing the thermal comfort of firefighters while protecting them from high external heating from fire environments is being explored. One such approach is to use passive cooling with phase change materials (PCM) [23, 24]; however, this has not been used broadly. Wearable devices to support monitoring firefighter's vital signs have also been used to collect data on firefighter and worker response [25, 26]; however, these are not broadly used. These wearable devices are becoming more popular in watches and rings, which makes it more likely that the cost will decrease and the technology will be improved.

Another aspect of situational awareness is understanding the current and future conditions of the fire hazard. One hazard that is difficult to predict is when the fire thermal environment is escalating to a point where a phenomenon known as flashover occurs. Flashover is when the gases inside of the space become sufficiently hot that combustibles inside the space begin igniting. This results in a rapid increase in fire size, gas temperatures, and toxic products which can lead to fatalities and severe injuries. Firefighters need to have some warning that conditions may be leading to flashover. Current research has used thermo-acoustics [27] as well as machine learning and body camera images to predict the onset of flashover [28].

A key component of situational awareness is understanding the location of the hazard and the firefighter within the structure. Unfortunately, visibility in a burning structure is reduced which can make it difficult for firefighters to keep track of their location within a structure. Integration of cameras and thermal imagers in the firefighting equipment has the potential to provide several benefits to a firefighter and their situational awareness. Handheld thermal imaging cameras (TIC) have been broadly used to support finding fires, especially deep-seated fires that are not visible. Thermal imagers have also been shown to be able to image through smoke making it possible for firefighters to better navigate through structures. Both thermal images and cameras have also been used to create maps and identify features (see Fig. 7.5) [29], which can be used to prevent firefighters from getting lost in structures. As these cameras become smaller, integrating these cameras as wearable devices becomes more practical.

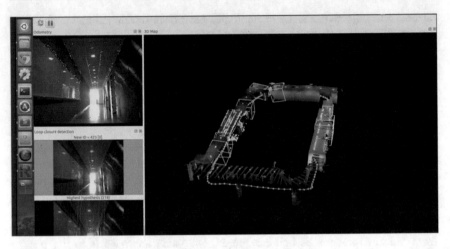

Fig. 7.5 Mapped environment using images and features of objects

7.4 Firefighting Activities

The activities that firefighters perform once a plan has been developed can be broad and may include suppression, protecting other potential hazards, and locating people. As shown in Fig. 7.6, future intelligent firefighting will likely include technology, robots, and firefighters working with each other to control and suppress fires. One example of this was the fire that occurred in Notre Dame on April 15, 2019. In this fire, a tracked robot worked alongside firefighters to control and suppress the fire. The tracked robot Colossus was moved into the structure by an operator and was used to cool critical parts of the structure while the firefighters focused on more complex tasks. Though the representation in Fig. 7.6 is still not fully realized, progress on these robotic systems is changing how firefighters are performing their activities.

7.4.1 Suppression

There are two general types of robotic systems that have been developed for firefighting: fixed systems and mobile systems. The fixed systems, such as automated fire monitors, are being used in applications where there is a significant fire hazard and the fire needs to be extinguished rapidly. Some example applications include aircraft landing areas, warehouse storage [30, 31], and tunnels [32]. These systems have UV and/or IR sensors to assist with fire localization to target the suppression onto the fire. Mobile systems have more advanced features to assist the operator in navigation and perform a wider range of tasks.

There are two types of fixed systems that are being increasingly considered for suppression applications in areas where there is a desire to limit the amount

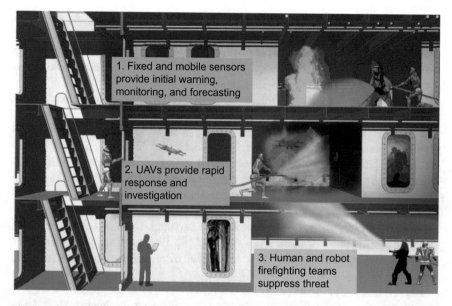

Fig. 7.6 Graphical representation of suppression activities with intelligent firefighting in the future

of damage due to the suppression agent, which is typically water. The first type is an automated monitor which contains a visible and IR vision system to target the fire location and direct the water spray onto the base of the fire. These have been considered for large spaces such as aircraft hangars or rack storage in warehouses. Simultaneous Monitoring, Assessment and Response Technology (SMART) sprinkler heads are the second type of fixed suppression technology that includes advanced processing to limit water damage and efficient suppression [33, 34]. In a SMART water sprinkler, the sprinkler and the fire detection systems are co-located. Through analysis of the smoke and thermal detection systems, sprinkler activation can be controlled. This has been found to result activating the correct sprinklers to efficiently suppress the fire while not activating sprinklers that may not assist in extinguishing the fire. Fixed suppression systems, such as automated monitor [35] and SMART sprinkler head [36], with advanced intelligence and vision systems to increase suppression efficiency while reducing the amount of water used to extinguish the fire.

Outdoor ground-based mobile robotic systems are predominately vehicles with onboard suppression systems that are remote controlled by an operator. Examples of ground-based mobile robotic systems that have been developed for outdoor firefighting are seen below. These robots travel 2.4–20 km/h (1.5–12.4 mi/h) using wheels or tracks, weigh 450–9300 kg (990–20,450 lbs), and have suppression capabilities onboard the robot. The robots are powered by batteries or a diesel engine. Suppression systems mounted onto the robots include water-based fire monitors, foam nozzles, nozzles on articulating arms for more range of motion, and a water fog system. In addition to the remote-control operation, these systems use a wireless connection to transmit information from sensors onboard the robot to the operator for assisting in navigation and fire suppression. Sensors on the robots include visual cameras, IR cameras, gas concentration sensors, and rangefinders to assist in avoiding obstacles.

A wide variety of robotic designs are being pursued for indoor mobile firefighting robots due to the confined, complex, and cluttered environments required for navigation. These include aerial vehicles (primarily quad or hex rotors), track/wheeled ground vehicles, biomimetic type robots (snake-like [37] and bug [38]), and humanoids [29]. Robots are being considered both as a fire watch as well as an assistant to firefighters. In these roles, the robots are being designed for detecting fires, sizing up the hazards inside a structure, locating and suppressing fires, and search-and-rescue. A recent high hazard fire has been exterior combustible façade fires, which rapidly spread across the exterior surface. Teams of UAVs have been developed to suppress of these types of fires [39]. For example, groups of UAVs have been demonstrated for use in suppressing an exterior façade fire on a structure [39].

Legged humanoid robots have also been used to perform suppression activities. During the demonstration on the ship, THOR walked on the heat warped decks while holding a water nozzle and worked with a human to suppress a compartment fire using the ship's water nozzle connected to a hose reel. An image of the robot putting out the fire with a person is shown in Fig. 7.7a. In addition, feedback control vision systems were developed to support autonomous suppression of fires as shown in Fig. 7.7b [40, 41]. This was able successfully suppress fires based on tracking where the fire is located as well as the water and making sure that the water is hitting the base of the fire. Suppression was also capable in dense smoke environments with reduced visibility [41].

7.4.2 Search and Rescue

Search for fires and other hazards as well as finding people that may be trapped inside of a structure with a fire requires vision systems with multiple spectra and image processing. A particular issue is imagers that are affected by the harsh fire environments such as smoke and elevated temperature. For example, THOR maintains perception in harsh environments using sensors with multiple modalities including stereoscopic IR thermal imagers for range finding through smoke and fire environment classification, a rotating laser rangefinder (LIDAR) to create a 3D point cloud of obstacle locations in unobscured environments, and stereoscopic RGB cameras to create a color point cloud of obstacles (i.e., obstacle locations over a color image of the scene) and for object recognition.

An example of a vision system to map, identify objects, and find fires is shown in Fig. 7.8. Although advances in using robots in confined, cluttered indoor environments has been accelerating, the use of robots to navigate through unknown spaces is challenging and still requires some level of human operation. In addition,

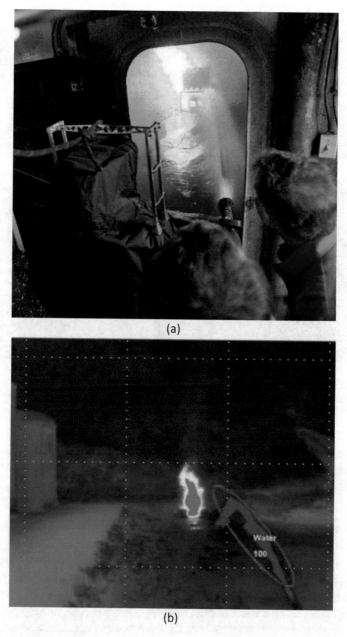

(a)

(b)

Fig. 7.7 (**a**) THOR robot suppressing a fire onboard a ship [29] and (**b**) a vision system to support autonomous suppression [40]

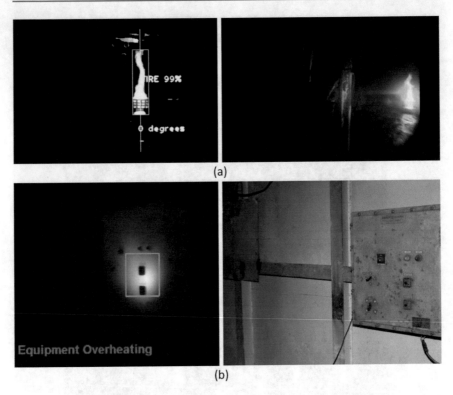

Fig. 7.8 Using advanced vision systems to (**a**) identify and classify the fire environment and (**b**) identify emerging hazards

identifying, localizing, and manipulating objects is a complicated task which still necessitates a human operator and significant computing power, especially for performing tasks on unknown objects.

7.4.3   Strength Augmentation

There are many aspects of a firefighter's job that require significant strength or endurance to perform the activity. One of the highest incidents of firefighter injury and fatality is due to overexertion in an emergency. This includes lifting and dragging hoses, using a water nozzle, carrying injured people, moving objects, and removal of structure. As previously seen, some robotic systems have been developed to support some of these activities such as dragging hoses and using a water nozzle. In addition, other wearable devices called exoskeletons have been created to support people in performing strenuous activities including lifting and scenarios where weights must be supported over long periods of time. An exoskeleton is a structure that is worn by a person that may include electronic actuators to support the upper body, lower body, or whole body. There are two types of exoskeletons: passive and active. Passive exoskeletons have no electronic parts but instead use leverage to augment strength. Active exoskeletons use actuators along with cables and/or bars to improve strength. Passive exoskeletons have the advantage that they can be quickly put on and have less parts that may fail. However, active exoskeletons have the potential to more dramatically impact strength. In interviews with firefighters on this technology, being reliable and able to quickly put on the device had high importance. Examples of passive [42] and active exoskeletons [43] used to augment strength and endurance can be found elsewhere.

7.5   Summary

Future use of technology and robots in firefighting will depend on the robot durability, sufficient sensors for environment monitoring and perception, task capabilities, cost, level of autonomy, and movement speed. Many of the robots being designed for firefighting applications are lacking in some or all of these areas. For firefighters, cost is a significant consideration and is currently driving the broader use of technology and robotics in firefighting. It is likely that a team of different types of robots will be required to assist firefighters in their widely varying and complex tasks on the fireground. As these robots become more effective at conducting firefighting tasks while firefighters monitor their performance at safe locations, robots will be used more routinely to support firefighters. Technology integrated onto the firefighter or within the building to assist in situational awareness continues to expand. Using this data to broadly understand the current state of the fire and firefighters remains a challenge. Predictions to understand the current state and forecast what is expected to happen continue to expand. A significant challenge will be to efficiently visualize the most necessary information to support incident command and firefighters.

References

1. R. Campbell and B. Evarts, "United States Firefighter Injuries in 2019," *NFPA Res. Rep.*, no. November, p. 15, 2020.
2. R. F. Fahy, J. T. Petrolli, and J. L. Molis, "Firefighter Fatalities in the US-2019," *NFPA Res. Rep.*, no. June, pp. 1–26, 2019.
3. C. Grant, A. Hamins, N. Bryner, A. Jones, and G. Koepke, "Research Roadmap for Smart Fire Fighting," *Summ. Report, NIST Spec. Publ. 1191*, p. 247, 2015.
4. G. Marbach, M. Loepfe, and T. Brupbacher, "An image processing technique for fire detection in video images," *Fire Saf. J.*, vol. 41, no. 4, pp. 285–289, 2006.
5. K. Muhammad, J. Ahmad, I. Mehmood, S. Rho, and S. W. Baik, "Convolutional Neural Networks Based Fire Detection in Surveillance Videos," *IEEE Access*, vol. 6, pp. 18174–18183, 2018.
6. A. E. Çetin *et al.*, "Video fire detection - Review," *Digit. Signal Process. A Rev. J.*, vol. 23, no. 6, pp. 1827–1843, 2013.

7. J. L. Hodges, B. Y. Lattimer, and K. D. Luxbacher, "Compartment fire predictions using transpose convolutional neural networks," *Fire Saf. J.*, vol. 108, no. November 2018, 2019.
8. W. C. Tam *et al.*, "Generating Synthetic Sensor Data to Facilitate Machine Learning Paradigm for Prediction of Building Fire Hazard," *Fire Technol.*, 2020.
9. B. Y. Lattimer, J. L. Hodges, A. M. Lattimer, "Using Machine Learning in Physics-based Simulation of Fire," *Fire Saf. J.*, 2020.
10. FLIR, "Packbot - Robots: Your Guid to the World of Robotics," *IEEE*, 2020.
11. G. Lab, "x-RHex - Robots : Your Guid to the World of Robotics," *IEEE*, 2020.
12. B. Dynamics, "BigDog - Robots: Your Guide to the World of Robotics," *IEEE*, 2020.
13. J. W. Starr and B. Y. Lattimer, "Application of thermal infrared stereo vision in fire environments," *2013 IEEE/ASME Int. Conf. Adv. Intell. Mechatronics Mechatronics Hum. Wellbeing, AIM 2013*, pp. 1675–1680, 2013.
14. J. W. Starr and B. Y. Lattimer, "A comparison of IR stereo vision and LIDAR for use in fire environments," *Proc. IEEE Sensors*, pp. 3–6, 2012.
15. J. W. Starr and B. Y. Lattimer, "Evaluation of Navigation Sensors in Fire Smoke Environments," *Fire Technol.*, vol. 50, no. 6, pp. 1459–1481, 2014.
16. J. H. Kim, S. Jo, and B. Y. Lattimer, "Feature Selection for Intelligent Firefighting Robot Classification of Fire, Smoke, and Thermal Reflections Using Thermal Infrared Images," *J. Sensors*, vol. 2016, 2016.
17. J.-H. Kim, Y. Sung, and B. Y. Lattimer, "Bayesian estimation based real-time fire-heading in smoke-filled indoor environments using thermal imagery," in *Proceedings - IEEE International Conference on Robotics and Automation*, 2017.
18. J. H. Kim and B. Y. Lattimer, "Real-time probabilistic classification of fire and smoke using thermal imagery for intelligent firefighting robot," *Fire Saf. J.*, vol. 72, pp. 40–49, 2015.
19. N. Kerle, F. Nex, M. Gerke, D. Duarte, and A. Vetrivel, "UAV-based structural damage mapping: A review," *ISPRS Int. J. Geo-Information*, vol. 9, no. 1, pp. 1–23, 2019.
20. F. Nex, D. Duarte, A. Steenbeek, and N. Kerle, "Towards real-time building damage mapping with low-cost UAV solutions," *Remote Sens.*, vol. 11, no. 3, pp. 1–14, 2019.
21. N. 2400, "Standard for Small Unmanned Aircraft Systems (sUAS) Used for Public Safety Operations," *NFPA*, 2019.
22. T. Chung, "DARPA Subterranean (SubT) Challenge," https://www.darpa.mil/program/darpa-subterranean-challenge, 2018.
23. A. Fonseca, T. S. Mayor, and J. B. L. M. Campos, "Guidelines for the specification of a PCM layer in firefighting protective clothing ensembles," *Appl. Therm. Eng.*, vol. 133, no. March 2017, pp. 81–96, 2018.
24. H. L. Phelps, S. D. Watt, H. S. Sidhu, and L. A. Sidhu, "Using Phase Change Materials and Air Gaps in Designing Fire Fighting Suits: A Mathematical Investigation," *Fire Technol.*, vol. 55, no. 1, pp. 363–381, 2019.
25. D. Dias and J. P. S. Cunha, "Wearable health devices—vital sign monitoring, systems and technologies," *Sensors (Switzerland)*, vol. 18, no. 8, 2018.
26. S. Rodrigues, J. S. Paiva, D. Dias, G. Pimentel, M. Kaiseler, and J. P. S. Cunha, "Wearable Biomonitoring Platform for the Assessment of Stress and its Impact on Cognitive Performance of Firefighters: An Experimental Study," *Clin. Pract. Epidemiol. Ment. Heal.*, vol. 14, no. 1, pp. 250–262, 2018.
27. Z. Deng, Y. Yu, D. Zou, W. Guan, and L. Yang, "OPTIMIZATION AND IMPLEMENTATION OF A THERMOACOUSTIC FLASHOVER DETECTOR," *MS Thesis, Univ. Maryland, Dep. Fire Prot. Eng.*, vol. 6, no. 3, 2013.
28. K. Yun, J. Bustos, and T. Lu, "Predicting rapid fire growth (flashover) using conditional generative adversarial networks," *arXiv*, pp. 10–13, 2018.
29. B. Lattimer *et al.*, "Humanoid Firefighting Robot for Structure Fires," *Interflam 2016*, 2016.
30. T. Chen, H. Yuan, G. Su, and W. Fan, "An automatic fire searching and suppression system for large spaces," *Fire Saf. J.*, vol. 39, pp. 297–307, 2004.

31. C. Yuan, Y. Zhang, and Z. Liu, "A survey on technologies for automatic forest fire monitoring, detection, and fighting using unmanned aerial vehicles and remote sensing techniques," *Can. J. For. Res.*, vol. 45, no. 7, pp. 783–792, 2015.

32. A. De Santis, B. Siciliano, and L. Villani, "A unified fuzzy logic approach to trajectory planning and inverse kinematics for a fire fighting robot operating in tunnels," *Intell. Serv. Robot.*, vol. 1, no. 1, pp. 41–49, 2008.

33. Y. Xin, K. Burchesky, J. de Vries, H. Magistrale, X. Zhou, and S. D'Aniello, "SMART Sprinkler Protection for Highly Challenging Fires—Part 2: Full-Scale Fire Tests in Rack Storage," *Fire Technol.*, vol. 53, no. 5, pp. 1885–1906, 2017.

34. Y. Xin, K. Burchesky, J. de Vries, H. Magistrale, X. Zhou, and S. D'Aniello, *SMART Sprinkler Protection for Highly Challenging Fires—Part 1: System Design and Function Evaluation*, vol. 53, no. 5. Springer US, 2017.

35. R. James, "Fire Fighting News: Automatic Fire Monitor Systems for Recycling and Water to Energy Plants," *Unifire.com*, 2020.

36. G. Malik, "What if your sprinkler system could do think?," *Sprink. Age*, 2018.

37. P. Liljebäck, Ø. Stavdahl, and A. Beitnes, "SnakeFighter - Development of a water hydraulic fire fighting snake robot," *9th Int. Conf. Control. Autom. Robot. Vision, 2006, ICARCV '06*, no. 7465, 2006.

38. J. Hong, B. Min, J. Taylor, V. Raskin, and E. Matson, "NL-Based Communication with Firefighting Robots," *2012 IEE Int. Conf. Syst. Man, Cybern. Oct. 14-27, COEX, Seoul, Korea*, pp. 1461–1466, 2012.

39. P. Strauss, "Using Drones to Fight Hi-Rise Fires," *Technabob*, vol. April, 2020.

40. J. G. McNeil and B. Y. Lattimer, "Robotic Fire Suppression Through Autonomous Feedback Control," *Fire Technol.*, vol. 53, no. 3, pp. 1171–1199, 2017.

41. J. G. McNeil and B. Y. Lattimer, "Autonomous Fire Suppression System for Use in High and Low Visibility Environments by Visual Servoing," *Fire Technol.*, vol. 52, no. 5, pp. 1343–1368, 2016.

42. T. Kevan, "Exoskeletons on the Move," *Digit. Eng.*, vol. December, 2018.

43. E. Kulisch, "Exoskeleton Could Give Delta Cargo Workers Superhuman Strength (with video)," *Air Cargo*, vol. January, 2020.

The Role of Artificial Intelligence in Firefighting

8

Jonathan L. Hodges, Brian Y. Lattimer, and Vernon L. Champlin

8.1 Introduction

Structural fires represent a significant hazard to the life safety of civilians and emergency responders, resulting in 3000–4000 annual fatalities, 1,000,000–2,000,000 annual injuries, and greater than $10 billion in annual property damage in the USA [1]. The total annual cost of fire in the USA has been estimated at approximately $300 billion, see Fig. 8.1 [2]. The objective of fireground operations is to reduce the impact of an on-going fire on the life safety of civilians and the property without sacrificing the safety of emergency responders.

The National Fire Protection Association (NFPA) provides an overview of fireground operations in Chap. 1, Section 13 of the Fire Protection Handbook. Tasks within fireground operations can be generally categorized into one of two main areas including pre-incident planning and incident response [3]. Pre-incident planning includes standard operating procedures which are applicable for most structures in a jurisdiction as well as specific plans tailored to individual structures or groups of structures. Incident response includes all activities associated with rescue and firefighting operations prior to extinguishment. Some of these tasks include sizing-up the incident, developing the incident action plan, and deploying emergency responders to execute the incident action plan. After the incident has been controlled, remaining actions after extinguishment include verifying structural integrity, minimizing damage to property, and investigating the origin and cause of the fire.

J. L. Hodges (✉) · V. L. Champlin
Jensen Hughes, Baltimore, MD, USA
e-mail: jhodges@jensenhughes.com

B. Y. Lattimer
Mechanical Engineering, Virginia Tech, Blacksburg, VA, USA

© Springer Nature Switzerland AG 2022
M. Z. Naser, G. Corbett (eds.), *Handbook of Cognitive and Autonomous Systems for Fire Resilient Infrastructures*, https://doi.org/10.1007/978-3-030-98685-8_8

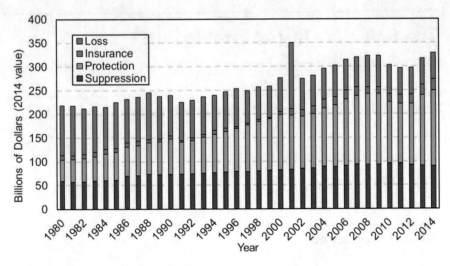

Fig. 8.1 Annual cost of fire in the USA [2]

There are numerous decisions which need to be made on the fire ground during structural fire suppression operations. What specific hazards exist, where are the exposures, what is the most effective method of extinguishment, and does the survivability profile warrant interior rescue operations to name a few? Successful firefighting operations require the incident commander (IC) and other emergency responders have an accurate understanding of the current state of the building and occupants and the likely progression of the fire. Unfortunately, due to the restrictive timeline of a fire, these decisions are often based on limited information and expert judgment.

An extensive survey of incident commanders with experience in fireground operations was conducted in 2008 to identify the primary barriers to situational awareness and decision-making on the fireground [4]. The top four categories of barriers identified by the responders accounted for 74.3% of all barriers and were related to staffing issues (understaffed, inexperienced staff, and response time delays), communication issues (incomplete communications loop, multiple radio channels being used, and missed communication/progress reports), physical and mental stress (feelings of stress, excitement, anger, frustration, and being over-whelmed), and workload management (multiple command roles and multitasking. Technology and applied artificial intelligence (AI) can be used to alleviate many of these barriers by improving training of staff, better identify the hazard based on available data, and optimize workload management strategies.

The decision-making process of emergency responders has been a focus of research for many years. Human decision-making paradigms generally fall into two categories: classical and naturalistic. Classical decision-making uses decision trees, multiattribute utility analysis (MAUA), and Bayesian statistics to identify optimal decisions. Klein provides a useful overview of the difficulties associated with this form of decision-making in the fireground [5], where incomplete data and short timelines make it difficult to apply this type of decision-making in an operational environment. Naturalistic decision-making (or intuitive decision-making) does not weigh all alternatives to make an optimal decision; rather, the first decision which is expected to be good enough is selected and acted upon. Klein et al. conducted a survey of qualified Fire Ground Commanders (FGCs) which primarily focused on the decision process used in critical incidents that were nonroutine [6]. The authors found that these experienced responders used their experience (intuition) to classify a nonroutine incident as its closest standard prototype and apply the standard course of action for that prototype to the new situation. AI systems generally follow the classical decision-making paradigm; however, complex models can often seem naturalistic due to the complex interactions embedded in the black box.

A key difficulty with naturalistic decision-making is the reliance on the experience of the individual. This leads to inherent uncertainties and potential inconsistencies in decision-making across individuals. For example, treating a warehouse, factory, or low-rise structure fire like a single-family residential fire has killed and injured many firefighters. A review of the National Institute for Occupational Safety & Health (NIOSH) Fire Fighter Fatality Investigation and Prevention (FFFIP) investigation reports indicates up to 67% of traumatic injury related line of duty deaths at structural fires could have been avoided if firefighters or incident commanders had conducted a more extensive risk/benefit analysis [7]. Researchers have concluded that a combination of naturalistic and classical decision-making can provide the best results, and it can be improved through training [8]. Time in the classroom and experience in the fireground are the traditional approaches to improve naturalistic decision-making [8]. However, researchers have recently shown that working with virtual training tools, such as fire simulations [9], can lead to improved decision-making in the physical environment [10]. Recent virtual reality tools have achieved acceptable levels of realism and presence to provide a similar experience to supervised live-fire simulations [11]. While these research efforts have shown that computer-based tools can improve the consistency of human decision-making, the role of AI in making these decisions is still an open question.

While most studies to date have been limited to traditional fireground operations where there is little data available to the incident commander, there has been a recent push to increase the data available to emergency responders. The National Institute of Standards and Technology (NIST) recently published a report summarizing a roadmap to more intelligent firefighting operations [12]. The focus of this report was to identify the current state-of-the-art and identify research needs to enhance data gathering, data processing, and targeted communications. Given the increasing amount of data which will be available on the fireground, there is a significant risk that key insights which could save lives may be overlooked due to information overload. As more data is available to incident commanders, AI will serve a critical role assisting incident commanders receive, analyze and apply available data to decision-making on the fireground.

AI as a discipline focuses on developing computer systems that interpret raw data to make actionable observations, interpretations, and decisions. AI systems address several fundamental issues encountered in traditional decision-making. AI systems can identify complex relationships in real-life scenarios which may historically have been overlooked by pairing prior knowledge with detailed probability inference computations [13]. Recent advancements in computational hardware and algorithms have led to the development of faster-than-real-time AI tools. These tools can make decisions significantly faster than humans who require time to evaluate data from a variety of sources. In addition, AI systems provide permanency which does not exist in traditional applications where past experience is lost when subject matter experts retire or leave an organization [14].

While AI can provide significant value over traditional decision-making, there are several roadblocks preventing its widespread use in applications involving life safety. Many of the highest performing state-of-the-art deep learning techniques operate as black boxes, where the specific rationale behind a decision made by the AI system may not be clear to a human reviewer [15]. However, people are less likely to develop trust in decisions made by black box models [16]. In applications involving life safety, such as autonomous driving or fireground operations, trusting a black box model to make a decision adds the additional roadblock of liability concerns when the model fails [17]. One of the greatest difficulties introducing AI into the fire service culture will be aligning new science with the high value placed on traditional approaches.

This chapter provides an overview of decisions which are made in the fireground, and how AI can be used as a tool to improve these decisions. The chapter begins with an overview of AI and provides some example tools. The chapter then presents specific tasks and decisions which are associated with each stage of fireground operations and discusses the current state-of-the-art research in using AI in each area. The final section of the chapter summarizes the current role of AI in fireground operations and recommends future areas of research which could result in a larger role for AI systems in these applications.

8.2   Artificial Intelligence

AI is a genre of computer programming that focuses on generating insights and making decisions based on data, see Fig. 8.2. There are many different types of data which can be used by a computer to make decisions. Some examples include drawings and models of a structure, databases of historic fire events, images from surveillance footage, observations from fire department inspections, text documents such as pre-incident plans and inspection reports, real-time sensor measurements such as air quality sensors or vibration sensors in the floor, predictions from simulated fires, and surveys. This data can be used to better predict the development of a fire, rapidly analyze the risk-benefit of different action plans, and inform the operations of autonomous robotic systems.

There are a variety of machine learning tools which are used in AI applications. Perhaps the technique most like the classical decision-making approach used by humans is the decision tree, see Fig. 8.3 for an example. Decision trees evaluate

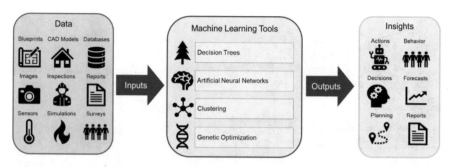

Fig. 8.2 Overview of artificial intelligence

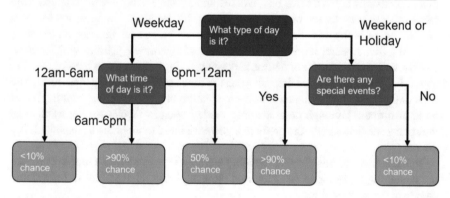

Fig. 8.3 Example decision tree

input data using simple logical functions such as splits along classification types or numeric values. The highest node, often called the root node, is generally based on the variable which has the highest discriminating power. Subsequent nodes are branches connecting the root to lower layers which are continually split on variables. End nodes with class labels or numeric values are generally called leaves. When used as a machine learning tool, the optimal composition of the tree is learned by evaluating data with known inputs and outputs. While the example shown in Fig. 8.3 is simple, these models can represent complex decisions through combinations of input parameter in each branch. A key advantage of this class of machine learning approach is the ease at which a human can interpret the decision made by the computer.

Recent developments in computational hardware and algorithms has led to significant advancements in deep learning tools [18]. These models often operate more like the naturalistic decision-making approach used by humans due to the high interconnectivity between input variables and learned parameters. As a result, these

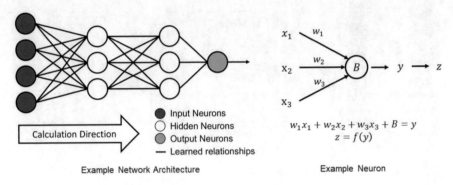

Fig. 8.4 Example neural network [19]

models can often act as a black box, where humans may not be able to understand the exact process used by the computer to develop its decision. The most well-known of these tools is the artificial neural network (ANN), see for Fig. 8.4 an example [19]. At a fundamental level ANNs are system of algebraic equations which can be rapidly solved. An individual neuron represents a single equation in the system. The network architecture can be thought of as the functional form of the equations to be solved. Optimal free parameters for each equation are solved during training based on known data. Although deep learning models such as ANNs are not as clear to interpret as decision trees, the predictive power of these models can be exponentially greater.

In the context of this chapter, machine learning is presented as a tool to correlate input and output data, and AI is focused on using those relationships to inform and empower decision-making.

8.3 Pre-incident Planning

An overview of the national standard for pre-incident planning is documented in NFPA 1620 [20]. Pre-incident planning is constructed on two foundational principals: fire prevention and preparedness. The objective of fire prevention is to reduce the risk of a fire occurring. This can be accomplished through effective application of construction codes and standards, periodic fire inspections, public education, and community risk reduction efforts. The objective of preparedness is to ensure that the response to an emergency will be streamlined and effective. This includes sufficient resource allocation, developing standard and site-specific operating procedures, adequate personnel training, and periodic testing of fire protection systems. Most of the research developing AI technologies for pre-incident planning have concentrated on the fire prevention side through prioritizing inspections; however, AI can aid in both areas.

8.3.1   Data Collection

The field of fire prevention has a history of using data to inform municipalities where to focus efforts to reduce the overall fire risk in communities. Historic fire incident data is often collected on the State and National levels, such as in the National Fire Incident Reporting System (NFIRS) in the USA [21]. These data sets typically include basic information about the property where the fire occurred, total losses and damages, fire injuries/deaths, and additional summary information about the event. The Society of Fire Protection Engineers (SFPE) Handbook of Fire Protection Engineering provides an overview of how this type of data has been used to quantify risk for individual structures in fire safety applications [22]. Researchers have also used this type of data to examine the impact of fire risk in communities. Some approaches have included considering the socioeconomic factors of local communities [23–26], local government policy [23], physical construction details [23], neighborhood vacancies [23, 27], and non-compliance with fire code [28]. These data sources provide valuable insight into the fire risks of specific communities and can be used by local governments and fire departments to develop risk reduction efforts within their communities. In recent years, these observations have been used to develop AI-based models to prioritize pre-incident planning efforts [29–36]. These models typically use machine learning to learn relationships between each of the risk factors and fire outcome data.

8.3.2   Risk-Informed Inspection Prioritization

The first of these efforts to be applied in the field was the FireCast software which weighs 60 factors and provides the basis for the Risk-Based Inspection System (RBIS) employed by the Fire Department of New York (FDNY) to better identify where fires may ignite [36]. The FDNY has adequate staff to inspect one tenth of properties requiring inspection each year. As a result, each property may only be inspected once every 10 years. Unfortunately, this rate of inspection has resulted in fires occurring which may have been prevented by a timely inspection. The historic model which operated under assumptions based on experience rather than data resulted in 1.9% of properties being inspected within three months of a fire occurring. The data-driven FireCast model increased this percentage to 16.5%, indicating a significant improvement in targeting properties for inspection, see Fig. 8.5 [37].

A similar system to the FireCast engine was developed and theoretically evaluated based on data available in British Columbia [31]. Interestingly, the researchers found that the time between inspection and a fire's occurrence was not highly indicative of fire outcomes with respect to fire spread and fire related casualties [29]. The researchers focused on the interconnection between level of risk and level of compliance for each property, leading to four categories of properties (i.e. high risk + low compliance, high risk + high compliance, low risk + low compliance, low

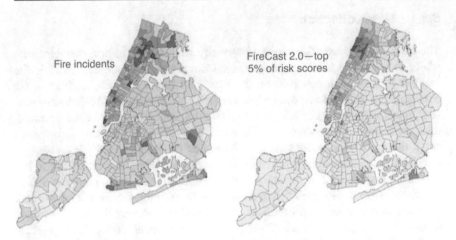

Fig. 8.5 Comparison of FireCast risk scores with actual fire incidents [37]. Republished courtesy of the National Institute of Standards and Technology

risk + high compliance). The results of this study indicated that coupling risk and compliance can provide a more targeted approach to inspection; however, additional research is needed to evaluate the long-term impacts of this type of procedure.

The Firebird initiative grew out of a similar need in the Atlanta Fire Rescue Department. The first goal of this initiative was to rank properties which require inspection by their fire risk, like the objective of the FDNY and BC approaches. However, this project added an additional objective to identify properties throughout the city which should be inspected but were not currently targeted as priorities for inspection. The researchers used occupancy classification data from properties currently being inspected as well as requirements established in the Fire Code to develop a model to determine if a business should be inspected or not, see Fig. 8.6 [32]. Each business within the Atlanta Business License dataset was then evaluated based on the model to determine if it should be inspected. The raw outputs from the AI-based model indicated that the 2573 currently inspected properties represented 11.7% of the total properties which may require inspection.

8.3.3 Inspection Assistance

Developing ranked prioritization lists of properties can provide significant value to fire departments by focusing their efforts on high-risk properties. Unfortunately, this may result in an increase in workload for the department should additional properties be identified, as was revealed in Atlanta's Firebird study. AI-based tools could alleviate these issues for fire departments by assisting in streamlining, crowdsourcing, and/or conducting inspections. Unmanned Aircraft System (UAS) based inspections of structures have gained traction in recent years. The primary difficulty has been the limited ground truth information available to the system to use

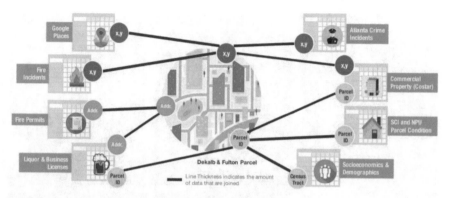

Fig. 8.6 Data used to identify new properties which need inspection [32]. Republished courtesy of ACM

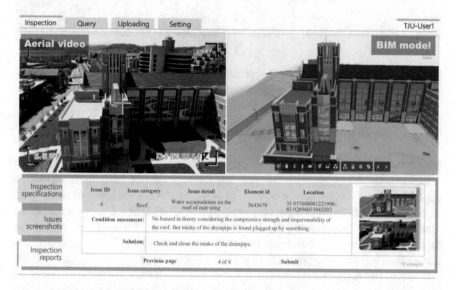

Fig. 8.7 BIM assisted aerial inspection [38]. Republished courtesy of the American Society of Civil Engineers

in identifying defects. Integrating building information models (BIM) of existing structures into the inspection process can assist autonomous systems in performing these inspections, see Fig. 8.7 [38]. While a similar AI-based approach to fire safety inspection has not been developed, Zhang et al. recently presented an alternative approach to fire safety inspection using BIM informed virtual reality (VR) and a crowdsourcing application with financial incentives, see Fig. 8.8 [39]. Data collected using this approach could be used to train an AI-based model to perform initial fire safety inspections using autonomous systems.

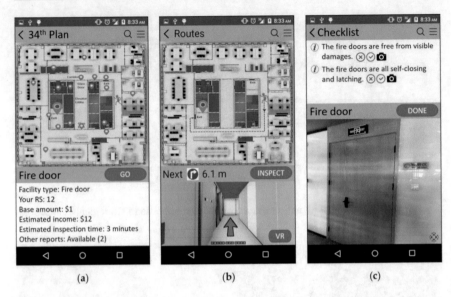

Fig. 8.8 Example crowdsourcing fire inspections using BIM, virtual reality, and financial incentives [39]. Republished courtesy of MDPI

8.3.4   Pre-incident Planning Prioritization

While each of these AI-based techniques focused on fire prevention by improving prioritization and application of fire inspections, similar techniques could be applied to increase preparedness. Kapalo et al. conducted an international survey of firefighters which focused on the benefits and challenges of using pre-incident plans during an emergency [40]. The authors found simplified maps of the site to be the most useful documents to emergency responders. These maps include general site information indicating locations of each structure as well as hydrants, access and fire department connection locations. In addition, simplified floor plans for each structure including fire alarm panel locations, key-box information, utility disconnects, indications of hazardous materials, and emergency contact information were helpful. The authors found two primary challenges associated with using pre-incident plans in responding to an incident. Many fire departments have mixed database storage for these plans, with some hard copy plans located in three-ring binders and other electronic plans stored on a Mobile Data Computer. This can make it difficult to locate the plans during an event. In addition, many plans are not regularly updated, which can lead to emergency responders working with out-of-date information or forgoing the use of the plans all together.

Unfortunately, fire departments may not have the resources required to develop and update site-specific plans for all buildings in their jurisdictions. However, AI approaches could be developed to identify which properties would benefit the most from developing or updating site-specific pre-incident response plans. Similarly,

Fig. 8.9 Example virtual reality training: incident commander view of a townhome fire. Image courtesy of XVR Simulation [41]

BIM of existing structures could be used to generate and evaluate sketches contained in the pre-incident response plans for a structure. While there may be a potential benefit to developing these AI-based approaches, this has not been a focus of research in recent years.

8.3.5   Training Personnel

Another key aspect of pre-incident preparedness is training personnel. Researchers have recently shown that working with virtual training tools, such as fire simulations [9], can lead to improved decision-making in the physical environment [10]. Although not a direct replacement for live-fire training exercises, virtual reality tools [41] have achieved acceptable levels of realism and presence to provide a similar experience [11], see Fig. 8.9 [41]. Advancements in computational hardware and machine learning-based fire simulation have made it computationally feasible to evaluate thousands of detailed fire scenarios [42–44]. Research is needed to develop an approach to evaluate the impact of different incident response strategies on the detailed fire behavior. After this hurdle is overcome, it will be possible to train an AI-based model to quantitatively evaluate a series of incident response tactics in real-time to assist Incident Commanders in developing a strategy for incident management.

8.4   Incident Response

Responding to an active fire is a multi-step process. The first step after detection is to mobilize resources to the site. After arriving on scene, the specific fire conditions at the site need to be evaluated (also known as sizing-up the situation). A specific incident action plan must then be developed based on observations from size-up, pre-incident plans, and standard operating procedures. At this stage, the incident commander assigns roles to each firefighting crew and deploys personnel to each task. Specific tasks may include establishing a water supply, forcible entry, search and rescue, opening horizontal or vertical ventilation, extinguishment (such as applying water or dry powder, or covering with a blanket), and overhaul (locate and extinguish hidden fires) [45]. Most of the research developing AI technologies for incident response have focused on situational awareness activities such as identifying where occupants are located or understanding the fire origin and growth. However, AI has the potential to aid fireground operations in each of these tasks.

8.4.1   Early Detection

The first stage of the incident response process where AI-based tools can have benefit is in early detection. Many structures will include some form of fire detection and/or suppression system. These systems are typically designed in accordance with guidance in standards such as NFPA 72 [46], NFPA 13 [47], the SFPE Handbook of Fire Protection Engineering [48], or informed through performance-based fire modeling [49]. Researchers have developed several approaches which are able to detect the present of fire and smoke more quickly than systems recommended in these standards [50]. Several approaches to early fire and smoke detection have been developed using combinations of video cameras, computer vision, and machine learning [51–63]. Other researchers have focused on minimizing false detections through multi-sensor fusion at individual locations [64] or through the internet of things throughout a structure [65]. While current buildings are instrumented with less sensors than are suggested in this research, the potential for early fire detection using these types of algorithms will increase as more structures are designed with intelligent systems and wireless sensor networks.

Similar data sources could also be used to identify and report the presence of unwanted accelerants prior to ignition. For example, anticipated fuel loads within a space should be identified during pre-construction fire hazard analysis and evaluated during fire inspections. Object recognition algorithms [66, 67] could be combined with fuel load data from fire testing [68, 69] and rapid real-time fire modeling [42, 43] to identify if the fire hazard has increased beyond design considerations.

8.4.2   Deployment

Little research has been conducted into how AI-based tools can assist emergency responders during the deployment phase; however, there are several tasks where AI could provide benefit in the future. Data on current weather conditions and event timing (such as day, night, holiday) could be automatically compared with statistical loss information from previous fires to identify the presence of higher risk conditions. In addition, similar historic fires could be identified, and lessons learned from those events compared to the pre-incident plan available for the site to recommend real-time modifications to the operation. Information from the pre-incident plan on the available water supplies, hazards, and occupancy could be used to identify the number and type of apparatus and personnel required. In addition, as mobile sensor platforms and autonomous robots are more widely available, these tools can be mobilized to size-up response tasks prior to emergency responders arriving on scene.

8.4.3   Size-Up/Data Collection and Analysis

Initial objectives upon arrival is to evaluate the specific conditions at the scene by gathering information related to life safety, extinguishment, property conservation, structure stability, and resource availability [3]. Table 8.1 presents a summary of the goals and data needed by emergency responders in a structural fire [70]. Information regarding the occupancy classification, floor plan, and locations of potential hazards and available water supplies should be included within the pre-incident plan. Key additional information which must be gathered onsite includes actual number of occupants and their location, fire location, size, and predicted growth, as well as structure stability.

8.4.3.1 Occupant Load Estimation

Actual occupant loads vary with time of day and season (for example, an office high-rise may meet or exceed approved occupant loads during the day but be nearly unoccupied at night [71]). There are two main approaches to estimate the occupant load of a structure. Statistical approaches can take into account the impacts time of day, season, and weather conditions have on the actual occupant loads [72–74]. Unfortunately, detailed data needed for these statistical analyses may not exist for all occupancy classifications, and data that exists may not be available in the fireground due to the rapid timeline of emergencies. Future AI-based tools will be able to provide this detailed information on statistical occupant load based on all available data and present conditions.

An alternative approach to estimating the occupant load is to use real-time surveillance and tracking algorithms based on sensor networks [75–77]. Researchers have developed algorithms to track the movement of people through video cameras [78–80], accelerometers in the floor [81–83] (see Fig. 8.10 [83]), and other

Table 8.1 Data requirements of emergency responders

	Firefighter	Company officer/division or group supervisor	Incident commander
Goals	Rescue victims Contain and extinguish fire Stay alive	Crew safety Resource management Communication with IC Accountability of firefighters assigned Situational awareness of other crews, groups, and divisions	Structural integrity Fire dynamics, growth, development Assigned crew/group/division location and health
Data needs	Victim location Fire location Conditions Interior layout information	Structural integrity information by proximity Information on the location of other firefighters Fire location Victim location	Structural integrity Changes in fire dynamics Fire crew location information Exposure threat information
Data sources	Caller indicators Scene indicators Maps/visual size-up	Sensors on scene Sensors on fire fighters	Same as fire fighter Same as first responder Pre-event building data Real-time building sensors

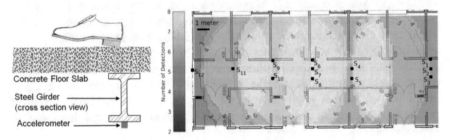

Fig. 8.10 Seismic sensor configuration and estimated footstep detection [83]. Republished courtesy of Elsevier

point sensors (such as light activations, ultrasonics, temperature, humidity, gas species concentrations, mechanical switches in chairs, or Wi-Fi traffic) [79, 80, 84–92], see Fig. 8.11 [92]. While advanced sensor networks may not be widely available in structures at this time, the prevalence of these systems in commercial structures is expected to increase in the future. AI is going to play a key role in gathering, processing, and presenting this data in a meaningful manner to emergency responders.

8.4.3.2 Fire Origin and Size Estimation

Fire detection and suppression systems within a structure should be documented in the pre-incident plan. Details regarding activation of these systems is typically available to emergency responders once they arrive on-site. While the general location of a fire's origin can typically be inferred from this data, it may not be

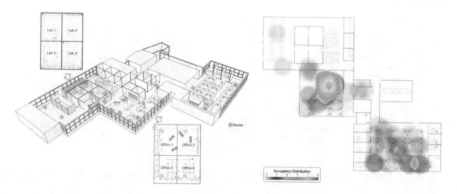

Fig. 8.11 Wi-Fi signal based occupant detection [92]. Republished courtesy of Elsevier

accurate and dependable for a number of reasons. In addition, these sensors on their own do not provide detailed indications to the Incident Commander about the size of the fire or how quickly it will grow. Researchers have developed several machine learning-based systems to provide scenario estimates of fire location, size, and growth. The general premise of these approaches is to conduct an inverse analysis where all available sensor data is compared to predictions of fire behavior from a fire model. This process is performed iteratively in an optimization framework until the most probable fire location, size, and growth based on the available data is determined. Some of these models use an additional machine learning component which is typically trained based on detailed predictions from high-fidelity computational fluid dynamics (CFD) fire models.

Similar to occupant load estimation, researchers have developed several approaches to the inverse analysis problem using multi-sensor fusion. Yuen et al. used a neural network to predict flow velocity and temperature through the door of a single room using flow velocity and temperature measurements to train the network [93]. The FireGRID project developed a framework which would assimilate all available sensor information to steer sensor driven forecasts of fire growth [94, 95]. The primary focus of this project was to determine if super-real-time forecasting of a fire scenario was possible using data assimilation and Monte Carlo style zone fire model simulations. Good forecasts were achieved with positive lead times for the simple geometry presented, see Fig. 8.12 [94]. However, a significant number of sensors were used which are unlikely to be present in a typical building (such as gas velocity measurement through doors, thermocouple trees in every room, surface thermocouples on the interior and exterior of every wall, and species concentrations of O_2, CO, and CO_2 in the center of multiple rooms). Ryu expanded the set of sensors applied in the inverse analysis using an Internet-of-Things-based approach to good effect [96]. Rasouli et al. developed a similar methodology for fire data analysis using pattern recognition on simulated sensor readings from fire dynamics simulator (FDS) [97]. The pattern recognition used temperature, smoke concentration, and heat sensors in each room to determine the fire size and growth

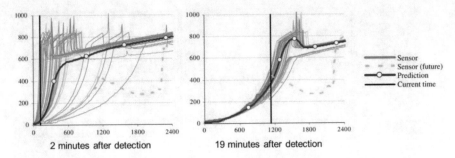

2 minutes after detection   19 minutes after detection

Fig. 8.12 Example inverse analysis and forecast from FireGRID [94]. Republished courtesy of Elsevier

rate. Wang et al. used an artificial neural network algorithm with thermocouple trees, smoke obscuration, and gas species concentration measurements to good effect [98]. While each of these sensors is more likely to be present in a structure, the sensor density is high.

Other researchers have developed approaches to the inverse analysis problem focusing primarily on dense temperature measurements. Jahn et al. [99, 100] and Overholt et al. [101, 102] demonstrated the potential effectiveness of dense thermocouple measurements in a single room to evaluate fire size and growth. Guo et al. and Wu et al. used a similarly dense matrix of thermocouples in hallways with Markov Chain Monte Carlo (MCMC) Bayesian Inference to consider fire location, growth rate, and maximum fire size for a single story 35 room building [103, 104]. Each of these groups presented good estimates of fire growth parameters, see Fig. 8.13 [104]; however, the density of thermocouple measurements required for the inverse analysis limits the implementation of these systems to laboratory settings or potentially unoccupied full-scale buildings. Lin et al. has had success in predicting fire size using less dense thermocouples for small geometries by using an ensemble Kalman filter [105–108]. However, this research still utilizes a thermocouple tree in each room of the building during test scenarios, which is unlikely to be available in actual occupied structures.

Each of the previous studies has presented inverse analysis approaches based on dense sensor networks which are unlikely to be widely available in the short-term. Similar approaches have been developed based on limiting the available data to typical sensors available now or those likely to be available in the future, such as smoke detector and sprinkler activation times. Jahn et al. evaluated the potential to determine fire location and growth parameters using these sensors by assimilating the sensors in a full CFD fire model using a coarse mesh [109]. The key limitation to this approach is the computational requirements of the full CFD model. This limitation can be alleviated through intelligent grid sizing of the CFD model [110], coupling the detailed CFD model with a more coarse zone fire model [111–118], or approximating the detailed CFD model using machine learning [42–44].

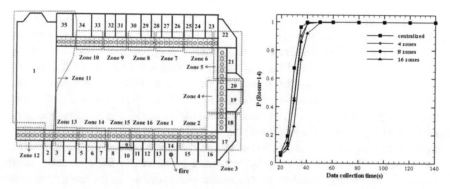

Fig. 8.13 Fire origin location determination through inverse analysis. From [104]; reprinted by permission of the American Institute of Aeronautics and Astronautics, Inc

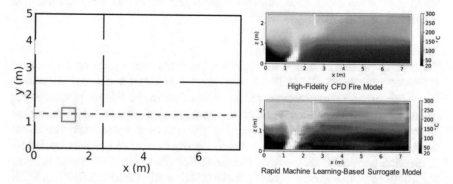

Fig. 8.14 Example spatially resolved thermal flow field predicted using a machine learning-based surrogate model compared with the CFD fire model predictions [42]. Republished courtesy of Elsevier

8.4.3.3 Occupant Survivability Profile Assessment

The previous studies showed that AI-based approaches have potential to evaluate the occupant load, as well as identify the fire location, size, and predicted growth rate based on sensor data (Fig. 8.14). The raw data from this analysis provides significant value for post-fire reconstruction and analysis; however, additional assessment is needed to provide the Incident Commander with actionable information. Once the inverse analysis is complete, faster-than-real-time modeling approaches can be used to provide additional detail about the conditions in the space. This information can be used to identify if occupants in the structure could survive existing and forecasted conditions, as well as understand the risk that a structure will collapse. Hodges et al. [42] and Lattimer et al. [43] recently presented an approach to couple coarse zone fire model predictions with a detailed machine learning model to predict the spatially resolved thermal flow field in a compartment fire 1000 to 10,000× faster than the high-fidelity fire model used to develop the training data, see Fig. 8.2 [42].

Combining the observed data and model predictions with detailed egress modeling [119, 120] could be used to develop personalized egress plans for occupants [121, 122], and inform the Incident Commander of egress paths which are likely to be uninhabitable due to heat and smoke. Detailed predictions of fire growth could be used to estimate the time until the room of origin or adjacent spaces reaches flashover conditions. In addition, coupling the fire behavior predictions with a thermal model of structural elements from a BIM model could provide a time-resolved risk potential of structural collapse. Probabilistic analyses could be used to evaluate the impact of different offensive and defensive extinguishment strategies. For example, Danzi et al. recently published a parametric fire risk assessment approach to quantify the risk to occupants and property based on fire growth categorization and expected firefighting expediency based on structure class and available suppression means [123]. While there is significant potential for these AI-based approaches to improve situational awareness on the fireground, additional research is needed [70].

8.4.3.4 Additional Fact Finding

The size-up process continues throughout a fire event which requires the incident commander to periodically reallocate resources towards fact finding [3]. An advantage of these AI-based processes for analyzing data in the size-up process is the capability to provide continual updates as more data is analyzed. While individual structures may not have enough sensors to perform these analyses, data from in situ sensors in the structure could be augmented through additional sensors carried by firefighting personnel, autonomous mobile sensor platforms, and robotic systems (e.g., Tethered Unmanned Aircraft Systems (UAS) equipped with infrared or FLIR cameras). Quantifying the uncertainty of the AI-based system in real-time could then be used to recommend optimal locations for fact finding teams and autonomous systems to further investigate and monitor. When the uncertainty in the analysis is low, the incident commander could reduce and reassign the firefighting teams which are allocated towards fact finding.

8.4.4   Incident Action Planning

Once an incident commander understands the conditions on the fireground, a decision must be made to pursue either offensive fire extinguishment, defensive fire extinguishment, or non-intervention strategies [3]. This decision is critical to the life safety of occupants and firefighters as well as the effectiveness of fireground operations. The Incident Commander weighs the risk to firefighters from smoke and structural collapse against the potential benefits of saving occupants and preserving property. The Incident Commander will also compare the estimated number of resources needed to conduct offensive operations with the firefighting resources available at the scene. The addition of autonomous robotic systems to the firefighting toolbox will adjust these calculations in the future. For example, autonomous systems could be assigned to evacuate occupants within a structure which is at risk of collapse, providing the benefit to life safety of the occupants while not increasing the risk to emergency responders. AI tools have the potential to assist Incident Commanders in making this decision in several ways.

8.4.4.1 Offensive Extinguishment

Once the decision is made to enter the structure, firefighters focus on search and rescue operations, extinguishment, and property conservation. Search and rescue operations include identifying where occupants are located, directing occupants towards egress paths, and assisting mobility limited occupants with evacuation. Occupant detection and tracking systems discussed earlier in this chapter can provide firefighters with a real-time occupant load map to better focus their efforts; however, the performance of these systems is likely to degrade throughout a fire as the presence of a fire smoke environment occludes optical sensors and erratic movement of occupants and firefighters confuses vibration-based sensors. Autonomous robotic systems can reduce the uncertainty during on-going evacuation by navigating through the structure to verify the location of occupants. In addition, networked autonomous robotic systems can develop a detailed real-time map of the structure including identifying egress paths which are blocked or likely to be blocked in a short time. Future robotic systems may be able to physically assist mobility limited occupants in evacuation.

The primary objective of interior extinguishment is to apply water to the burning fuel until it stops burning. The specific tactics associated with extinguishment operations vary from structure to structure depending on the available suppression systems. Structures with no fire protection systems require firefighters to manually fight the fire with water pumped through hoses from an exterior source. Firefighting activities necessitate enough water in order to be successful. Incident Commanders estimate the amount of water supply needed to suppress a fire, and subsequently the number and size of hose lines that need to be deployed. While this information should be contained within the pre-incident plan, changes in the building and community may impact the effectiveness of the plan. For example, changes in water supply demand around the building (such as from additional development constructed around the building), changes in fuel load in the building or occupancy, or new construction since development of the pre-incident plan. Detailed information provided by AI-based systems during size-up on the predicted heat release rate profile of the fire and fire location could be used to provide a more accurate estimate of water supply demands, and the hose deployment configuration needed to extinguish the fire. In addition, an AI-based system could provide an early warning that available water resources may be insufficient to suppress the fire, which would allow the Incident Commander to develop alternate strategies or request additional resources earlier in the event.

Interior extinguishment tactics in structures protected with automatic fire protection systems differ slightly from manual fire suppression operations. When a fully automated fire protection system is present, AI systems could be used to continuously monitor the automated system to ensure each sprinkler nozzle near the fire is operating and maintaining desired flow rates. If the automated system is not operating properly, AI-based tools can provide this information to the Incident Commander who can allocate resources to investigate the issue or provide localized manual suppression activities to augment the system.

In addition to directly supporting offensive extinguishment, AI tools can be used to continuously monitor conditions around firefighters. For example, sensors integrated into firefighter personal protection equipment (PPE) can monitor key aspects of the thermal environment, such as gas temperatures and presence of contaminants/gas species concentrations. AI-based systems can continuously compare these measurements to the predicted state of the fire from the inverse analysis to understand if actual conditions deviate from the expected value. These insights could provide early detection of untenable conditions, and early warning from the onset of flashover. Providing these measurements from all firefighters as feedback to the AI-based system could further improve the accuracy of the inverse analysis of the fire behavior, resulting in improved situational awareness for the IC and all firefighters.

8.4.4.2 Defensive Extinguishment

If an Incident Commander determines there is too much risk to enter a structure, defensive strategies may be employed. Incident Commanders evaluate the occupant survivability profile, severe structural damage, limited water supplies, and extenuating circumstances (saving high value contents). AI tools can provide a better understanding of the spread of the fire and smoke throughout the structure, forecasted fire growth/intensity, and the degradation of the structural integrity. This information can be used to assist the Incident Commander in determining the potential for structural survivability. AI systems can assist firefighters by identifying the locations on the structure which will benefit the most from suppression actions.

Different approaches can be used to protect adjacent structures. One approach focuses on containing the flame volume by targeting the building of origin. The primary consideration in this approach is to reduce the heat release rate of the fire while not pushing the flame volume towards an adjacent structure. AI-based tools could provide rapid estimates on the impacts of weather (e.g. relative humidity, ambient temperatures, and wind speed/direction) and high-pressure hose streams on the movement of the flame volume. Another approach focuses on removing heat from the adjacent buildings by directly targeting the adjacent structures with water, gel, or foam. Coating the adjacent structures with water will extract heat from the surface by increasing the temperature of the water until it undergoes a phase change to steam. In addition, firefighting in other areas has shown application of gels and class A foam to be effective at saving adjacent structures [124, 125]. AI-based systems could couple predictions of thermal exposure with a community layout and BIM of adjacent structures to understand the impact of different strategies on the thermal degradation of adjacent structures. This information could be used to assist Incident Commanders in developing optimal defensive exposure protection strategies for a specific fire event.

8.4.4.3 Salvage and Overhaul

After the incident has been controlled, remaining actions include verifying structural integrity, minimizing damage to property, and investigating the origin and cause of the fire. While little research has been conducted into using AI in this area, the same tools which have been studied in other applications may provide significant benefit to response personnel during the salvage and overhaul phase.

The first consideration during the salvage and overhaul is to determine if the structure is safe to enter. The previous sections have discussed how AI-based approaches can provide a better picture of the degradation of a structure by coupling the results of an inverse analysis of sensor data with detailed modeling of fire-structure interactions. This analysis can provide an overall evaluation of collapse potential before personnel enter the structure. If the risk is sufficiently low, the resulting analysis can provide personnel with a prioritized list of structural members to physically inspect to verify structural integrity.

The objective of overhaul is to find and extinguish hidden fire, remove smoldering materials, and remove entrapped smoke from the structure [3]. Detailed fire predictions can be used to understand the risk of enclosed areas containing ventilation limited, sufficiently heated pyrolyzed fuel which may re-ignite if ventilation conditions change. In addition, the fire behavior information provided by the AI-based system can be used to develop a prioritized list of areas of the structure to check for remaining fires.

Similar predictions could be used to understand which areas within the structure are likely to contain toxic materials and combustion products behind walls, ceilings, and other structural components which will need to be opened. In addition to overhaul operations, an investigation into the origin and cause of the fire may be conducted at this time. If sufficient sensors are available to conduct the inverse analysis, AI-based systems should be able to provide details on the room of origin and the early growth of the fire during the incipient stage. This information can be used to indicate the presence of accelerants in the space and/or identify likely fuel sources.

8.5   Summary

Key decisions on the fireground impact the life safety of occupants, firefighters, and the preservation of property must be made quickly and are often made on limited data. The increase in smart infrastructure and sensor networks in structures presents a key opportunity to improve this decision-making using AI. Researchers have recently started investigating how these tools can improve pre-incident planning through prioritizing inspections. While similar techniques could be used to prioritize the development and auditing of site-specific pre-incident plans, this has not been investigated. Several researchers have presented approaches to assist emergency responders in understanding the current state of the fire and occupants by analyzing on-site sensors. However, additional research is needed to better understand the optimal types, quantity, and locations of sensors required for this analysis. In addition, while researchers have developed tools which can predict the detailed thermal conditions faster-than-real time based on this inverse analysis, additional work is needed to integrate these systems to evaluate the risk to emergency personnel. Continued development of AI technologies will lead to improved decision-making in the fireground which may save lives and reduce the damage to property.

References

1. J. R. J. Hall and E. R. Twomey, "Challenges to Safety in the Built Environment," in *Fire Protection Handbook: Safety in the Built Environment*, 20th ed., R. E. Solomon, Ed. Quincy, MA: National Fire Protection Association, 2008, pp. 3–29.
2. NFPA FAR Division, "The total cost of fire in the United States," no. March, pp. 2–3, 2014.
3. B. J. Klaene and R. Sanders, "Fireground Operations," in *Fire Protection Handbook: Organizing for Public Sector Emergency Response*, 20th ed., Quincy, MA: National Fire Protection Association, 2008, pp. 3–11.
4. R. B. Gasaway, "Fireground Command Decision Making: Understanding the Barriers Challenging Commander Situation Awareness," Capella University, 2008.
5. G. A. Klein and R. Calderwood, "Decision Models: Some Lessons From the Field," *IEEE Trans. Syst. Man Cybern.*, vol. 21, no. 5, pp. 1018–1026, 1991.
6. G. a Klein, R. Calderwood, and A. Clinton-cirocco, "Rapid Decision Making on the Fire Ground," 1988.
7. R. J. Looby, "Fire Fighter Understanding and Application of NIOSH Recommendations," Grand Canyon University, 2020.
8. R. B. Gasaway, "Making Intuitive Decisions Under Stress: Understanding Fireground Incident Command Decision-Making," *Int. Fire Serv. J. Leadersh. Manag.*, vol. 1, no. 1, pp. 8–18, 2007.
9. K. A. Hall, "The effect of computer-based simulation training on fire ground incident commander decision making," 2010.
10. S. Gillespie, "Fire Ground Decision-Making: Transferring Virtual Knowledge to the Physical Environment," 2013.
11. C. H. Wijkmark, M. M. Metallinou, and I. Heldal, "The Role of Virtual Simulation in Incident Commander Education – A field study," in *NIK 200x*, 2020.
12. C. Grant, A. Hamins, N. Bryner, A. Jones, and G. Koepke, *Research Roadmap for Smart Fire Fighting*. National Institute of Standards and Technology, 2015.
13. D. Patterson, *Introduction to Artificial Intelligence*. Prentice Hall, Inc., 1990.
14. M. Chowdhury and A. W. Sadek, "Advantages and Limitations of Artificial Intelligence," in *Artificial Intelligence Applications to Critical Transportation Issues*, Transportation Research Circular, 2012, pp. 6–8.
15. Y. Lecun, Y. Bengio, and G. Hinton, "Deep learning," *Nature*, vol. 521, no. 7553, pp. 436–444, 2015.
16. K. Siau and W. Wang, "Building trust in artificial intelligence, machine learning, and robotics," *Cut. Bus. Technol. J.*, vol. 31, no. 2, pp. 47–53, 2018.
17. G. E. Marchant and R. A. Lindor, "The Coming Collision Between Autonomous Vehicles and the Liability System," *Santa Clara Law Rev.*, 2012.
18. A. C. Ian Goodfellow, Yoshua Bengio, "The Deep Learning Book," *MIT Press*, vol. 521, no. 7553, p. 785, 2017.
19. J. L. Hodges, "Predicting Large Domain Multi-Physics Fire Behavior Using Artificial Neural Networks," Virginia Polytechnic Institute and State University, 2018.
20. NFPA, *NFPA 1620: Standard for Pre-Incident Planning*. 2020.
21. J. Burris, "NFIRS: Better Data for Better Decisions," *Fire Eng.*, vol. 153, no. 5, 2000.

22. J. M. J. Watts, "Fire Risk Indexing," in *SFPE Handbook of Fire Protection Engineering, Fifth Edition*, 2016, pp. 3158–3182.
23. C. R. Jennings, "Urban Residential Fires: An Empirical Analysis of Building Stock and Socioeconomic Characteristics for Memphis, Tennessee," p. 286, 1996.
24. F. USFA National Fire Data Center, "Socioeconomic Factors and the Incidence of Fire," no. June, pp. 1–35, 1997.
25. Y. Lizhong, C. Heng, Y. Yong, and F. Tingyong, "The effect of socioeconomic factors on fire in China," *J. Fire Sci.*, vol. 23, no. 6, pp. 451–467, 2005.
26. P. Chhetri, J. Corcoran, R. J. Stimson, and R. Inbakaran, "Modelling potential socio-economic determinants of building fires in South East Queensland," *Geogr. Res.*, vol. 48, no. 1, pp. 75–85, 2010.
27. S. E. Schachterle, D. Bishai, W. Shields, R. Stepnitz, and A. C. Gielen, "Proximity to vacant buildings is associated with increased fire risk in Baltimore, Maryland, homes," *Inj. Prev.*, vol. 18, no. 2, pp. 98–102, 2012.
28. L. Garis, L. Thomas, S. Robinson, and A. Tyakoff, "Recovery Houses: Non-Compliance with the British Columbia Fire Code and Implications for Life Safety," no. May, 2016.
29. L. Garis and J. Clare, "Examining 'regular' fire-safety inspections: the missing relationship between timing of inspection and fire outcome," 2012.
30. E. Copeland, "Big Data in the Big Apple," 2015.
31. L. Garis and J. Clare, "A Dynamic Risk Based System for Scheduling Inspections A Dynamic Risk," no. September, 2015.
32. M. Madaio *et al.*, "Firebird: Predicting Fire Risk and Prioritizing Fire Inspections in Atlanta," in *Proceedings of the 22nd ACM SIGKDD International Conference on Knowledge Discovery and Data Mining*, 2016, pp. 185–194.
33. B. S. Walia *et al.*, "A dynamic pipeline for spatio-temporal fire risk prediction," *Proc. ACM SIGKDD Int. Conf. Knowl. Discov. Data Min.*, pp. 764–773, 2018.
34. D. Liu, Z. Xu, Y. Zhou, and C. Fan, "Heat map visualisation of fire incidents based on transformed sigmoid risk model," *Fire Saf. J.*, vol. 109, no. January, p. 102863, 2019.
35. J. Anderson-Bell, C. Schillaci, and A. Lipani, "Predicting non-residential building fire risk using geospatial information and convolutional neural networks," *Remote Sens. Appl. Soc. Environ.*, vol. 21, no. January, p. 100470, 2021.
36. J. C. Chen and J. P. Gore, "Real-Time Data Analytics," in *Research roadmap for smart fire fighting*, National Institute of Standards and Technology, 2015.
37. J. C. Chen and J. P. Gore, "Real-Time Data Analytics," in *Research Roadmap for Smart Fire Fighting*, C. Grant, A. Hamins, N. Bryner, A. Jones, and G. Koepke, Eds. National Institute of Standards and Technology, 2015, pp. 123–128.
38. D. Liu, X. Xia, J. Chen, and S. Li, "Integrating Building Information Model and Augmented Reality for Drone-Based Building Inspection," *J. Comput. Civ. Eng.*, vol. 35, no. 2, p. 04020073, 2021.
39. D. Zhang, J. Zhang, H. Xiong, Z. Cui, and D. Lu, "Taking advantage of collective intelligence and BIM-based virtual reality in fire safety inspection for commercial and public buildings," *Appl. Sci.*, vol. 9, no. 23, 2019.
40. K. A. Kapalo and J. J. LaViola, "Failing to Plan is Planning to Fail: Capturing the Pre-incident Planning Needs of Firefighters," *Proc. Hum. Factors Ergon. Soc. Annu. Meet.*, vol. 63, no. 1, pp. 612–616, 2019.
41. J. Beerthuis, "The Added Value of Virtual Reality Simulation for Safety and Security," 2021.
42. J. L. Hodges, B. Y. Lattimer, and K. D. Luxbacher, "Compartment fire predictions using transpose convolutional neural networks," *Fire Saf. J.*, vol. 108, no. November 2018, pp. 1–22, 2019.
43. B. Y. Lattimer, J. L. Hodges, and A. M. Lattimer, "Using Machine Learning in Physics-based Simulation of Fire," *Fire Saf. J.*, 2020.
44. T. Buffington, J. M. Cabrera, A. Kurzawski, and O. A. Ezekoye, "Deep-Learning Emulators of Transient Compartment Fire Simulations for Inverse Problems and Room-Scale Calorimetry," *Fire Technol.*, 2020.

45. J. Norman, *Fire Officer's Handbook of Tactics*, 4th ed. Fire Engineering, 2005.

46. NFPA, *NFPA 72: National Fire Alarm and Signaling Code*. 2019.

47. *NFPA 13 Standard for the Installation of Sprinkler Systems*. 2019.

48. R. P. Schifiliti, R. L. P. Custer, and B. J. Meacham, "Design of Detection Systems," in *SFPE Handbook of Fire Protection Engineering, 5th Edition*, Springer, 2016, pp. 1314–1377.

49. D. W. Stroup and D. D. Evans, "Use of computer fire models for analyzing thermal detector spacing," *Fire Saf. J.*, vol. 14, no. 1–2, pp. 33–45, 1988.

50. A. Gaur *et al.*, "Fire Sensing Technologies: A Review," *IEEE Sens. J.*, vol. 19, no. 9, pp. 3191–3202, 2019.

51. B. Ko, K. H. Cheong, and J. Y. Nam, "Early fire detection algorithm based on irregular patterns of flames and hierarchical Bayesian Networks," *Fire Saf. J.*, vol. 45, no. 4, pp. 262–270, 2010.

52. B. C. Ko, K. H. Cheong, and J. Y. Nam, "Fire detection based on vision sensor and support vector machines," *Fire Saf. J.*, vol. 44, no. 3, pp. 322–329, 2009.

53. F. Yuan, "A double mapping framework for extraction of shape-invariant features based on multi-scale partitions with AdaBoost for video smoke detection," *Pattern Recognit.*, vol. 45, no. 12, pp. 4326–4336, 2012.

54. P. Santana, P. Gomes, and J. Barata, "A vision-based system for early fire detection," *Conf. Proc. - IEEE Int. Conf. Syst. Man Cybern.*, pp. 739–744, 2012.

55. S. Verstockt *et al.*, "A multi-modal video analysis approach for car park fire detection," *Fire Saf. J.*, vol. 57, pp. 44–57, 2013.

56. D. K. Appana, R. Islam, S. A. Khan, and J. M. Kim, "A video-based smoke detection using smoke flow pattern and spatial-temporal energy analyses for alarm systems," *Inf. Sci. (Ny).*, vol. 418–419, pp. 91–101, 2017.

57. J. Rong, D. Zhou, W. Yao, W. Gao, J. Chen, and J. Wang, "Fire flame detection based on GICA and target tracking," *Opt. Laser Technol.*, vol. 47, pp. 283–291, 2013.

58. H. J. Zhang, N. Zhang, and N. F. Xiao, "Fire detection and identification method based on visual attention mechanism," *Optik (Stuttg).*, vol. 126, no. 24, pp. 5011–5018, 2015.

59. K. Muhammad, J. Ahmad, I. Mehmood, S. Rho, and S. W. Baik, "Convolutional Neural Networks Based Fire Detection in Surveillance Videos," *IEEE Access*, vol. 6, pp. 18174–18183, 2018.

60. S. M. Nemalidinne and D. Gupta, "Nonsubsampled contourlet domain visible and infrared image fusion framework for fire detection using pulse coupled neural network and spatial fuzzy clustering," *Fire Saf. J.*, vol. 101, no. August, pp. 84–101, 2018.

61. F. Yuan, J. Shi, X. Xia, Y. Fang, Z. Fang, and T. Mei, "High-order local ternary patterns with locality preserving projection for smoke detection and image classification," *Inf. Sci. (Ny).*, vol. 372, pp. 225–240, 2016.

62. G. Xu, Y. Zhang, Q. Zhang, G. Lin, and J. Wang, "Deep domain adaptation based video smoke detection using synthetic smoke images," *Fire Saf. J.*, vol. 93, pp. 53–59, 2017.

63. W. S. Qureshi, M. Ekpanyapong, M. N. Dailey, S. Rinsurongkawong, A. Malenichev, and O. Krasotkina, "QuickBlaze: Early Fire Detection Using a Combined Video Processing Approach," *Fire Technol.*, vol. 52, no. 5, pp. 1293–1317, 2016.

64. F. Derbel, "Performance improvement of fire detectors by means of gas sensors and neural networks," *Fire Saf. J.*, vol. 39, no. 5, pp. 383–398, 2004.

65. T. Listyorini and R. Rahim, "A prototype fire detection implemented using the Internet of Things and fuzzy logic," *World Trans. Eng. Technol. Educ.*, vol. 16, no. 1, pp. 42–46, 2018.

66. Y. Guo, Y. Liu, A. Oerlemans, S. Lao, S. Wu, and M. S. Lew, "Deep learning for visual understanding: A review," *Neurocomputing*, vol. 187, pp. 27–48, 2016.

67. Z. Q. Zhao, P. Zheng, S. T. Xu, and X. Wu, "Object Detection with Deep Learning: A Review," *IEEE Trans. Neural Networks Learn. Syst.*, vol. 30, no. 11, pp. 3212–3232, 2019.

68. NIST, "NIST Fire Calorimetry Database," 2021.

69. V. Babrauskas, "Heat Release Rates," in *SFPE Handbook of Fire Protection Engineering, Fifth Edition*, 2016, pp. 799–904.

70. P. Siebert and N. Venkatasubramanian, "Use of Data During an Emergency Event," in *Research roadmap for smart fire fighting*, C. Grant, A. Hamins, N. Bryner, A. Jones, and G. Koepke, Eds. National Institute of Standards and Technology, 2015, pp. 139–159.

71. NFPA, *NFPA 101: Life Safey Code*. 2021.

72. G. De Sanctis, J. Kohler, and M. Fontana, "Probabilistic assessment of the occupant load density in retail buildings," *Fire Saf. J.*, vol. 69, pp. 1–11, 2014.

73. Z. Chen, J. Xu, and Y. C. Soh, "Modeling regular occupancy in commercial buildings using stochastic models," *Energy Build.*, vol. 103, pp. 216–223, 2015.

74. Z. Li and B. Dong, "A new modeling approach for short-term prediction of occupancy in residential buildings," *Build. Environ.*, vol. 121, pp. 277–290, 2017.

75. T. D. Räty, "Survey on contemporary remote surveillance systems for public safety," *IEEE Trans. Syst. Man Cybern. Part C Appl. Rev.*, vol. 40, no. 5, pp. 493–515, 2010.

76. Z. Chen, C. Jiang, and L. Xie, "Building occupancy estimation and detection: A review," *Energy Build.*, vol. 169, pp. 260–270, 2018.

77. J. Ahmad, H. Larijani, R. Emmanuel, M. Mannion, and A. Javed, "Occupancy detection in non-residential buildings – A survey and novel privacy preserved occupancy monitoring solution," *Appl. Comput. Informatics*, vol. 17, no. 2, pp. 279–295, 2018.

78. J. Zou, Q. Zhao, W. Yang, and F. Wang, "Occupancy detection in the office by analyzing surveillance videos and its application to building energy conservation," *Energy Build.*, vol. 152, pp. 385–398, 2017.

79. M. Amayri, A. Arora, S. Ploix, S. Bandhyopadyay, Q. D. Ngo, and V. R. Badarla, "Estimating occupancy in heterogeneous sensor environment," *Energy Build.*, vol. 129, pp. 46–58, 2016.

80. F. Wang *et al.*, "Predictive control of indoor environment using occupant number detected by video data and CO2 concentration," *Energy Build.*, vol. 145, pp. 155–162, 2017.

81. S. Pan, A. Bonde, J. Jing, L. Zhang, P. Zhang, and H. Y. Noh, "BOES: Building Occupancy Estimation System using sparse ambient vibration monitoring," *Sensors Smart Struct. Technol. Civil, Mech. Aerosp. Syst. 2014*, vol. 9061, no. April 2014, p. 90611O, 2014.

82. R. Bahroun, O. Michel, F. Frassati, M. Carmona, and J. L. Lacoume, "New algorithm for footstep localization using seismic sensors in an indoor environment," *J. Sound Vib.*, vol. 333, no. 3, pp. 1046–1066, 2014.

83. J. D. Poston, R. M. Buehrer, and P. A. Tarazaga, "Indoor footstep localization from structural dynamics instrumentation," *Mech. Syst. Signal Process.*, vol. 88, pp. 224–239, 2017.

84. L. M. Candanedo and V. Feldheim, "Accurate occupancy detection of an office room from light, temperature, humidity and CO2 measurements using statistical learning models," *Energy Build.*, vol. 112, pp. 28–39, 2016.

85. Z. Chen, M. K. Masood, and Y. C. Soh, "A fusion framework for occupancy estimation in office buildings based on environmental sensor data," *Energy Build.*, vol. 133, pp. 790–798, 2016.

86. S. H. Ryu and H. J. Moon, "Development of an occupancy prediction model using indoor environmental data based on machine learning techniques," *Build. Environ.*, vol. 107, pp. 1–9, 2016.

87. S. Zikos, A. Tsolakis, D. Meskos, A. Tryferidis, and D. Tzovaras, "Conditional Random Fields - Based approach for real-time building occupancy estimation with multi-sensory networks," *Autom. Constr.*, vol. 68, pp. 128–145, 2016.

88. J. Chaney, E. Hugh Owens, and A. D. Peacock, "An evidence based approach to determining residential occupancy and its role in demand response management," *Energy Build.*, vol. 125, pp. 254–266, 2016.

89. T. Labeodan, K. Aduda, W. Zeiler, and F. Hoving, "Experimental evaluation of the performance of chair sensors in an office space for occupancy detection and occupancy-driven control," *Energy Build.*, vol. 111, pp. 195–206, 2016.

90. C. Jiang, M. K. Masood, Y. C. Soh, and H. Li, "Indoor occupancy estimation from carbon dioxide concentration," *Energy Build.*, vol. 131, pp. 132–141, 2016.

91. M. Gruber, A. Trüschel, and J. O. Dalenbäck, "CO2 sensors for occupancy estimations: Potential in building automation applications," *Energy Build.*, vol. 84, pp. 548–556, 2014.

92. H. Zou, H. Jiang, J. Yang, L. Xie, and C. J. Spanos, "Non-intrusive occupancy sensing in commercial buildings," *Energy Build.*, vol. 154, pp. 633–643, 2017.
93. R. K. K. Yuen, E. W. M. Lee, S. M. Lo, and G. H. Yeoh, "Prediction of temperature and velocity profiles in a single compartment fire by an improved neural network analysis," *Fire Saf. J.*, vol. 41, no. 6, pp. 478–485, 2006.
94. S. H. Koo, J. Fraser-Mitchell, and S. Welch, "Sensor-steered fire simulation," *Fire Saf. J.*, vol. 45, no. 3, pp. 193–205, 2010.
95. L. Han *et al.*, "FireGrid: An e-infrastructure for next-generation emergency response support," *J. Parallel Distrib. Comput.*, vol. 70, no. 11, pp. 1128–1141, 2010.
96. C. S. Ryu, "IoT-based intelligent for fire emergency response systems," *Int. J. Smart Home*, vol. 9, no. 3, pp. 161–168, 2015.
97. S. Rasouli, O.-C. Granmo, and J. Radianti, "A methodology for fire data analysis based on pattern recognition towards the disaster management," *2015 2nd Int. Conf. Inf. Commun. Technol. Disaster Manag.*, pp. 130–137, 2015.
98. X. G. Wang, S. M. Lo, and H. P. Zhang, "Influence of feature extraction duration and step size on ANN based multisensor fire detection performance," *Procedia Eng.*, vol. 52, pp. 413–421, 2013.
99. W. Jahn, G. Rein, and J. L. Torero, "Forecasting fire growth using an inverse zone modelling approach," *Fire Saf. J.*, vol. 46, no. 3, pp. 81–88, 2011.
100. W. Jahn, G. Rein, and J. L. Torero, "Forecasting fire dynamics using inverse computational fluid dynamics and tangent linearisation," *Adv. Eng. Softw.*, vol. 47, no. 1, pp. 114–126, 2012.
101. K. J. Overholt and O. A. Ezekoye, "Characterizing Heat Release Rates Using an Inverse Fire Modeling Technique," *Fire Technol.*, vol. 48, no. 4, pp. 893–909, 2012.
102. K. J. Overholt, "Forward and Inverse Modeling of Fire Physics Towards Fire Scene Reconstructions," University of Texas at Austin, 2013.
103. S. D. Guo, R. Yang, H. Zhang, and X. Zhang, "New Inverse Model for Detecting Fire-Source Location and Intensity," *J. Thermophys. Heat Transf.*, vol. 24, no. 4, pp. 745–755, 2010.
104. N. Wu, R. Wang, H. Zhang, and L. Q. (United T. R. Center-China), "Decentralized Inverse Model for Estimating Building Fire Source Location and Intensity," vol. 27, no. 3, 2012.
105. C.-C. Lin and L. Wang, "Applications of data assimilation to forecasting indoor environment," *IEEE Int. Conf. Autom. Sci. Eng.*, vol. 2014-Janua, pp. 1097–1102, 2014.
106. C. C. Lin and L. Wang, "Forecasting smoke transport during compartment fires using a data assimilation model," *J. Fire Sci.*, vol. 33, no. 1, pp. 3–21, 2015.
107. C. Lin, G. Zhao, and L. L. Wang, "Using Real - Time Sensing Data for Predicting Future State of Building Fires," *IEEE Int. Conf. Autom. Sci. Eng.*, pp. 1313–1318, 2015.
108. C. C. Lin and L. (Leon) Wang, "Real-Time Forecasting of Building Fire Growth and Smoke Transport via Ensemble Kalman Filter," *Fire Technol.*, vol. 53, no. 3, pp. 1101–1121, 2017.
109. W. Jahn, "Using suppression and detection devices to steer CFD fire forecast simulations," *Fire Saf. J.*, vol. 91, no. January, pp. 284–290, 2017.
110. W. Jahn, F. Sazunic, and C. Sing-Long, "Towards Real-Time Fire Data Synthesis Using Numerical Simulations," *J. Fire Sci.*, 2021.
111. L. Wang, Q. Chen, and Q. Chen, "THEORETICAL AND NUMERICAL STUDIES OF COUPLING MULTIZONE AND CFD MODELS FOR BUILDING AIR DISTRIBUTION SIMULATIONS," pp. 348–361, 2007.
112. Y. Ishida, "Method for Coupling Three-Dimensional Transient Pollutant Transport into One-Dimensional Transport Simulation Based on Concentration Response Factor," *ASHRAE Trans.*, vol. 114, 2008.
113. M. Bartak, I. Beausoleil-morrison, J. A. Clarke, J. Denev, F. Drkal, and M. Lain, "Integrating CFD and building simulation," vol. 37, pp. 865–871, 2002.
114. W. Zhang, K. Hiyama, S. Kato, and Y. Ishida, "Building energy simulation considering spatial temperature distribution for nonuniform indoor environment," *Build. Environ.*, vol. 63, pp. 89–96, 2013.
115. F. Colella, G. Rein, V. Verda, and R. Borchiellini, "Multiscale modeling of transient flows from fire and ventilation in long tunnels," *Comput. Fluids*, vol. 51, no. 1, pp. 16–29, 2011.

116. J. Floyd, "Coupling a Network HVAC Model to a Computational Fluid Dynamics Model Using Large Eddy Simulation," *Fire Saf. Sci. 10*, 2011.
117. I. Vermesi, G. Rein, F. Colella, and M. Valkvist, "Reducing The Computational Requirements for Simulating Tunnel Fires by Combining Multiscale Modelling and Multiple Processor Calculation," pp. 1–16.
118. A. Haghighat, K. Luxbacher, and B. Lattimer, "Development of a Methodology for Interface Boundary Selection in the Multiscale Road Tunnel Fire," *Fire Technol.*, vol. 54, no. 4, pp. 1043–1080, 2018.
119. M. Aleksandrov, C. Cheng, A. Rajabifard, and M. Kalantari, "Modelling and finding optimal evacuation strategy for tall buildings," *Saf. Sci.*, vol. 115, no. December 2018, pp. 247–255, 2019.
120. M. Choi and S. Chi, "Optimal route selection model for fire evacuations based on hazard prediction data," *Simul. Model. Pract. Theory*, vol. 94, no. December 2018, pp. 321–333, 2019.
121. M. Gamaleldin, Z. Liao, M. Asfour, and L. Zhao, "Optimizing the Egress Route Using a New Smoke Emulator IoT System," *IEEE Internet Things J.*, vol. 8, no. 11, pp. 9373–9382, 2021.
122. H. Jiang, "Mobile Fire Evacuation System for Large Public Buildings Based on Artificial Intelligence and IoT," *IEEE Access*, vol. 7, pp. 64101–64109, 2019.
123. E. Danzi, L. Fiorentini, and L. Marmo, *FLAME: A Parametric Fire Risk Assessment Method Supporting Performance Based Approaches*, vol. 57, no. 2. Springer US, 2021.
124. J. Urbas, "Effectiveness of Pre-Applied Wetting Agents in Prevention of Wildland Urban Interface Fires," *Fire Mater.*, vol. 37, pp. 563–580, 2013.
125. R. W. Gorte and K. Bracmort, "Wildfire protection in the Wildland-Urban interface," *Congr. Res. Serv. Rep. Congr.*, pp. 99–104, 2012.

Implementing AI to Assist Situation Awareness: Organizational and Policy Challenges

9

Charles R. Jennings

9.1 Introduction

Achieving situation awareness (SA) among first responders is an enduring challenge. As illustrated perhaps most starkly in the 9/11 response to the World Trade Center, the consequences of separate command posts and inability to utilize radio interoperability led to a breakdown of SA which exacerbated the toll in terms of deaths, primarily among first responders ([1]; [2], p. 9).

Many existing efforts to advance SA have consisted of aggregating textual and visual information such as dispatch or unit status, resources, and video and audio feeds from source such as in-building or public fixed or deployable assets. While these efforts have proliferated under private and quasi-public leadership, no true standard has emerged, and such systems have gained limited utilization in field settings.

The proliferation of data streams and sensors have exponentially increased the sources of data, and the challenges of making sense of such information in a way that can be leveraged by incident commanders in real time have increased apace. An emergent challenge is *how to make sense and impart actionable intelligence* to support decision-making during an emergency incident. We term this utilization of AI to assist in producing SA "AI assisted SA" [3]/AI in and of itself is an instrument for achieving SA, which is a meaningful, defined end state.

The advances in Artificial Intelligence (AI) offer tremendous potential to enhance safety by making sense of complex and disparate data streams to improving SA. While its potential application is widespread, this chapter will focus on the built environment, and the case of assisting with evacuation, and needs of incident

C. R. Jennings (✉)
Department of Security, Fire, and Emergency Management, John Jay College of Criminal Justice, The City University of New York, New York, NY, USA
e-mail: cjennings@jjay.cuny.edu

© Springer Nature Switzerland AG 2022
M. Z. Naser, G. Corbett (eds.), *Handbook of Cognitive and Autonomous Systems for Fire Resilient Infrastructures*, https://doi.org/10.1007/978-3-030-98685-8_9

commanders in managing emergencies. AI can assist in providing earlier and more definitive detection of hazardous conditions, assist in reporting them, and in leveraging the building and its systems to facilitate evacuation of occupants. While our analysis is generally applicable to any structure, our discussion will be focus primarily on the high-rise building, which has received research attention for many years, and is in many ways a likely leading case for application of this technology.

Indeed, the closely-related *smart city* movement encompasses this challenge, and cites public safety as one of its key benefits ([4], p. 225; [5, 6]). The International Standards Organization's *ISO 37120 Sustainable cities and communities* standard explicitly calls out numbers of firefighters, response times, and fire deaths per capita as elements of its indicators for quality of life [7]. The smart building, then, would serve as the "building block" of the smart city [8].

Smart cities are not merely products of technological innovation. Context, to include "complexities and interconnections among social and technical factors" must be considered [9]. Allam and Dhunny further emphasize the need to consider the societal integration of these concepts [10]. This context is critical to the implementation of AI to achieve improved SA for building evacuation and incident management.

Significant challenges remain to realization of the potential of AI. Amid the hype and hyperbole, much of it promoted by vendors, start-ups, and researchers with prototypes still in the lab, there are real organizational, policy, and implementation challenges. These challenges are generally ill-defined, and the streams of funding to address these issues is wholly inadequate. Furthermore, there is little recognition of these needs and consequently, we are inundated with rosy predictions and inadequately-vetted solutions.

The path to implementation of AI as an aid to evacuation and to enable this potential requires that *analytics* be developed for interpretation of various sensor outputs, signals, and merged with reports from humans via traditional emergency reporting networks. The emergency reporting networks themselves are in transition, moving from analog technology to digital-capable systems that can accept inputs from emerging and currently available technologies [11–13].

As a consequence, there is a mismatch between the readiness of emergency services for utilization of AI assisted incident management, and the capacity of private sector interests who are attempting to develop and promote availability of sensor outputs for decision support. Mediating the delivery of this information is the public emergency reporting system—namely the 9-1-1 system. The 9-1-1 system is itself undergoing transition, and is largely unable to manage or utilize these prospective streams of information.

This chapter will define key terms relevant to this problem, cite current examples in production or in advanced testing. We will then define some conceptual use cases based on humans as information sources, building infrastructure, and finally emergency reporting. We then discuss barriers to implementation and utilization of AI assisted SA, ethical implications of AI for SA, and the outlook for the near future. This chapter will focus on the United States, although the concepts are generally applicable.

9.2   Definitions

The following definitions may be helpful in understanding the chapter.

Algorithm—An unambiguous specification of a process describing how to solve a class of problems that can perform calculations, process data, and automate reasoning [14].

Artificial Intelligence—A methodology used in machine learning to determine which one of several used models have the highest performance [14].

IoT—The Internet of Things (IoT) is the network of physical objects that contain embedded technology to communicate and sense or interact with their internal states or the external environment [15, 16].

Machine Learning—The subfield of Artificial Intelligence that often uses statistical techniques to give computers the ability to "learn," i.e., progressively improve performance on a specific task, with data, without being explicitly programmed [14].

Real-time analytics—It is the discipline that applies logic and mathematics to data to provide insights for making better decisions quickly. For some use cases, real time simply means the analytics is completed within a few seconds or minutes after the arrival of new data. *On-demand real-time analytics* waits for users or systems to request a query and then delivers the analytic results. *Continuous real-time analytics* is more proactive and alerts users or triggers responses as events happen [15, 16].

9.3   Conceptual Use Cases

The use of AI assisted SA can be represented In a number of domains. For our purposes, we can consider two frames of reference—human-based sensors and AI and physical infrastructure-based AI. Many conceptions of smart cities and theorized evacuation or incident management SA systems straddle these two domains. Considering the two domains as distinct is important because of the privacy, information sharing, and funding or resource bases required for each.

While there is considerable work on conceptual use cases integrating these domains, we will focus on building evacuation and incident management. In reality, the range of sensors can be brought to bear on helping to build SA for fire and evacuation incidents. In practice, the stages of everyday building operations, management, and emergency incident management should be tightly integrated.

The Internet of Things (IoT) has been used to describe various arrangements of sensors arrayed to produce a desired functionality or service. While no consensus has emerged on the structure of IoT networks, the consumer market has shown potential, while also confirming the challenges of adapting mass consumer-driven products to mission critical public safety uses [17].

9.3.1   Humans as Sensors and AI Inputs

Humans as sensors is actually a leading area for development of AI assisted SA. Traditionally, humans serve as a valuable intelligence tool for incident detection and emergency reporting. The vanguard of sensors for humans is the proliferation of mobile phones and affiliated technologies such as smart watches and fitness monitors. These technologies are pervasive, and exist irrespective of the building stock.

9.3.1.1 Wearables
The rise of wearables is an important development. Even in buildings where no smart infrastructure is installed, the mobile sensors associated with occupants will provide an opportunity, with data aggregation and analytics, for improved SA.

Using the sensors embedded within mobile phones can be a critical data source in managing an evacuation. The ability to track position and movement using the in-phone accelerometer and gyroscope is a key necessary capability [18]. In addition, the presence of people, details of the quality of an evacuation such as congestion or free flow down stairs can be inferred from mobile phone sensors [19].

There is considerable research ongoing on the use of smart phone sensors to infer a number of valuable characteristics of both individual users as well as the ability to aggregate data for inferring the status of groups or sub-groups such as occupants of upper floors of a building, for example [20]. Such data can include physiological information, movement, behaviors, and even the presence of or exposure to communicable diseases. There are significant concerns in regard to privacy and data access to permit the structured use of this data for incident management purposes [21].

9.3.2   Physical Infrastructure-Based AI

The second main source of inputs for AI assisted SA is through the use of sensors and analytics applied to the built environment. The built infrastructure can be considered both from the scale of the "smart city" as well as the individual "smart building," with the potential for sharing data and jointly contributing to SA [22].

The scale of intelligence can work from the individual building up to the city scale or work from the city scale down to the individual building.

9.3.2.1 Buildings
Examples of physical infrastructure sensors would include monitoring and evalua-tion of evacuation routes, sensing presence of hazardous chemicals, GPS location of vehicles, and weather [6]. Others have suggested using computer vision, RFID, LIDAR, footpads, gas analyzers, and Bluetooth [23, 24].

The importance of being able to use BIM as a basis for both pre-event fire safety engineering is an important and implied if not explicitly necessary for detection of adverse events and management of evacuation [25]. Siddiqui et al. have identified the need for data sharing to accomplish these goals, which is an important beginning, even as they acknowledge that connections and coverage of data within BIM models does not necessarily support the detail needed for fire safety engineering.

Vandercasteele, Merci, and Verstockt link SA with BIM using compute vision to detect and characterize fire detection and extent of progress, but acknowledge that most BIM do not contain sufficient detail to enable such efforts generally [26].

Some have proposed integrating evacuation modeling with general occupant comfort and energy efficiency using BIM and agent-based models [27, 28]. Others have explicitly proposed the use of BIM as a basis for assessing evacuation capabilities of tall buildings [29], or fire safety generally [30].

Other studies do not explicitly reference BIM, but discuss more targeted monitoring of fire protection equipment using IoT and aided by AI. The use of IoT can monitor the status of fire protection systems and features as well as hazards within the building, and such a system is being actively deployed in Seoul, Korea [31], and suggested for use elsewhere [32]. Jiang [33] proposes a model for fire evacuation for large buildings integrating fire modeling, fire protection systems, and evacuation in real time.

As an example of interim steps toward using sensors to enhance SA are efforts to use video or other monitoring of high-rise building stairwell discharge (to enable counts of persons exiting or flow rates) or common areas to assess the progress of egress and the tenability of exits (Fig. 9.1) [34] which was incorporated in the NFPA 101, *Life Safety Code*® [35] for high-rise buildings with an occupant load of more than 4000 persons.

9.3.2.2 Highways and Streetscape

Highways, streets, and the outdoor built environment are the venue for installation of numerous sensors and devices. These sensors are useful when in proximity to a building of interest, and can provide information such as best access routes for emergency responders, presence of environmental hazards, and exterior views of structures involved in emergencies.

9.3.3 Emergency Reporting

Perhaps the most intuitive (but hardly the least complex) use case for AI assisted SA is through the emergency reporting system. The use of AI to reduce false alarms, increase the likelihood that a reported emergency is actually happening, and transmit this information to emergency responders is a fundamental capability. AI offers the potential to identify problems more quickly, with greater certainty, and notify responders with more definitive information.

Of course, underlying the adoption of AI and related technologies is the interface with emergency responders. This interface can occur through electrical or electronic signaling mechanisms. The most common of these is interface with the emergency dispatch systems. This emergency dispatch system is not a unitary entity, and in spite of efforts to impose a digital, integrated system, considerable work remains.

Fig. 9.1 Video monitoring
(not recorded) of common
hallways in high-rise
apartment building arranged
for fire service use (White
Rock, BC, Canada)

The existing infrastructure for emergency reporting in the United States is highly decentralized, with some centers serving individual communities or villages of a few thousand people, all the way to major city or county-wide systems servicing millions. The technical sophistication of these centers also varies (see Table 9.1). This interface typically occurs through a complex topology of 9-1-1 centers, dispatch centers, serving a checkerboard of jurisdictional entities of varying size, scope, and complexity. These centers also vary with regard to the emergency response disciplines they serve. It is important to distinguish between 9-1-1 call receiving and the dispatch function, which may be performed by different personnel, working at different agencies and at different locations. These arrangements vary from community to community.

9.3.3.1 NG9-1-1

NG9-1-1, or next-generation 9-1-1, describes a plan to migrate the emergency reporting 9-1-1 system from the 1960s vintage analog, copper wire system built for the land-line phone to a digital system. The 9-1-1 system has struggled for years with the advent of new technologies for reporting emergencies. The rise of mobile phones undermined the cornerstone of the 9-1-1-system's ability to identify the location of callers. The original system relied on a database that linked each land-line phone number to a specific address. The rise of cell phones and the advent of voice over internet protocol (VoIP) frustrated the ability to locate callers, and the Federal Communications Commission reacted with several upgrades to accommodate location features for wireless callers.

Table 9.1 Overview of 9-1-1 center and dispatch center structure

	Single-discipline client, e.g., fire and emergency services only	Multiple-discipline public safety clients, e.g., fire and emergency services and police services	Multi-discipline clients, including non-emergency sevices such as parks or pulic works or transportation
Single agency provider			
Municipality as provider			
County as provider			
Regional (more than one country) center as provider			

Source: International City/County Management Association. *Managing Fire and Emergency Services*. Washington, DC: 2012, p. 455

Implementation of these and other upgrades is managed locally, and funded with local government resources, meaning that the rollout of such "fixes" can take years to become ubiquitous. While widely adopted, they are still not universal [36].

Even with these refinements, most 9-1-1 centers have little capacity to natively receive and interact with rich content including video, sensor data, images, and other sources of field-based intelligence and data.

There is a long (if not tortuous) path to implementation of NG911 ([37], pp. 28–29) and implications on staffing levels needed, alignment of policies, and other critical steps remain to be done. These challenges are both managerial and organizational. Current funding is clearly inadequate ([38], p. 200). Local government structure in many parts of the United States empowers local governments, and requires cumbersome and lengthy processes for communities to provide shared services. As a consequence, numerous states have launched programs and incentives to encourage communities to employ regional service delivery, and pool their resources to operate larger, better-staffed and equipped dispatch centers. These efforts are moving at a deliberate pace, as the origins of NG9-1-1 stretch back to 2001 [11–13].

9.3.3.2 FirstNet and Wireless Broadband Data

FirstNet is a nationwide broadband data network designed to be used for emergency responders. Inspired by the communication failures in New York City on 9/11, FirstNet was developed as a result of federal legislation [39] which raised funds through an auction of public frequencies to fund construction of the network. Construction and operation of the network the $46.5 billion network was awarded through a competitive process to a major telecommunications provider in the United States [40].

FirstNet is poised to provide the potential for field access to services such as Automatic Vehicle Location, shared video, automatic crash detection, patient tracking, and evacuee tracking. Also included in its feature set if the use of long-

Public Safety Communications Evolution

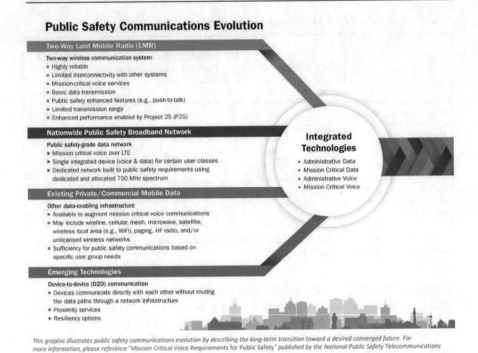

Two-Way Land Mobile Radio (LMR)

Two-way wireless communication system
- Highly reliable
- Limited interconnectivity with other systems
- Mission-critical voice services
- Basic data transmission
- Public safety enhanced features (e.g., push-to-talk)
- Limited transmission range
- Enhanced performance enabled by Project 25 (P25)

Nationwide Public Safety Broadband Network

Public safety-grade data network
- Mission critical voice over LTE
- Single integrated device (voice & data) for certain user classes
- Dedicated network built to public safety requirements using dedicated and allocated 700 MHz spectrum

Existing Private/Commercial Mobile Data

Other data-enabling infrastructure
- Available to augment mission critical voice communications
- May include wireline, cellular, mesh, microwave, satellite, wireless local area (e.g., WiFi), paging, HF radio, and/or unlicensed wireless networks
- Sufficiency for public safety communications based on specific user group needs

Emerging Technologies

Device-to-device (D2D) communication
- Devices communicate directly with each other without routing the data paths through a network infrastructure
- Proximity services
- Resiliency options

Integrated Technologies
- Administrative Data
- Mission Critical Data
- Administrative Voice
- Mission Critical Voice

This graphic illustrates public safety communications evolution by describing the long-term transition toward a desired converged future. For more information, please reference "Mission Critical Voice Requirements for Public Safety" published by the National Public Safety Telecommunications Council at http://www.npstc.org/npstcReports.jsp.

Fig. 9.2 Public Safety Communications Evolution (Source: U.S Cybersecurity and Infrastructure Agency)

term evolution (LTE) for use of mission critical voice over broadband ([41], p. 2). The network is also designed to provide for enhanced reliability through public safety priority pre-emption.

One of the resultant benefits of FirstNet will be the prospect of greatly enhanced interoperability—meaning that disparate agencies will be able to communicate during a major incident using existing communications hardware. This interoperability will extend beyond voice communication to include exchange of data. Interoperability is not solely a function of technology, and requires administrative and policy direction to be effective.

In the long term, FirstNet aims to provide a convergence of land mobile radio and voice over broadband which could further enable enhanced, incident-specific voice interoperability between disparate users including not just traditional emergency responders, but also data sharing that can enable a common SA between the field and dispatch (Fig. 9.2).

FirstNet could provide a near-ubiquitous broadband access for first responders, which could enable transmission of rich data between incident scenes, private networks, building and infrastructure, mobile responders, and dispatch centers. At present, connectivity and reliable data access are serious challenges in many communities.

9.4   Barriers to Utilization

There are many barriers to utilization of potential AI in evacuation and incident management. These barriers begin within the built environment, and include the divided responsibility between the private and public sectors, limitations on analytics and display for first responder utilization, a failure to elaborate a concept of operations across these sectors, challenges of organization scale among first responders, market failure for realization of first responder needs, and difficulties in allocating and measuring costs across the various sectors and organizations involved in utilization of AI supported SA.

9.4.1   Emergent Standards for BIM Data Exchange

There is a need for technologies to tie together the various sensors and actuators that can automatically manage building functions to achieve greater efficiency, user comfort, and reduced waste. While tremendous progress has been made, currently the country lacks mandated or widely used national BIM standards. Instead BIM approaches are haphazard, varying by company, public agency, or client requirements" [42, 43]. The National Institute of Building Sciences' Building Information Management (BIM) Council heads a consensus process for development and promulgation of a national BIM Standard for the United States [44].

One of the most significant challenges is the fragmented nature of Building Information Models (BIM) Systems, and the multiple standards in place. The purpose of BIM is to enable architects, engineers, and construction specialists to virtually "view and test [a digital representation of a building], revise it as necessary, and then output various reports and views for purchasing, fabrication, assembly, and operations" [45].

BIM systems form the basis for harnessing the in-building sensors and even human sensors within the building to enable collection, processing, and production of meaningful intelligence in emergencies.

9.4.2   Analytics and Display

The use of information by incident commanders can quickly surpass their ability to utilize it [46]. The mere existence of information purported to promote SA does not assure its acquisition and usage by those who need it ([47], p. 6). This information is often acquired via some interface, particularly for large or complex structures.

An important initial step in deciding what information should be displayed is informed by research by Groner and others which suggests that discipline and role are critical to understanding information needs. Also critical is understanding the time frame in which information is most critical for sensemaking and ultimately SA. Existing research on first responder information needs also support this approach, specifically for the case of fire in a tall building [48]. Emergency incidents generally begin with incomplete, vague, and sometimes wrong information.

A research study examining the information needs of police, fire, and EMS responders to a series of reported emergencies set in a high-rise building showed clearly that different roles required different information, and that the information need was most critical in the early stages of an incident—from initial report, to the first unit arriving on scene, and the period from arrival to forming an accurate understanding of the nature of the incident to include such characteristics as type of incident, location within the building, extent, and property or people threatened [49].

Groner and others applied a structured interviewing technique known as *Situation Awareness Information Requirements Analysis* (SAIRA). Using a negotiated text approach [50], SAIRA offers promise as a simple and effective means for eliciting the information critical to decision-making and avoids the common pitfall of providing undifferentiated information to all responders (irrespective of role) and asking for extraneous information that is not directly relevant to decisions (wish lists) [49]. Most importantly for AI assisted SA, these roles should be considered, which will in turn determine the information needed, its form, and at what stage of the incident it is needed. The authors also used interviews to identify potential sensors and information needs at several representative emergency scenarios.

9.4.3 CONOPS

A concept of operations is a term originally developed in the military to describe a commander's intent and the plan for achieving this.

> The concept of operations is a statement that directs the manner in which subordinate units cooperate to accomplish the mission and establishes the sequence of actions the force will use to achieve the end state. The concept of operations describes how the commander sees the actions of subordinate units fitting together to accomplish the mission [51].

The term is also widely used to describe technical systems, taking into account the social and organizational dimensions of such systems. These dimensions include organizational leadership's intent on how the system should operate to meet the desired goals and objectives. Such information should be consistent with both operational and long-range strategic plans ([52], p. 87). Indeed, with the deep integration of technical systems into military planning, considerable attention is paid to concept of operations as it pertains to electronic command, control systems (later expanded to C5I, representing command, control, communications, computers, combat systems, and intelligence) [53].

The MITRE Handbook of Systems Engineering defines CONOPS as:

the full range of factors that are needed to support the mission (i.e., doctrine, organization, training, leadership, materiel, personnel, facilities, and resources). Post-fielding life cycle costs often dwarf those of the development effort. Therefore, it is critical that the CONOPS provide sufficient information to determine long-term life cycle needs such as training, sustainment and support throughout capability fielding and use. [54]

No similar structured effort is currently imposed on the task of developing AI supported SA for emergency response and evacuation. This approach requires establishment of a "common set of operating principles … relevant to all network stakeholders operating in a 'shared' system" [55]. On the absence of such an approach, disparate parties and players work independently, or in incomplete integration, on aspects of the problem. Some have even suggested that such work represents a new branch of engineering [56]. Creating a venue and assembling the stakeholders for development of CONOPS is a major hurdle that is yet to be addressed. It is complicated by the many other barriers identified here.

9.4.4 Organization Scale (Command Overhead)

Organization scale is important to understand in terms of ability to utilize and interact with SA information, even when mediated by AI. The operating environment varies greatly between emergency service organizations. While metropolitan fire departments may dispatch dozens of personnel on an initial alarm in a high-rise building, other agencies may send a handful of personnel, who must later call for help if an emergency is confirmed. Similarly, the command element that is sent to an incident varies similarly from one officer to a team of officers assisted by command support personnel.

Serious utilization of AI supported SA requires some minimum personnel complement devoted to command in order to be effective. Many agencies are simply too small to have sufficient personnel available on the initial alarm in order to enable a command structure that can effectively consume and utilize this information. A sizable number of first response agencies simply do not respond to emergencies with a command cadre sufficient to engage with, process, and utilize this information in decision-making during initial stages of an incident.

Existing tools and demands such as operating a vehicle, interacting with existing land mobile radio, and physical movement to observe the incident scene and interact with initial responders and the public will fully occupy incident commanders.

A similar concern arises in the dispatch function. While there are roughly 5800 9-1-1 centers today [11–13], the number of dispatch centers (who actually mobilize resources and manage the tactical communications) are many, many more. For example, during the wildfire in Paradise, California in 2018, the Town's dispatch staff was overwhelmed with the rapidly escalating incident. Their information technology was limited, and relied mainly on voice communication. This limited SA [57].

Simply put, there is a need for wholesale consolidation of the 9-1-1 and dispatch functions of public safety agencies to enable a scale at which sufficient staff are on duty, economic sufficiency of acquiring necessary technology is achieved, and an adequate resource base of response resources is served to make the investment worthwhile. A dispatch center staffed by one person, serving an agency with a small number of personnel on duty is not able to participate in AI assisted SA in a meaningful way.

9.4.5 Market Failure Due to Variation

Emergence of industry standards have been slow. As a consequence, much work on analytics is considered proprietary, and remains restricted to customers of specific software or sensor hardware. Indeed, private companies seek competitive advantage, and defend their intellectual property and seek to recoup their investments in analytics. This is antithetical to the emergence of a common standard for such systems.

We must also remember that the sensor industry is not vertically-integrated. Multiple manufacturers, and system integrators collaborate and compete, meaning that proprietary solutions will be embedded in solutions far into the future.

Within emergency services, market failure has been observed as a condition that hindered research since the 1961 National Research Council report on fire research. The report observed "areas of economic interest to the whole nation are often of insufficient interest, to any one group, to produce a desirable level of attention ..." [58].

The number of first responder organizations in the United States is staggering. There are an estimated 18,200 local EMS agencies in the United States [59]. Law enforcement agencies number over 18,000 [60] and there are local and over 29,000 fire departments [61]. This large group of small agencies is unable to identify or advocate for their needs. Beyond the large numbers of agencies, regulatory and legal structures vary from state to state ([38], p. 214).

The "one-off" nature of first responder organizations, policies, and settings defies an easily-marketed solution that can be widely adopted. Faced with a highly differentiated marketplace, vendors see no profit in developing products. Even large organizations face challenges in designing and implementing requisite SA technologies. As an example, the concept of an electronic "command board," originally proposed following 9/11 in New York ([62], p. 35) was to replace a dry-erase board with magnetic markers designed to track resources and provide accountability. Planned for a wide rollout in 2012 [63], The concept has not yet been widely implemented.

Table 9.2 Conceptual benefits and costs of AI assisted SA

Benefits	Costs
Building owner	
Reduced direct loss (earlier notification and mitigation)	Installation coasts
Reduced indirect loss (less building disruption)	Maintenance costs
Avoidance of liability from tenants/employees	
Possible insurance benefits	
Tenants/Occupants/Employees	
Lower risk of injury or death	
Employers	
Reduced business interruption	
Reduced liability from employees/other tenants	
Public sector emergency services	
Reduced time on scene (reduced overtime, reduced need for excess staff to manage the incident)	Staffing at alarm center/technology to interface with AI
Reduced injury to workers and the public	Added cost of technology in the field to interact with AI for SA/incident management

9.4.6   Cost Effectiveness/How to Measure

As with many complex problems spanning multiple domains, measuring the cost effectiveness of AI assisted SA will be difficult. In addition, attributing costs requires accounting for costs and benefits that accrue across both public and private domains, and across a building's life cycle.

As an example, the building owner bears the cost of installing and maintaining the sensors and necessary upkeep. Benefits of AI assisted SA will accrue only in relatively limited cases (in the event of a fire or other emergency), but these benefits apply to different groups, as shown in Table 9.2.

This simplistic accounting of benefits and costs is designed to illustrate the complex attribution of across the public, building owners, and tenants or occupiers of these buildings. Notably, while costs are borne by the building owner, secondary costs are borne by the public sector in the form of additional staff and technology both at dispatch and in the field. Benefits accrue much more widely, with building owners seeing reduced direct losses (assumed due to more efficient and more rapid incident management), occupants/tenants see reduced injury and death, while employers see reduced business interruption and reduced liability. The public sector sees benefits in terms of lower demands on emergency services (through earlier notification and increased efficiency), reduced injury to first responders, and reduced costs of injury among members of the public. The cost of injury and various litigations following a consequential fire or even in the event of a serious injury can greatly surpass the costs of direct loss.

Of course at the individual building level, emergencies are thankfully not the norm, and while estimates are that the risk of a fire in a high-rise building is non-negligible from the perspective of the building as a whole, these costs are divided over a relatively small set of incidents. This suggests that the incentives for installation of such systems may be aided by insurance, taxes, and other incentives.

9.5   Ethical Implications of AI in Situation Awareness

While we have discussed the promise of AI assisted SA, the ethical implications of adopting such technologies must be considered ([5], p. 115). The use of AI can have dire impacts on reinforcing the dominant social order, including attendant discriminatory or adverse effects [64, 65]. While disparate impact, bias, or reinforcement of harms is not foremost in the realm of AI for SA and evacuation, we must remain cautious about unintended effects in implementation and practice.

We must remember that the driving force behind innovations in this realm is market forces, led by consumer demand in various forms. Those who will get this technology are those who can afford to pay. Of course this lays bare the dilemma—the innovations and harnessing of sensors in built infrastructure to inform emergency response will begin in newly-constructed buildings—those with the ability to pay for such innovation. Such buildings will tend to install such measures not because they are required by building codes, but because the developers seek to attract tenants who value such measures, or to lower the cost of operating the building.

The fire problem generally is one that is overwhelmingly in the existing building stock, and fire deaths are highly negatively correlated with household income [66]. This creates the potential for added stratification of safety—the affluent get enhanced safety, while the poor remain at elevated risk.

Simple numerical analysis of the existing building stock makes clear that any solution incorporated into new construction will have only marginal impact on the overall scope of any particular emergency such as fires. This suggests that opportunities to develop solutions affordably, and designed explicitly for retrofit will be important to meaningfully impact the fire problem, for example, sensor systems to provide early detection of fires and alerting of occupants in existing high-rise buildings could make a substantial contribution to safety, but would ideally offer additional benefits to offset the costs of their installation.

A second concern with regard to ethics is data privacy. Particular concerns arise around access to sensitive information which could include location, health, and other records and data streams. Who will own the data leveraged by these systems? Such data likely falls under the realms of both corporate and personal ownership, with attendant challenges of data sharing and permission. Subsequent access to the data by public entities will similarly require forethought and legal resolution.

Wearable technologies that render humans as sensors should have adequate safeguards to assure anonymity for all but situations of individual distress. For example, location data could be generalized during routine operations such as energy efficiency, particularized as counts or estimates of persons for purposes of building security and evacuation, and made specific only under circumstances of an emergency, For example, real-time or periodic estimates of the number of occupants on a floor could be available to emergency responders and building staff, and individual-level data could be revealed in the case occupants were trapped or needed assistance in evacuation during an emergency.

Care should also be taken that such information cannot be accessed by potentially hostile parties. For example, knowing the location of building occupants is highly desirable during a fire, but such information should not be accessible to others in the event of a hostile party or targeted attack (ASHER) event.

Ensuring accountability for AI systems must be considered throughout design. "The use of AI technologies ... requires attribution of responsibility within complex systems in which responsibility is distributed among numerous agents" ([67], p. 28).

9.6   Future Outlook

Perhaps a paramount purpose in leveraging AI is to produce improved SA for both building occupants and emergency services. While individual efforts to develop AI, based upon analytics of sensor outputs and human-driven information are advancing, there are significant challenges and barriers to adoption of AI supported SA. While the technical feasibility of communicating sensor and other rich data to first responders has been demonstrated, gaps remain in making the leap to using AI for purposes of sensor analytics and display to emergency responders.

In the area of emergency reporting, even assuming the widespread implementation of NG911, significant challenges remain in aligning the current topology of 9-1-1 centers and their articulation with dispatch centers across the country. Despite years of efforts to standardize quality of service, limited progress has been made, and a large number of dispatch and 9-1-1 centers remain too small to effectively utilize AI in the incident management realm due to limited staffing and equipment.

The lack of a CONCOPS, frustrated by the structure of the sensor and system integration industry as well as the myriad emergency service organizations and their vast differentiation, further complicates the ability to build widely-applicable sets of analytics that can be used to improve SA. Meaningful efforts to gather emergency services, NG9-1-1 developers, and BIM and developers will be necessary to forge ahead.

In spite of the challenge, high-rise buildings remain a leading building type for development of sensor analytics, followed by meaningful display and ultimately AI assisted SA. The extension of these efforts to existing buildings offers promise to make larger and more equitable reductions in fire losses than focusing on only on new construction. A vigorous engagement with the organizational dimension of the emergency response community is a necessity.

Acknowledgements This work was performed with support from National Science Foundation grant# 1934968, Growing Converege Research: Emotionally Responsive Computation and Communication.

References

1. Dwyer, J., & Flynn, K. (2002, July 7). Fatal Confusion: A Troubled Emergency Response; 9/11 Exposed Deadly Flaws in Rescue Plan. *New York Times*. Retrieved from https://www.nytimes.com/2002/07/07/nyregion/fatal-confusion-troubled-emergency-response-9-11-exposed-deadly-flaws-rescue.html
2. McKinsey and Company. (2002, October). *McKinsey Report – Increasing FDNY's Preparedness*. Retrieved from FDNY: https://www1.nyc.gov/assets/fdny/downloads/pdf/about/mckinsey_report.pdf
3. Kokar, M. M., & Endsley, M. R. (2012, May/June). Situation Awareness and Cognitive Modeling. *IEEE Intelligent Systems*, 91-96.
4. Pereira, G. V., Macadar, M. A., Luciano, E. M., & Testa, M. G. (2017). Delivering public value through open government data initiatives in a SMart City context. *Information System Frontiers*, 213-229.
5. Halegoua, G. (2020). *Smart Cities*. Cambridge, MA: MIT Press.
6. Lacinak, M., & Ristvej, J. (2017). Smart city, Safety and Security. *Procedia Engineering*, pp. 522-527.
7. International Standards Organization. (2018). ISO 37120 briefing note: the first ISO International Standard on city indicators. Geneva, Switzerland.
8. Shastri, P. (2015, September 21). Smart cities would require smart buildings for fire safety. *The Times of India*.
9. Pardo, T. A., Gil-Garcia, J. R., Gasco-Hernandez, M., Cook, M. E., & Choi, I. (2021). Creating Public VAlue in Cities: A Call for Focus on Context and Capability. In E. Estevez, T. Pardo, & H. J. Scholl (Eds.), *Smart Cities and Smart Governance* (pp. 119-139). Cham, Switzerland: Springer Nature Switzerland.
10. Allam, Z., & Dhunny, Z. A. (2019). On big data, artificial intelligence and smart cities. *Cities*, 80-91.
11. National Emergency Number Association. (2021a, November 11). *9-1-1 Statistics*. Retrieved from National Emergency Number Association: The 91-1- Association: https://www.nena.org/general/custom.asp?page=911statistics
12. National Emergency Number Association. (2021b, September 22). *NG9-1-1 Project*. Retrieved from NENA: The 9-1-1 Association: https://www.nena.org/general/custom.asp?page=NG911_Project
13. National Emergency Number Association. (2021c, July 21). *NG-911 Project*. Retrieved from National Emergency Number Association: https://cdn.ymaws.com/www.nena.org/resource/resmgr/ng9-1-1_project/whatisng911.pdf
14. Appen. (2021, September 4). *The AI Glossary*. Retrieved from Appen Artificial Intelligence Glossary: https://appen.com/ai-glossary/
15. Gartner. (2021a, September 2021). *Internet of Things*. Retrieved from Gartner Glossary: https://www.gartner.com/en/information-technology/glossary/internet-of-things
16. Gartner. (2021b, September 9). *Real-time analytics*. Retrieved from Gartner Glossary: https://www.gartner.com/en/information-technology/glossary/real-time-analytics
17. Fraga-Lamas, P., Fernandez-Carames, T. M., Suarez-Albela, M., Castedo, L., & Gonzalez-Lopez, M. (2016). A Review on Internet of Things for Defense and Public Safety. *Sensors*.
18. Li, Y.-S., & Ning, F.-S. (2018). Low-Cost Indoor Positioning Application Based on Map Assistance and Mobile phone data. *Sensors*.
19. Ouchi, K., & Doi, M. (2012). Living activity recognition using off-the-shelf sensors on mobile phones. *Annals of Telecommunications*, 387-395.

20. Nuttaki, C., Mathew, R. J., Sureth, A., Vijay, A. R., Krishna, S., Babu, A., & Diwakar, S. (2020). Classification and kinetic analysis of healthy gait using multiple accelerometer sensors. *Procedia Computer Science*, 395-402.
21. Psychoula, I., Amft, O., & Chen, L. (2020). Privacy RIsk Awareness in Wearables and the INternet of Things. *IEEE Pervasive Computing*, 60-66.
22. Wan, S., Lu, J., Fan, P., Lataief, & A., K. (2018). To Smart City: Public Safety Network Design for Emergency. *IEEE Access*, 1451-1460.
23. Gomaa, I., Gwynne, S., Spencer, B., Zalok, E., & Kinateder, M. (2021). A framework for intelligent fire detection and evacuation system. *Fire Technology*.
24. Kulkarni, S. W., & Agashe, S. D. (2016). Study of Intelligent Evacuation Systems of high=Rise Buildings in India – A Review. *2016 International Conference on Computing, Analytics, and Security Trends (CAST)*, 190-194.
25. Siddiqui, A. A., Ewer, J. A., Lawrence, P. J., Galea, E. R., & Frost, I. R. (2021). Building Information Modelling for performance-based Fire Safety Engineering Analysis – A strategy for data sharing. *Journal of Building Engineering*.
26. Vandercasteele, F., Merci, B., & Verstockt, S. (2017). Fireground location understanding by semantic linking of visual objects and building information models. *Fire Safety Journal*, 1026-1034.
27. Bishop, C., He, Y., & Magrabi, A. (n.d.). *Situation Awareness and Occupants' Pre-Movement Times in Emergency Evacuations*. Retrieved from Fire Protection Association Australia: http://www.fpaa.com.au/media/230358/d2-fse-p11-bishop.pdf.pdf
28. Micolier, A., Tallandier, F., Tallandier, P., & Bos, F. (2019). Li-BIM, an agent-based approach to simulate occupant-building interaction from the Building-Information Modelling. *Engineering Applications of Artificial Intelligence*, 44-59.
29. Lofti, N., Benham, B., & Peyman, F. (2021). A BIM-based framework for evacuation assessment of high-rise buildings under post-earthquake fires. *Journal of Building Engineering*.
30. Wang, S.-H., Wang, W.-C., Wang, K.-C., & Shih, S. Y. (2015). Applying building information modeling to support fire safety management. *Automation in COnstruction*, 158-167.
31. Seoul Metropolitan Government. (2019, November 26). *Seoul Metropolitan Fire & Disaster Headquarters Introduces IoT-Based Firefighting Facility Management System*. Retrieved from Seoul Metropolitan Government : http://english.seoul.go.kr/iot-based-firefighting-facility-management-system/
32. Goldstein, P. (2020, August 31). *Fire Technology in Smart Cities and Beyond: How IoT Helps Fight Fires*. Retrieved from State Tech Magazine: https://statetechmagazine.com/article/2020/08/fire-technology-smart-cities-and-beyond-how-iot-helps-fight-fires-perfcon
33. Jiang, Huixian. Mobile fire evacuation system for large public buildings based on artificial intelligence and IoT. IEEE Access May 29, 2019.
34. Bigda, K. (2012, April 18). *NFPA 1010 Public Input #1 – Video Monitoring and Situational Awareness*. Retrieved from National Fire Protection Association.
35. NFPA 101 Life Safety Code. (National Fire Protection Association, 2021).
36. National 9-1-1 Program. (2018). *Review of Nationwide 911 Data Collection*. Office of Emergency Medical Services. National Highway Traffic Safety Administration. Retrieved from https://www.911.gov/pdf/National_911_Program_Review_of_Nationwide_Data_Collection_2017.pdf
37. Task Force on Optimal Public Safety Answering Point Architecture (TFOPA) Working Group 2. (2016). *Task Force on Optimal Public Safety Answering Point Architecture (TFOPA)*. Retrieved from Federal Communications Commission: https://transition.fcc.gov/pshs/911/TFOPA/TFOPA_WG2-120216.pptx
38. Kemp, B., & Lovett, B. (2021). *Handbook of Next-Generation Emergency Services*. Norwood, MA: Artech House.
39. 112th Congress. (2012, February 22). Middle Class Tax Relief and Job Creation Act of 2012. *Public Law 112-96*. Washington, DC: United States Government Printing Office. Retrieved from https://www.congress.gov/112/plaws/publ96/PLAW-112publ96.pdf

40. FirstNet. (2021, September 22). *FirstNet: The History of our Nation's Public Safety Network*. Retrieved from FirstNet Authority: https://www.firstnet.gov/about/history
41. Cybersecurity and Infrastructure Security Agency. (2019, January). *Public Safety Communications Evolution*. Retrieved from https://www.cisa.gov/sites/default/files/publications/Public_Safety_Communications_Evolution_FINAL_01222019_508C.pdf
42. Criminale, A., & Langar, S. (2017). Challenges with BIM Implementation: A Review of Literature. *53rd Associated Schools of Construction Annual Conference Proceedings*, 329-335.
43. Peters, T. (2021, January 27). Renewed Efforts for a Universal BIM Standard. *Architect Magazine*. Retrieved from https://www.architectmagazine.com/practice/renewed-efforts-for-a-universal-bim-standard_o
44. National Institute for Building Sciences. (2017). *National BIM Guide for Owners*. Washington, DC: National Institute of Building Sciences. Retrieved from https://www.nibs.org/files/pdfs/NIBS_BIMC_NationalBIMGuide.pdf
45. NBIMS-USv3. (2021, August 9). *FREQUENTLY ASKED QUESTIONS ABOUT THE NATIONAL BIM STANDARD-UNITED STATES™*. Retrieved from National BIM Standard: https://www.nationalbimstandard.org/faqs
46. Huang, X. (2021, June 29). AI in Smart Firefighting and the Future of Fire Safety Engineering Performance Based Design. (W. Wegrzynski, Interviewer) Retrieved from https://www.firescienceshow.com/007-ai-in-smart-firefighting-and-the-future-of-fse-pbd-with-xinyan-huang/
47. Endsley, M. (2021). *Situation Awareness Measurement: How to Measure Situation Awareness in Individuals and Teams*. Washington, DC: Human Factors and Ergonomics Society.
48. Nunavath, V., Prinz, A., & Comes, T. (2016, January-March). Identifying FIrst Responders Information Needs: Supporting Search and Rescue Operations for Fire Emergency Response. *International Journal of Information SYstems for Crisis Response and Management*, pp. 25-46.
49. Groner, N., Jennings, C. R., & Robinson, A. (2011). *Situation Awareness Requirements Analysis for Sensor Analytics Display to First Responders*. John Jay College of Criminal Justice (CUNY). New York: Christian Regenhard Center for Emergency Response Studies. Retrieved from http://christianregenhardcenter.org/wp-content/uploads/2017/04/racers-wp1101.pdf
50. Fontana, A., & Frey, J. (2000). The Interview from Structured Questions to Negotiated Text. In N. Denzin, & Y. Lincoln (Eds.), *Handbook of Qualitative Research*. Thousand Oaks, CA: Sage.
51. United States Army. (31 July 2019). *Army Doctrine Publication (ADP) 5-0: The Operations Process*. Washington, DC, USA: United States Amy.
52. Institute of Electrical and Electronics Engineers. (2018). *Systems and software engineering – Life cycle processes – Requirements engineering IEEE 29148* (Second ed.). New York, NY, USA: Institute of Electrical and Electronics Engineers, Inc.
53. Jackson, K. (2015, March 4). *Technology Drives C5I Interoperability Testing*. Retrieved September 20, 2021, from CHIPS, The Department of the Navy's Information Technology Magazine: https://www.doncio.navy.mil/CHIPS/ArticleDetails.aspx?id=6095
54. MITRE Corporation. (2021, September 22). *Concept of Operations*. Retrieved from Systems Engineering Guide: https://www.mitre.org/publications/systems-engineering-guide/se-lifecycle-building-blocks/concept-development/concept-of-operations
55. Harrington, T. S., & Srai, J. S. (2016, May 17). Designing a 'concept of operations' architecture for next-generation multi-organizational service networks. *AI and Society*.
56. Jordan, M. I. (2019). Artificial Intelligence – the Revolution Hasn't Happened Yet. *Harvard Data Science Review*(1.1).
57. Taddonio, P. (2019, October 28). *"No Danger to Paradise," 911 Callers Were Told as the Deadly Camp Megafire Approached*. Retrieved from PBS Frontline: https://www.pbs.org/wgbh/frontline/article/no-danger-to-paradise-911-callers-were-told-as-the-deadly-camp-fire-approached/#:~:text=In%20the%20upcoming%20FRONTLINE%20documentary%20Fire%20in%20Paradise%2C,2018%2C%20speaks%20out%20publicly%20for%20the%20firs

58. National Research Council. (1961). *A Study of Fire Problems.* Washington: National Academy of Sciences.
59. NHTSA Office of EMS. (2021, August 16). *NATIONAL EMS ASSESSMENT DESCRIBES CURRENT STATE OF EMS IN THE UNITED STATES.* Retrieved from ems.gov: https://www.ems.gov/newsletter/fall2020/national_ems_assessment_describes_current_state_of_ems_in_the_united_states.html
60. U.S. Department of Justice. (2021). *Agencies.* Retrieved from United States Department of Justice: https://www.justice.gov/agencies/list
61. United States Fire Administration. (2021, October 6). *U.S. Fire Statistics.* Retrieved from U.S. Fire Administration: https://www.usfa.fema.gov/data/statistics/
62. Fire Department of New York. (2004). *Goal 6: Advance Technology.* Retrieved from FDNY Strategic Plan: http://www.nyc.gov/html/fdny/pdf/pr/2004/strategic_plan/goal_6.pdf
63. Raytheon Technologies. (2012). *Improving New York City Firefighters' Command Response and Safety.* Retrieved from https://www.raytheon.com/sites/default/files/capabilities/rtnwcm/groups/iis/documents/content/rtn_iis_fdny_case_study.pdf
64. Kontokosta, C. E., & Hong, B. (2021). Bias in smart city governance: How socio-spatial disparities in 311 complaint behavior impact the fairness of data-driven decisions. *Sustainable Cities and Society.*
65. Wallace, R. (2021). "The names have changed, but the game's the same": artificial intelligence and racial policy in the USA. *AI and Ethics.*
66. Jennings, C. R. (2013). Social and economic determinants of residential fire risk in urban neighborhoods: a review of the literature. *Fire Safety Journal*, 13-19.
67. World Health Organization. (2021). *Ethics and governance of artificial intelligence for health: WHO guidance.* Geneva: World Health Organization.

Probabilistic Reliability Analysis of Steel Mezzanines Subjected to the Fire

10

Wojciech Kowalski, Adam Krasuski, Andrzej Krauze, and Adam Baryłka

10.1 Introduction

The fire-resistance testing has its long history. The early attempt of fire testing was based on exposure of the buildings assembles on fire. It was generally large-scale fire conditions in which full-size buildings elements were burned and evaluated [1]. One of the first furnace tests is credited to T. Hyatt [2, 3]. He conducted tests in a wood-fired furnace, over which a specimen was tested. The test specimen represented sections of a floor slab. The section was loaded with a mass $1500 \, \text{kg/m}^2$ for testing its bearing behavior. Then after 10 h, a hose stream test was conducted. Hyatt was unable to measure and control the temperature of furnace. However, with using the immersion calorimetry the results were surprisingly repeatable in obtaining the iron temperature.

Until the 1903, the fire-resistance testing was based on ad hoc tests where the temperature of furnace or temperature of buildings assembly during the tests varied [2]. The general assumption was saying that the temperature should be maintained on average at certain level. In 1903 British Fire Prevention Committee proposed standard testing method, with three endurance classes: full protection, partial protection, and temporary protection. The full protection was sufficient to endure a burn out of the content of the building, while temporary would not be sufficient [3].

In 1917 American Society for Testing and Materials (ASTM) along with other testing institutes and bureaus issued C19 standard. The strongest innovation of the new standard was its prescribed time–temperature curve. It is important to stress

W. Kowalski (✉) · A. Krasuski (✉) · A. Krauze
Institute of Safety Engineering, The Main School of Fire Service, Warsaw, Poland
e-mail: kowalskiwojciech986@gmail.com; akrasuski@sgsp.edu.pl

A. Baryłka
Centre of Construction Expertise, Lomianki, Poland

© Springer Nature Switzerland AG 2022
M. Z. Naser, G. Corbett (eds.), *Handbook of Cognitive and Autonomous Systems for Fire Resilient Infrastructures*, https://doi.org/10.1007/978-3-030-98685-8_10

that this time–temperature curve was proposed without the knowledge what actual temperatures in building fires can be [3], nor their time dependence. The curve just took into account that the furnaces do not heat up instantaneously, and for repeatable testing this heating period should be quantified.

The other countries followed the approach proposed by ASTM. The C19 standard curve (currently known as E119) is still in use today. ISO 834 standard temperature–time curve also known as the Cellulosic curve or the standard nominal fire curve is based on ASTM approach. ISO 834 curve is broadly used to test the fire resistance of materials subjected to a category of general combustible building materials fire and building contents.

Since its introduction, the test method has been updated, although its general idea remained unchanged. The test methods described in ASTM E 119 or ISO 834 prescribe a standard fire exposure for comparing the test results of building construction assemblies. For the tests of floors and roofs, a test assembly is structurally loaded, and the standard fire exposure is applied to the underside of the specimen. The assembly is evaluated for its ability to maintaining the applied load.

As the results of the test, a fire-resistance rating is attached to the building assemblies. The fire-resistance rating defines the duration for which a passive fire protection system can withstand the time–temperature curve test. The rating is quantified simply as a measure of time expressed in minutes or hours, and it may entail other qualitative criteria proving functionality or fitness for defined purposes.

The fire-rating is further used by national fire codes defining the specific value of fire-rating for the assemblies of given type of the buildings. For example, a medium rise office building a fire-resistance R60 can be required for main building construction assemblies.

The proposed approach is easy to apply and assure repeatability of the tests and comparability among specimen. During the design phase the constructor just selects the specimen that satisfies the fire-resistance posed by the regulations. In most cases, no prediction of the temperature of the actual fire in building is need nor the assemblies response to the temperature generated by fire.

However, the approach also has shortcomings. Excluding the physical factors of realistic fires causes under- or overestimation of passive fire protection measures reflected in fire-rating of building specimen. The influence of sprinkler system— that is very often present in buildings—is also not taken in account. The scientific records [4] show that, the method of testing is identical regardless of the sprinkler system assumed in building.

The collected data in [5] show that the temperature–time curves from natural fire and from ISO-Fire are substantially different (see Fig. 10.1). The collective knowledge in fire safety domain profession proved ISO 834 performance regarding the safety. The building specimen with fire-rating in most cases satisfies the condition of actual fires. However their cost–benefit is questionable. This is especially symptomatic for warehousing and assembly building where the fire-load is based on its heat release rate during fire, may be in wide spectrum of values. These data are not taken into account and even for building with no burning materials,

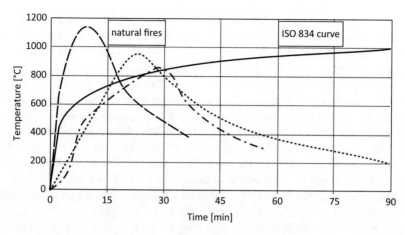

Fig. 10.1 The temperature–time curves from natural and ISO fires [6]

the specimen is tested regarding demanding ISO curve. This demands for more tailored performance-based design with a more realistic and more credible approach justifying investment in the safety.

However, the recent needs and changes related warehousing revealed that the safety can be also an issue with using current standards of testing. The movement to the e-commerce due to the Sars-Cov-2 pandemic caused that the goods are shipped to the end-client directly from the large warehousing centers. This raises a demand for storing a small piece of good ready to be sent. That form of logistic process does not allow to use rack storage operated by forklifts. Very few number of companies can afford the autonomous robotic system for operating these warehouses. In most cases these warehouses are operable by humans who access various levels of storage with using mezzanines, platforms or walkways, the so-called pick-towers. The specimens of these mezzanines are not fire-rated, while they are not main assembles of the building. The very high storage of goods with various fire characteristics may cause very fast developing fires. The safety of these systems is mostly assured by the sprinkler systems. However, these system may fail and there is a strong need to prove the safety of the people even in these situations.

The behavior of the fire in buildings remains outside deterministic quantification. Its stochastic nature means the fire is difficult to predict as it develops. In the case of such dynamic facilities as warehouses live load can also vary with time and location during its lifespan. Fire protection systems as sprinklers or Smoke and Heat Exhaust Ventilation Systems (SHEVS) are also burdened with a possibility of failure [7]. Chances of the fire outbreak also varies across the different types of facilities [8].

To cope with this problem, the collective experience of the profession or adequate level of conservatism [8] is used. The experience was gained by the application of longstanding approaches that proved their performance in multiple past fire events. Conservatism is introduced in the analysis by safety margin in the input parameters, fire scenario definition, or performance criteria. However, collective experience is limited to the typical fire safety engineering designs. Conservatism, on the other hand, requires a good understanding of fire phenomena and performance criteria. Moreover, it is criticized by unjustified costs, low flexibility, and preventing technological innovations and alternative solutions [9, 10].

Switching to the performance-based design [10] or attainment of adequate safety, defined by explicit safety targets for more sophisticated structures, requires more systematic engineering approaches. In this case, the acceptability of the design cannot be demonstrated using expert judgments and deterministic methods. The broad spectrum of possible fire scenarios should be covered by explicit calculation of risk associated with the design to demonstrate acceptability. The probabilistic risk assessment is considered state of the art among the methods for risk assessment.

In order to consider all these physical factors in a systematic way, a more realistic and more credible approach should be developed. The approach should analyze structural safety in the case of fire including active fire protection measures and real fire characteristics. The proposed methodology should be developed based on the statistical, probabilistic, and deterministic approaches and analysis. Moreover, the methodology should also take into account the cost–benefit analysis regarding the applied safety measures.

In this chapter, we present an approach for the performance-based design regarding safety of the structures subjected to fire. The approach was validated on the assembly and warehousing buildings with mezzanines. The buildings and mezzanines specimen have no requirements regarding fire rating. Nevertheless, they should prove their performance in the case of fire. The method is based on the *Natural Fire Safety Concept* that has been proposed through various research projects [11–13]. In order to analyze the problem comprehensively we proposed the probabilistic risk assessment defined by standard PD 7974-7:2019 instead of deterministic one.

The rest of the paper is structured as follows. In Sect. 10.2 we described the methods, models, software, and data we use in our approach. We discussed the natural fire safety concept, probabilistic risk assessment, and the software for deterministic analysis. This section also contains a workflow during the design phase of steel structures. In Sect. 10.3 we presented a case study that is the illustration of the application and verification of our approach. We use pick-tower type mezzanine placed in a modern logistic building. The conclusions and outlook for future work summarize our paper in Sect. 10.4.

10.2 Materials and Methods

The approach we proposed is based on the three primary concept: (a) natural fire concept defined in Eurocode, (b) probabilistic risk assessment defined in PD 7974-7:2019 supported by the multisimulation approach, and (c) deterministic engineering tools allowing calculation of thermal and mechanical response of the structure subjected to the fire.

10.2.1 Natural Fire Safety Concept

The application of the natural fire safety concept is provided in the current Eurocode generation as an alternative to the common ISO standard fire testing for fire safety design. The basic assumption of the methodology presented in the Eurocode is the estimation of the load capacity time with regard to a fire of the materials stored in a given building. Acceptance criteria should be based on functional goals such as safety of evacuation or safety of rescue operations. Further design goals are to ensure environmental protection, property protection, and the continuity of the building's operation.

In the case of engineering analysis based on the Eurocode, the first step is to determine the thermal exposition. For this purpose, fire models should be used. Figure 10.2 shows a diagram of the fire models listed in Eurocode 1 [14]. The diagram groups the fire models due to their complexity.

The next step is thermal analysis to determine the heating rate and temperatures of the structural elements. The Eurocode 3 [15] standard defines several methods for determining the behavior of the structure subjected to the fire. The commonly used methods are the verification of the load-bearing capacity of the element at specific temperatures or criterion whether and when the element made of carbon steel reaches critical temperature. The rules however, have their limitation, i.e.,

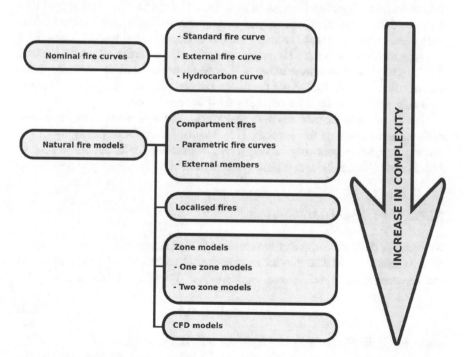

Fig. 10.2 Natural fire model for the structural fire design [14]

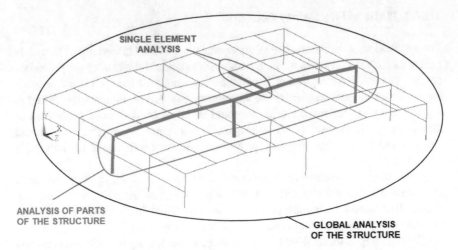

Fig. 10.3 Design methods for determining the mechanical response of structures under fire conditions [16]

cannot be applied along with deformation criteria or when instability of the structure should be analyzed. Hence, the other approach defined by Eurocode is to calculate the mechanical response of structural members. The approach is further split into analysis of the entire structure, subpart of structure, or finally single element. The mechanical response is more accurate while considers also mechanical interactions among structural elements. Moreover, analysis of the entire structure (global) enables a much better understanding of the overall behavior of the structure under fire conditions. Figure 10.3 shows the design methods that can be used to determine the mechanical response of the structure under fire conditions.

Non-linear finite element software enables detailed analysis of the thermal and mechanical response of the structure [17]. Detailed analysis of the safety of the structure consists in checking the occurrence of excessive deformations leading to the loss of stability, but also constituting an obstacle in the functioning of the system.

10.2.2 Risk Model and Matrices

We consider the performance of the structure in fire, relating it to the risk resulted from its failure. We define the risk as a product of probability of unwanted events and the consequences that it poses. Equation (10.1) quantifies the risk measure.

$$R(A) = P(A) \times S(A) \tag{10.1}$$

where P—probability of the event A; S—severity of the event A.

The unwanted event, we consider, is the fire in the proximity of the mezzanine. In the case of the analysis of the mezzanine occupied by the people, the most severe consequences is loss of their lives due to lack of the stability of the structure. The further consequences are injuries, damages to the environment, and monetary losses. The time horizon we consider is the averaged annual risk. The assessment of the total risk is based on the scenarios defined by the probabilistic model of the exposures, and failure of the safety measures.

We assess the structural performance by its responses divided into two domains consisting of desirable and undesirable states. The boundary between these states is called limit state. We define the failure as entering the undesirable domain [18]. In our analysis we consider two boundaries of limit state: ultimate limit (ULS) state and limit state of serviceability (LSS).

We define that the ULS is achieved when the following conditions are registered: (a) rupture or excessive deformation as an effect of the instantaneous attainment of the maximum capacity of cross-sections of the members, (b) instability of the structure or large part of it.

When the fire is detected, occupants of the mezzanine should be safely evacuated from the structure. Evacuees are traveling across the structure until they terminate on exit or fire-rated staircases. During the evacuation the platforms used for the evacuation are supported by structural elements of the mezzanine. Therefore the structure must attain its serviceability. We define LSS as the unacceptable deformations of the structural elements. The values of unacceptable deformation are defined on the basis of the EN-15512 standard.

We describe the performance of the structure by the limit state function (LSF), where the argument for this function is input parameters that can be random. We denote the LSF as $g(X)$, where $X = (X_1, X_2, \ldots)^T$ are input random variables. The input variable are further discussed in Sect. 10.2.2.1. The inequality (10.2) identifies the undesirable domain.

$$g(X) \leq 0 \qquad\qquad (10.2)$$

The formula (10.2) is calculated for each considered fire scenario. A fire scenario is defined as specific manifestation of the vector of input parameters X. The input parameters are random variables defined by probability distribution functions. To address the possible fire scenarios with the manifestation of their consecutive input parameters in accordance with their probabilities, we apply Monte Carlo simulations.

We express the consequences of the unwanted event both qualitatively as well as quantitatively. At the current state of the research we are more concentrated on the consequences to humans: fatalities and casualties. However, the output from the mechanical response of the structure may also serve as an input for the calculation of the loss in property.

We assume that, if the ultimate limit state will be achieved before successful evacuation from mezzanine it results with the loss of lives. There is no current model available for traumatic injury due to the collapse of the structure. Therefore, we assume these consequences qualitative. However, with using the evacuation model we are able to evaluate number of people at risk and also plot FN curves.

In the case of limit state of serviceability, we assume that it may result in serious injury. Achieving the limit state of serviceability of the structure, during the evacuation, the failure of the structural element may be expected. This failure may result with stumbles, bones fracture, or other type of traumatic injuries. This measure is also qualitative and we are unable to calculate the number of people affected. There is no current behavioral model that addresses people behavior crossing mezzanine platforms with the fire and smoke below. Therefore, the model we use is burdened with uncertainty. It should be stressed that the approach used for the consequences analysis for LSS, ULS, and traumatic injuries is based on the expert judgment. We found no justification for these assumptions in literature.

Our risk-informed decision making is based on the uncertainty model represented by stochastic variables addressed by Monte Carlo simulations, and the limit state approach for the identification of failure events. The candidate design is perceive reliable when the probability of the failure does not exceed a specified target value. We represent the probability of failure regarding the formula (10.3).

$$p_f = P[g(X) \leq 0] \tag{10.3}$$

In order to address the uncertainty and consequences comprehensively we use event tree analysis (ETA) technique. Figure 10.4 depicts the exemplary ETA.

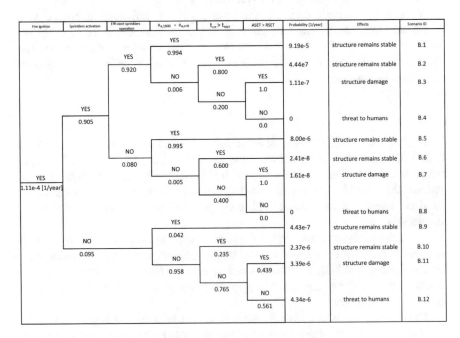

Fig. 10.4 Event Tree Analysis graph for the fire event in a warehouse

The initiating event for our ETA is fire in the proximity of a mezzanine. The value of the ignition probability is calculated according to the PD 7974-7:2019 [8]. The probability of the following branches: *sprinkler activation* and *effective sprinkler operation* are also taken from the statistics [7]. However, the consequences for these scenarios are modeled with using fire models, thermal response of the elements, and Monte Carlo simulations.

Failure on sprinkler activation is modeled by unlimited fire growth with αt^2 rate in our fire model. Non-efficient sprinklers operation is supposed to reduce fire development and limit the HRR to the constant value. Fire curve modified by effectively operating sprinklers considers the extinguishing phase. From the time of sprinklers' activation the HRR is shaped with Eq. (10.4) [8, 19]

$$Q(t) = Q_0 e^{-k(t-t_0)} \tag{10.4}$$

The definition of fire development defines the input for deterministic software for structure response OZone and SAFIR. Stochastic modeling checks the thermal response of the structure's elements and then compares it with the critical temperature. With such derived data the probability that the critical temperature was achieved is calculated as well as the time of the exceedance. Figure 10.5 shows the default CDF chart generated by the model.

We use evacuation model Pathfinder to calculate the evacuation time. The time when the critical temperature was achieved is compared with the evacuation time. This defines the another branch in the ETA. For the scenarios where the time of the evacuation was longer than the exceeding critical temperature of structural element, the mechanical response is calculated. The finite element analysis (FEA) provides some precise data as bending moments, stresses, displacements, or torsion. The mechanical response will allow for the calculation of LSS and ULS with relation to the evacuation.

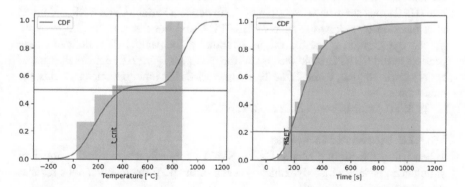

Fig. 10.5 The cumulative distribution function of the critical temperature's exceeding time with the RSET marked

10.2.2.1 Model Parameters

The model's geometry remains unchanged during the simulation process (at least for a given candidate design); hence it can be considered as a fixed set of parameters defining input Z. This set is then expanded by other invariants related to the physical properties of the building obstacles, environmental parameters (initial indoor temperature, humidity, and others), and physical parameters of safety measures, such as ventilation flow. Also the load on the structure is considered as fixed parameter. In the case of fire, according to [20], the accidental combination of actions is used (see Eq. 10.5).

$$E_d = \sum_{j \geq 1} G_{k,j} + A_d + \psi_{1,1} Q_{k,1} + \sum_{i > 1} \psi_{2,i} Q_{k,i} \tag{10.5}$$

Dead loads $G_{k,j}$ are introduced explicitly—without multiplying by partial safety factors γ_i. Live loads $Q_{k,i}$ are reduced with combination factors $\psi_{j,i}$. Similarly to the dead loads, safety factors γ_i are not considered in the case.

The accidental load A_d is calculated implicitly by the FEA software. This part of load combination is classified as dependent variable.

Global imperfection of frame as well as local imperfection of member can be also considered in analysis. Equivalent imperfection forces, calculated with methodology presented in Eurocode 3 [24], can be implemented as fixed punctual or line loads.

The second set of input parameters X is uncertain or variable and is drawn from the Monte Carlo sampler. We apply Simple Monte Carlo approach [21]. As an example may serve fire growth rate or fire coordinates. The parameters of the probability distributions are mostly based on standards such as [22, 23] or the scientific records [25–28].

The third type of the parameters comprises dependent variables. These variables are related to the fixed values or those drawn from the distributions. The heat release rate (HRR) may serve as an example of the dependent variable. The HRR is defined as a function of a drawn sample of HRRPUA and the fire area that depends on the storage location at the fire origin. Eventually, the peak HRR is defined as the product of the HRRPUA and the area of the racking. Figure 10.6 depicts the idea of Monte Carlo routine, while Table 10.1 summarizes the input parameter used in the simulations.

Table 10.1 consists of default input variables.

10.2.2.2 Stochastic Sampling

When using the Simple Monte Carlo (SMC) approach, the required number of simulations is determined dynamically. Each of the performed simulations reduces the approximation error [29]:

$$\hat{p}_n \pm \Phi^{-1} \cdot \sqrt{\frac{\hat{p}_n(1 - \hat{p}_n)}{n}} \tag{10.6}$$

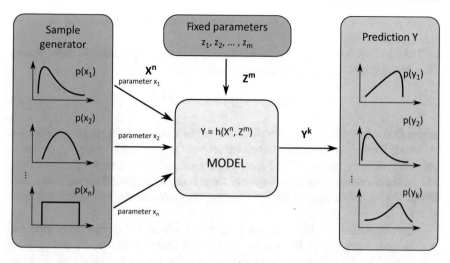

Fig. 10.6 Probabilistic Reliability Analysis with multisimulation method

Table 10.1 Default input variables in McSteel

Variables	
Random	Fixed
Built-in	
HRR(t) curve	maximum HRR
Fire growth rate	
(few occupation types)	
Fire coordinates	
Configuration files	
Sprinklers' reliability	Sprinklers' activation time
Fire area	Structure geometry
Fire growth rate *(user's)*	Enclosure geometry
HRRPUA	SHEVS
	Material properties
	Steel profile geometry
	Required Safe Egress Time
	Utilization factor
	Fuel distribution
	Mechanical loads

where $\hat{p}_n$ is the approximated probability, Φ is the cumulative distribution function of the standard normal distribution, and n is the number of simulations.

The root mean squared error (RMSE) of estimated expected value $\hat{\mu}_n$ can be calculated as follows [29]:

$$\text{RMSE} = \sqrt{\mathbb{E}((\mu - \hat{\mu}_n)^2)} = \frac{\sigma}{\sqrt{n}} \tag{10.7}$$

where μ is the mean of the expected value of some random variable, and σ its standard deviation.

The convergence rate of SMC estimation of μ is of the $\sqrt{n}$ order. In other words, for a given Y RMSE is a function with respect to n:

$$\text{RMSE}_Y = f_Y(n) = O(n^{-1/2}) \tag{10.8}$$

where O is a denotation used to characterize a function according to its growth rate.

Monte Carlo is a sampling method extensively used to model complex phenomena. The main statement of the Monte Carlo method is that it is possible to model a complex phenomena by sampling over large enough set of random variables. The higher the sample size, the higher convergence between the estimation and real value.

Multisimulation, which is a variation of the Monte Carlo has been used for probabilistic modeling of fire event [30]. Using a deterministic model and random input variables it is possible to receive the distribution of the result value. With appropriately large sample of input randoms it is possible to fairly estimate the shape of results [30]. In the case of modeling structures subjected to a fire results would be a probability distribution of failure in specified time.

The output from Monte Carlo analysis is clustered regarding the results: critical temperature was obtained or not. Then they are compared with the evacuation analysis calculating the consequences of given clustered results on evacuees and presented in a form of ETA. Finally, having the consequences and their probabilities risk is calculated.

10.2.2.3 Risk Ranking

The results from the ETA analysis are probability and for various consequences defining risk related to the fire. Once the probability and severity, hence the risk, are determined the next step of probabilistic reliability analysis can be made. The risk categorization is described in detail in the codes [8,31,32]. The two methods, chosen due to their applicability and versatility, are presented below. SFPE Handbook offers the common matrix approach and suggests to assign risk components to the one of five categories on each axis [31]. Property or human safety can be analyzed with the matrix shown in the Fig. 10.7.

British code for risk assessment presents quite a different approach. It specifies the critical values of the risk for fatal events. The results can be compared with such defined criterion easily. The design below acceptable risk should be done accordingly to As Low As Reasonably Possible (ALARP) method [8].

10.2.3 Deterministic Models

Multisimulation method uses deterministic models to conduct stochastic analyses. The main model proposed in this paper to assess the temperature in steel members is *OZone 3*. The tool is an effect of cooperation of ArcelorMittal Research and

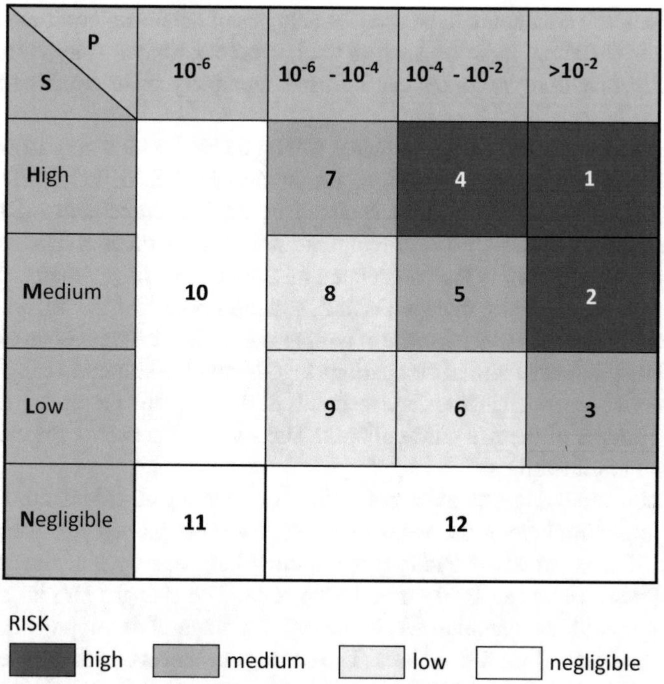

P⟍S	10^{-6}	10^{-6} - 10^{-4}	10^{-4} - 10^{-2}	$>10^{-2}$
High		7	4	1
Medium	10	8	5	2
Low		9	6	3
Negligible	11	12		

RISK

▮ high ▨ medium ☐ low ☐ negligible

Fig. 10.7 Risk matrix for human safety [31]

Development Esch and University of Liege. OZone 3 is an engineering tool aimed at aiding fire design of steel elements. The tool consists of the three main parts: *geometry, fire, and section.*

Users can define rectangle compartment with at last four walls and flat, single- or double-pitch roof. Walls consists of up to five layers, where specific heat, density, thickness, and conductivity have to be specified. Type of heating curve can be set in *fire* menu. There are available nominal (ISO 834, ASTM E119) and parametric fire curves. Moreover, one of natural fire models (zone, localized fire, user defined) can be also defined. The profile of the cross-sections has to be chosen from the OZone's library. There are numbers of pre-defined *I-profiles* attached (IPE, HE, HD, UB, W, and others), consistent with American, European, or Japanese standards. Heat transfer to the section is calculated according to the EN 1993:1-2.

The natural fire models, which are advocated to be used in this paper, implemented into OZone 3 can be described as simple fire models. Zone model includes one- (1Z) and two-zone (2Z) model.

University of Liege's SAFIR software is used to conduct two step numerical analysis of thermal and mechanical response of structure subjected to the fire. SAFIR is a computer program that models the behavior of building structures subjected to fire. The structure can be made of a 3D skeleton of linear elements such as beams and columns, in conjunction with planar elements such as slabs and walls. Volumetric elements can be used for analysis of details in the structure such as connections. Different materials such as steel, concrete, timber, aluminum, gypsum, or thermally insulating products can be used separately or in combination in the model.

The thermal exposure on the elements can be defined with using, frontier—ISO or user defined temperature, as well as fire models HASEMI, LOCAFI, or output from CFD simulations. From the calculated or set gas temperatures, SAFIR will first calculate the evolution of the temperatures in the structure. The mechanical behavior of the structure is then calculated on the basis of its geometry, its support conditions, the loads that it must withstand, and the strength of the materials, taking into account the progressive increase of temperature. The elevation of temperature in the materials produces thermal elongations together with a reduction of strength and stiffness. As a consequence, the displacements of the structure increase continuously during the course of the fire until collapse. The software accounts for material and geometrical nonlinearities.

Stochastic modeling characterizes with high amount of calculations. To cope with the multisimulations the versatile computer code named *McSteel* has been developed. Thanks to user-friendly programming language Python and as an open-source software users can freely modify the code. The changes and improvements can be shared with the community. On the primary stage of development, the tool is limited to analyses of steel structures. The core code, however, can be conveniently applied to operate on the different models in order to analyze wider variety of materials.

10.2.4 Workflow

The design process starts with the identification of the fire hazards. In this phase the probability of a fire ignition for given type of building is considered, as well as possible threats for people. In this phase, for example, the automatically operated areas by robots or conveyors, may be excluded from the safety analysis—if they do not have impact on people safety. We also collect the information regarding the safety systems installed in the building, such as: fire alarm system, sprinkler system, SHEVS, and safety management policy.

Then, a model of the steel structure, selected for the analysis is prepared. We use GiD software that is very well integrated with SAFIR. However, any type of CAD software, which is able to export DXF file and SAFIR inputs, can be used.

McSteel reads the two types of objects into its mapping function: *LINE* and *3DFACE*—linear elements (beams or columns) and shells (ceilings), respectively. The current version of McSteel supports only simple frame steel structures.

Shell elements allow the McSteel's script to consider ceiling lowering in a part of analyzed enclosure. The shell objects will not be taken into account in finite elements analysis. However, McSteel limits fire plume length and excludes elements above ceiling or beneath the floor level. In this phase also the loads on the structure are calculated and placed to the consecutive elements.

Next, the review of the list of storage goods is performed in order to define possible fire scenarios. In this phase we define the HRRPUA of possible fires as well as we select the probability distributions of fire growth rate.

Input variables required by the McSteel's script can be divided into stochastic (with default built-in distribution or specified by the user) and deterministic. Both, stochastic and deterministic input variables, are specified in the set of configuration files. They are ASCII files (txt, CSV, or JSON), with examples to be downloaded from the GitHub repository https://github.com/kowalskiw/mcsteel.

Then the first phase of the simulations is launched using OZone 3 and the bundle of scripts of McSteel software. The analysis is performed in the thermal response domain. To predict the effect of the fire event the Monte Carlo sampling method is used [21].

The results of the analysis are the probability distributions of the exceeding of critical temperature, as well as probability distributions of the time of this exceeding. The results allow drafting ETA and selection of most severe scenarios with the probability higher than 10^{-6} for the further analysis in SAFIR. McSteel handles choosing scenarios associated with the highest risk and runs other scripts automatically. User has to set input parameters properly, and the framework will start all necessary processes and calculations on its own. Also primary post-processing of multisimulation results is provided by the script.

This second step of the modeling process, Finite Elements Analysis (FEA) is handled by the McSteel's core script. The selection of scenarios to FEA is based on the severity criterion. Script iterates over multisimulation's results and chooses the simulations where severity, defined as the time to exceeding critical temperature of steel, is the shortest. Fire scenarios are selected from the multisimulation results database (1st percentile as default). Thereafter analyses are set up and run one by one. The results from the analysis allow for the definition of limit states and updating the ETA.

There are also three default charts available: probability density function (PDF) and cumulative density function (CDF) of maximum temperature in a section; and CDF of time to exceeding critical temperature (failure). Numerical analyses derive precise data about internal forces, displacements, and stability of the structure. Post-processing is possible via Diamond software by the University of Liege.

The final assessment of safety, and reliability, is quantified by risk categorization. The SFPE risk matrix as well as PD-7974-9 standard is used for this purpose. Final phase of the analysis is the risk ranking and the reporting. The workflow is presented graphically in the Fig. 10.8.

10.3 Case Study

The usability and functionality of the developed methodology and software has been checked with test analysis. The analyzed facility was a steel mezzanine in a multi-

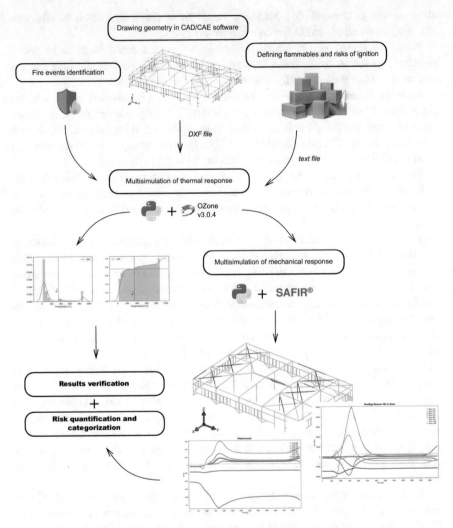

Fig. 10.8 Scheme of probabilistic fire risk analysis workflow

storey logistic, storage, and service hall. The working platform has been made of thin-walled steel profiles. The main superstructure consists of columns of a square or rectangular pipe section and beams made of double channel bars. The secondary structure meant to stabilize the working platforms in the transverse direction is C-section beams. Furthermore, the rigidity of the structure is assured by purlins and slabs forming the mezzanine floor. Table 10.2 presents basic information concerning the analyzed part of the building.

Evacuation to the outside of the building is feasible via twenty-six escape exits. Eight of them are on the ground floor. On floors 1st and 2nd there are seven exits each, and on the third floor of the mezzanine—four escape exits.

Table 10.2 Basic information concerning the analyzed part of the facility

Hall	
Floor area	48,286 m^2
Height	21.2–21.8 m
Mezzanine	
Mezzanine floor area	26,914 m^2
Height	9.6–14.4 m
Number of storeys	4

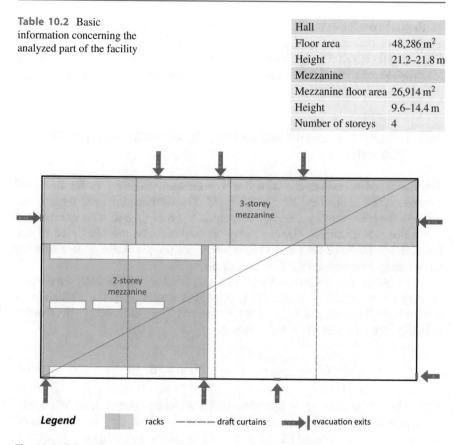

Fig. 10.9 Schematic view of the logistic, storage, and production hall

The main fire protection means installed in the analyzed facility are fire alarm system, acoustic signaling units, sprinkler system, mechanical smoke control system, hydrant installation, and emergency lighting system.

Figure 10.9 presents information concerning the facility of key importance for the analysis. The area where the mezzanine is situated has been marked, as well as a division into smoke control zones.

Analyzed structural parameters were derived from Eurocode standards [20, 24]. The ULS and LSS were considered, due to evacuation and structure safety. ULS and LSS were related to the Required Safe Egress Time (RSET).

To calculate RSET, a simulator for crowd behavior Pathfinder 2020 was used. Calculations were based on British standard [33] as regards RSET components and pre-evacuation time distributions accordingly M1 B1-B2 A1 category. The value of the detection time was assumed to be 30 s, basing on CFD simulations. Pre-evacuation time for Pathfinder was set as $t_{pre1\%} = 30$ s and $t_{pre99\%} = 90$ s.

To produce reliable results of RSET, multiple simulations were made. Location and pre-evacuation time of occupants were changed in every single iteration. Consequently the RSET was determined as $\mu + 2\sigma$ to obtain an error at the level of 5%. Time required for the evacuation from the mezzanine enlarged with the margin of safety was assumed to be 250 s.

10.3.1 Analysis of Load-Bearing Capacity of the Structure in Fire Conditions

The likelihood of occurrence of a fire was assumed similarly as for industrial facilities at the level of $9 \cdot 10^{-3}$ year $^{-1}$ [8]. The likelihood defined by standard refers to the entire facility. During the analysis a part of the hall with incorporated mezzanine was modeled. Hence it may be assumed that the likelihood of fire outbreak on the mezzanine area should be the surface share of the mezzanine to the whole logistic and storage floor area of the hall.

A fire initiated in the given smoke zone may become automatically suppressed or may be extinguished manually. In accordance with [34] the likelihood of automatic fire extinction of 0.29 has been assumed. The likelihood of effective fire initiation P_i on the area of the mezzanine is

$$P_i = 9 \cdot 10^{-3} \cdot \frac{26914 m^2}{48286 m^2} \cdot (1 - 0.29) = 3.56 \cdot 10^{-3} \tag{10.9}$$

The facility in question is furnished with a sprinkler system. A sprinkler may become activated to suppress the initiated fire provided that it is in a good working order. The standard PD 7974-7:2019 [8] defines the likelihood of technical efficiency of the sprinkler system in storage buildings as 0.86. According to [7] failure of sprinklers to function may have the following causes: (a) disabled system—62%; (b) damage to the system due to manual intervention—16%; (c) damage to elements of the sprinkler system—7%; (d) lack of inspections—5%; (e) inappropriate selection of the system to the type of fires—9%.

Based on this information we updated of sprinkler infallibility in the analyzed building as 0.905. Failure of the sprinklers to operate enables unconstrained development of the t-squared fire. The ultimate fire severity adopted in the scenario assuming a total failure of sprinklers is 300 MW. This arises from limitations of the model implemented in OZone 3.

Once sprinklers are activated, they may affect the fire in an effective or ineffective way. The likelihood of effective functioning of the sprinklers comes up to 0.92 [7]. In accordance with [7] the effectiveness of sprinklers is defined as having the fire controlled by a given number of sprinklers. This means that there is no further temperature increase and activation of further sprinklers. Pursuant to [7] the fire is controlled by a single sprinkler in 97%.

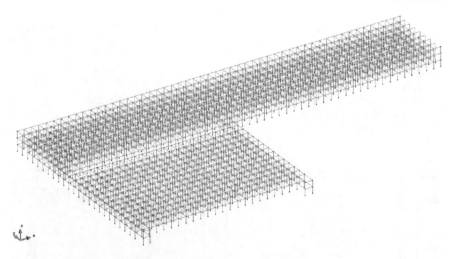

Fig. 10.10 Geometrical model of the mezzanine used in stochastic analysis

Consequently as regards effective functioning, an unconstrained development of the fire was assumed until activation of the sprinklers determined as taking 108 s (accordingly to the method derived from [31]).

The scenario of effective tripping of sprinklers assumes that after their activation the extinguishing phase would proceed in line with the empirical curve [19,35]. The present method of product storage in the analyzed facility may not allow complete suppression of the fire. It was assumed that in the suppression phase the fire severity would not fall below the level of 15% of power achieved by the fire until the moment of sprinkler triggering.

Ineffective triggering of the system has been modeled assuming maintaining the unchanged power severity from triggering of sprinklers until the end of simulations.

10.3.2 Stochastic Analysis of the Thermal Response

In order to assess the probability of element failure the analysis in terms of thermal response had been made. The model consisting linear elements of the structure had been prepared. Also taken into account was the impact of flooring tiles of the mezzanine above the fire on its further development. The vertical development of the fire was limited by the ceiling—a slab that constitutes the floor of the next storey. Pursuant to the LOCAFI model the spread of flames under the ceiling was modeled, along with their impact on structural elements [36]. It was necessary to add the sections' parameters and dimensions of key importance for heating. The heating rate of an element depends on the so-called exposure index, which may be defined as a ratio of the circumference of the section to its surface. Those values was added directly to the OZone's *Profiles.sys* file. Figure 10.10 shows the geometry of the mezzanine.

	Series	$P(\theta_a \geq \theta_{a,cr})$
Table 10.3 Partial probability of events—results of stochastic analysis of the mezzanine. Series 1—sprinklers are functioning in an effective way. Series 2—sprinklers are functioning in an ineffective way. Series 3—sprinklers are not functioning	Series 1	0.179
	Series 2	0.248
	Series 3	0.998

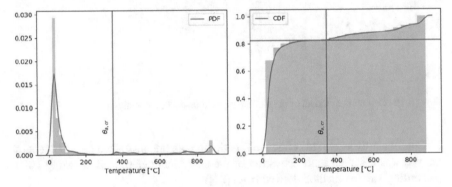

Fig. 10.11 Probability functions of exceeding the critical temperature for Series 1. Left diagram—probability density function; Right diagram—cumulative distribution function

It was assumed that flammable materials are uniformly distributed in each storey up to 2.5 m from the each of floor levels. The HRRPUA of stored materials varies from 2.1 to 2.6 MW/m^2 [37] and was implemented as the triangular distribution with mode 2.4 MW/m^2.

The fire growth factor α was derived from the empirical log-normal distribution adequate for storage buildings [38].

The utilization factor for the mezzanine structure was assumed at a level appropriate for thin-walled profiles (class IV) $\mu_0 = 1.0$. Thanks to adopting such a conservative approach a safety margin was achieved that enables safe assessment of stability of the analyzed structure. For such parameters the critical temperature pursuant to Eurocode equals to $\theta_{a,cr} = 340\,°C$ [15].

For the adopted input parameters of particular series of simulations, estimations were made as to the probability of exceeding critical temperature in the section of the structural element. Values obtained in subsequent simulation series were presented in a Table 10.3 and as likelihood distributions in Fig. 10.11). Subsequently the results of stochastic analysis were included as constituent probabilities in the tree of events (see: Fig. 10.12).

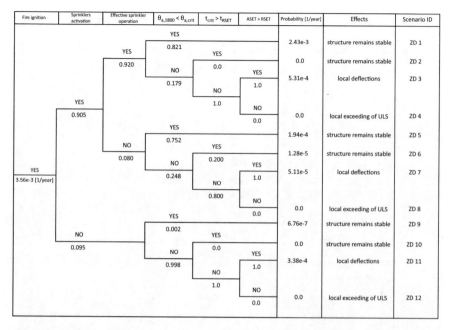

Fig. 10.12 Tree of events for an undesired event in the form of fire initiation in the mezzanine area

10.3.3 Numerical Analysis of Thermal and Mechanical Response

Five of the most severe scenarios were picked from each series of multisimulation to further analysis. Parts of the structure in explicit proximity of fire source were introduced into FEA software—SAFIR. Appropriate dead and live loads, reduced accordingly to accidental load combination had been attached to beam elements. Loads from upper storeys were attached to the column nodes as punctual. All nodes were assumed to be rigid. The structures were subjected to localized fires derived from stochastic analysis. The LOCAFI model was used to heat elements.

Results of each FEA were investigated to identify the time of structure's failure. Changes in deflections, bending moments, and shear forces allowed us to determine the moment of ULS exceedance. Table 10.4 contains the summary of FEAs. Additionally the time to exceeding $L/200$ deflection limit in beams was indicated.

As the conclusion, results of both, stochastic and finite element analyses, are served as input to Event Tree Analysis. Figure 10.12 shows the tree of events, which can be used to calculate the frequency of selected scenarios.

10.3.3.1 Estimation and Categorization of Risks

Table 10.5 contains a specification of likelihood values for particular undesired events, calculated physical effects, their impact on people, and effects arising from

Table 10.4 Results of FE analyses of scenarios, in which the estimated effects were the most pronounced

ID	$t_{L/200}$	t_{ULS}	P($t_{L/200}$<RSET)	P(ASET$_{ULS}$<RSET)
Sprinklers not working				
1.1	184	455		
1.2	176	450		
1.3	198	600	1.0	0.0
1.4	245	487		
1.5	164	397		
Sprinklers working ineffectively				
2.1	111	1800		
2.2	114	1800		
2.3	240	1800	0.8	0.0
2.4	131	1800		
2.5	365	1800		
Sprinklers working effectively				
3.1	165	1800		
3.2	134	1800		
3.3	100	1800	1.0	0.0
3.4	131	1800		
3.5	138	1800		

Table 10.5 Estimation of risks for people for defined undesired events, where: *negligible*—implies no hazard for people; *low*—implies minor injuries that do not require hospitalization; *high*—implies possible demise of one or more people

Scenario	Frequency [1/year]	RSET	ASET<RSET	Effects
ZD 3	$5.31 \cdot 10^{-4}$	250	LOCALLY	Low
ZD 4	0.0	250	NO	Negligible
ZD 7	$5.11 \cdot 10^{-5}$	250	LOCALLY	Low
ZD 8	0.0	250	NO	Negligible
ZD 11	$3.38 \cdot 10^{-4}$	250	LOCALLY	Low
ZD 12	0.0	250	NO	Negligible

them. The table does not comprise events, the effects of which did not generate risk for people.

Next a categorization of risks was conducted pursuant to SFPE Risk matrix [31]. The aim of this categorization is to allow the determination of negligible risk, acceptable risk, and risk that requires minimization. The risk matrix comprises verses that define losses (from negligible to high ones) and chances of their occurrence divided into intervals as regards frequency. Figure 10.13 presents a risk matrix applicable to people.

As regards the presented matrix, the risk of events ZD 4, ZD 8, and ZD 12 was classified as insignificant. In the facility in question the risk of death to people

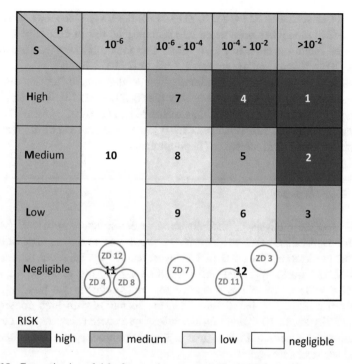

Fig. 10.13 Categorization of risks for people pursuant to the SFPE Risk matrix

belongs to the category of negligible risks, i.e., the likelihood is lower than 10^{-4}, considered as a threshold of death for natural causes.

The likelihood of occurrence of events ZD 3, ZD 7, and ZD 11 is within the range from 10^{-6} to 10^{-2}. Effects of occurrence of those events are low, which allows classifying the risk also as low. This risk will need to be minimized.

Next the risks were categorized with PD 7974-7 [8] methodology. For needs of the analysis the risk of death in a fire of a residential building was adopted as a threshold value of tolerable risk. According to statistical data of the Polish State Fire Service [39] residential buildings are characterized by the highest value of that risk. Given the fact that at present no changes are anticipated in fire regulations applicable to this group of building structures, it may be assumed that the present state as regards mortality in fires of residential buildings appears to be tolerable. The value of the risk of death in a fire of a residential building has been assumed on the basis of the study [30], and it amounts to $1.3 \cdot 10^{-5}$ 1/year, respectively. On the other hand, $1.0 \cdot 10^{-6}$ 1/year has been adopted as an acceptable risk value. This level is commonly accepted, because this is the value of a risk of death for natural reasons [31, 40].

On the basis of results obtained in the conducted multisimulations and numerical FEM calculations, a presumption may be made that the loss of the load-bearing capacity of a structure in a fire situation is most likely to happen after completion of evacuation (see scenarios ZD 4, ZD 8, ZD 12 in Table 10.4). Once the uncertainty of the Monte Carl model has been taken into account, the risk of death due to a fire in the analyzed facility is close to a zero value. This makes it lower than the tolerable risk ($1.3 \cdot 10^{-5}$ 1/year) and the acceptable risk ($1.0 \cdot 10^{-6}$ 1/year).

However, the significant deformations of single elements of the structure before the time of evacuation may not be ruled out (ZD 3, ZD7, ZD 11). The likelihood of occurrence of those events is higher than tolerable risk ($1.3 \cdot 10^{-5}$ 1/year). However, taking into account the consequences (injury), the minimization of the risk should be considered along with the cost of its minimization.

10.4 Conclusion

We presented the analysis of the mezzanine steel structure subjected to a fire. Using the probabilistic risk assessment we were able to assure adequate safety for the people keeping the costs of the investment in the safety at rational level. Applied Monte Carlo approach and simplified OZone model allow for the fast review of a wide range of fire scenarios that may occur in the building, and to assign them adequate probabilities. Then, the scenarios with high consequences and probability above 10^{-6} were further investigated with using more accurate but computationally expensive FEA software. The entire process was automated by the proposed McSteel software.

McSteel is still in the development phase. The essential functionalities to make the software stable have to be improved. OZone software is closed-code desktop application. The necessity to navigate through tabs and buttons of the graphic interface significantly slows down analyze process. Although works on implementing simplified LOCAFI model to a Python script directly are in advanced progress. Making this step will allow to speed up the process and have a full control over errors and warnings raised by the script. McSteel is developed as an open-source software to ensure full access and create a wide community of users. In the future it would be advisable to made the software independent on closed-code software. An alternative to zone fire model could be for example the *CFAST* model by NIST. Also the SAFIR has a strong, open-sourced competitor—*OpenSees for fire* by Polytechnica University of Hong Kong. Access to the models' source code would allow to parallelize some processes in the future. More research should be done to find reliable distributions of as much stochastic input data as possible and implement those values as the default. Detailed documentation will provide a guideline for adding stochastic variables to the code and make the tool more user-friendly and more versatile.

References

1. Lawson, J.R.A.: *A history of fire testing* (National Institute of Standards and Technology. Building and Fire Research, 2009)

2. Babrauskas, V.; Williamson, R.B.: *The historical basis of fire resistance testing—Part I* (1978)
3. Babrauskas, V.; Williamson, R.B.: *The historical basis of fire resistance testing—Part II* (1978)
4. Hietaniemi, J.: *Risk-based fire resistance requirements: contract no. 7210-PR/251, 1 July 2000 to 31 December 2003; final report* (Office f. Official Publ. of the Europ. Communities, 2005)
5. Vassart, O and Cajot, LG and Brasseur, M and Strejček, M: *Dissemination of Fire Safety Engineering Knowledge. DIFISEK. PART 1: Thermal & Mechanical Actions* (National Institute of Standards and Technology. Building614and Fire Research, 2009)
6. Lyzwa, J., Zehfuss, J.: *Thermal material properties of concrete in the cooling phase.*(ASFE conference 2017)
7. Ahrens, Marty: *US experience with sprinklers* (National Fire Protection Association. Research, Data and Analytics Division, 2017)
8. BSI. Published Document PD 7974-7:2019: *Application of fire safety engineering principles to the design of buildings - Part 7* (2019)
9. Alvarez, A. and Meacham, B. J. and Dembsey, N. A. and Thomas, J. R.: *A Framework for Risk-Informed Performance-Based Fire Protection Design for the Built Environment* (Fire Technology, 2014)
10. Hurley, Morgan J. and Rosenbaum, Eric R.: *Performance-based fire safety design* (CRC Press, 2015)
11. Schleich, J-B and Cajot, L-G: *Natural fire safety for buildings* (Revue de Métallurgie, EDP Sciences, 2002)
12. Schleich, J-B and Cajot, L-G and Pierre, M and Moore, D and Lennon, T and Kruppa, J and Joyeux, D and Holler, V and Hosser, D and Dobbernack, R and others: *Natural fire safety concept: full-scale tests, implementation in the Eurocodes and development of a user-friendly design tool* (European Commission, 2003)
13. LU, ProfilARBED SA and DE, Arbeitsgemeinschaft Brandsicherheit AGB: *Natural Fire Safety Concept–The Development and Validation of a CFD-based Engineering Methodology for Evaluating Thermal Action on Steel and Composite Structures* (2005)
14. EN 1991-1-2:2002 Eurocode 1: *Actions on structures - Part 1-2: General actions - Actions on structures exposed to fire* (2002)
15. Norma, Polska: *EN 1993-1-2:2005 Eurocode 3: Design of steel structures - Part 1-2: General rules - Structural fire design* (2005)
16. *Single-Storey Steel Buildings Part 7: Fire Engineering.* ArcelorMittal, Peiner Träger, Corus, CTICM, SCI.
17. Gernay, Thomas and Van Coile, Ruben and Elhami Khorasani, Negar and Hopkin, Danny: *Efficient uncertainty quantification method applied to structural fire engineering computations* (Engineering Structures, 2019)
18. Technical Committee: ISO/TC 98/SC 2: *ISO 2394:2015. General principles on reliability for structures* (2015)
19. Yu, Hong-zeng and Lee, James L and Kung, HSIANG-CHENG and Brown, WR: *Suppression of rack-storage fires by water* (Fire Safety Science, 1994)
20. *EN 1990:2002 Eurocode: Basis of structural design* (2002)
21. Metropolis, Nicholas and Ulam, S.: *The Monte Carlo method* (Journal of the American Statistical Association, 1946)
22. BSI, Standards Publication: *9999: 2008-Code of practice for fire safety in the design, management and use of buildings* (British Standards Institute, London, 2008)
23. PD 7974-6: *The application of fire safety engineering principles to fire safety design of buildings. part 6: Human factors: Life safety strategies- occupant evacuation, behavior and condition (sub-system 6)* (2004)
24. EN 1993-1-1:2005 Eurocode 3: *Design of steel structures - Part 1-1: General rules and rules for buildings* (2005)
25. Purser, David A.: *Toxicity assessment of combustion products* (National Fire Protection Association Quincy, MA, 2002)
26. Elicson, Tom and Harwood, Bentley: *Calculation of Fire Severity Factors and Fire Probabilities for a DOE Facility Fire PRA* (2011)

27. Frank, Kevin and Spearpoint, Michael and Weddell, Steve: *Finding the probability of doors being open using a continuous position logger* (Fire Safety Science, 2014)
28. Rahikainen, Jussi and Keski-Rahkonen, O.: *Statistical determination of ignition frequency of structural fires in different premises in Finland* (Fire Technology, 2004)
29. Owen, Art B.: *Monte Carlo variance of scrambled net quadrature* (SIAM Journal on Numerical Analysis, 1997)
30. Krasuski, Adam: *Multisimulation: Stochastic simulations for the assessment of building fire safety* (SGSP Publishing House, 2019)
31. Hurley, M. J.: *SFPE handbook of fire protection engineering* (Springer, New York, 2016)
32. Hurley, Morgan J and Gottuk, Daniel T and Hall Jr, John R and Harada, Kazunori and Kuligowski, Erica D and Puchovsky, Milosh and Watts Jr, John M and WIECZOREK, CHRISTOPHER J and others: *SFPE handbook of fire protection engineering* (Springer, 2015)
33. BSI Standards Publication: *PD 7974-6 Application of fire safety engineering principles to the design of buildings - Part 6: Human factors: Life safety strategies - Occupants evacuation, behaviour and condition* (2019)
34. Kati Tillander. *Utilisation of statistics to assess fire risks in buildings. VTT Publications*, (537): 3–224 (2004)
35. Klote, John H and Milke, James A: *Principles of smoke management* (American Society of Heating, Refrigerating and Air-Conditioning Engineers, 2002)
36. Vassart, O. and Hanus, F. and Obiala, R. and Brasseur, M. and Franssen, J.M. and Scifo, A. and Zhao, B. and Thauvoye, C. and Nadjai, A. and Sanghoon, H. and Zaharia, R. and Pintea D.: *Temperature assesment of a vertical steel member subjected to localised fire (LOCAFI)* (Directorate-General for Research and Innovation, 2017)
37. Sztarbała, G. and Węgrzyński, W. and Krajewski, G.: *Zastosowanie gorącego dymu do oceny skuteczności działania systemów bezpieczeństwa pożarowego podziemnych obiektów* (2012)
38. Deguchi, Yoshikazu: *Statistical Estimations of the distribution of fire growth factor study on risk-based evacuation safety design method.* (International Association for Fire Safety Science, Kyoto, 2011)
39. Leśniakiewicz, Wiesław: *Zasady ewidencjonowania zdarzeń w Systemie Wspomagania Decyzji Państwowej Straży Pożarnej* (Komenda Główna Państwowej Straży Pożarnej, 2014)
40. Colaborative work: *FRAME Fire Risk Assessment for Engineering. FRAME – Validation* (2014)

Autonomous Sensor-Driven Pressurization Systems: Novel Solutions and Future Trends

11

Wojciech Węgrzyński and Piotr Antosiewicz

11.1 Introduction

Smoke is considered the principal threat to building occupants in fires. To reduce this threat, technical solutions are used under a common name "smoke control systems." The role of such systems is to limit the spread of smoke in the building to a predefined area (e.g., room of fire origin, predefined smoke control zone, and compartment), which allows to maintain sufficient conditions in the rest of the building. This enables safe evacuation of building occupants and improves the rescue operations. The role of smoke control systems as a part of modern buildings and the fire safety strategies was described in [1].

Smoke control systems can be distinguished by their principle of operation—for this chapter we will consider two. The first group are systems that create a pressure difference between the protected space (e.g. stairwell) and a compartment with fire (e.g. corridor), and prevent smoke from infiltrating protected space. These systems will be referred to as the pressurization systems or pressure differential systems (PDSs). The second group are all solutions that remove the smoke and other combustion products from the building. These systems will be referred to as smoke exhaust systems. A similar classification is given by Klote [2], following the definitions of NFPA 92 [3]. In this view, the smoke control systems are "engineered system that includes all methods that can be used singly or in combination to modify smoke movement." These systems may be subdivided into two roles—(1) systems used to prevent people from coming into contact with the smoke and (2) systems that provide tenable conditions in the time necessary to leave the premise by its occupants. The pressurization solutions fall into the group (1) and smoke exhaust into the group (2). The physical mechanism of smoke control can be (1)

W. Węgrzyński (✉) · P. Antosiewicz
Fire Research Department, Building Research Institute (ITB), Warsaw, Poland
e-mail: w.wegrzynski@itb.pl

© Springer Nature Switzerland AG 2022
M. Z. Naser, G. Corbett (eds.), *Handbook of Cognitive and Autonomous Systems for Fire Resilient Infrastructures*, https://doi.org/10.1007/978-3-030-98685-8_11

compartmentation, (2) dilution, (3) pressurization, (4) airflow, and (5) buoyancy [2]. In this classification, the pressurization systems fall obviously into the group (3); however, important aspects of their operation are common for mechanisms (1) and (4).

A more thorough categorization of smoke control solutions is given BS 7346-4 [4], which defines that the smoke control may serve as:

1. means of protecting escape routes (keeping the escape and access routes free from smoke and radiant heat); or
2. means of protection of property (protecting equipment and furnishings by reducing the damage caused by thermal decomposition products, hot gases, and heat radiation); or
3. means of controlling the temperature of hot, smoky gases affecting, for example, the building's structure, façades or glazing; or
4. means of facilitating fire-fighting operations by creating a smoke-free layer; or
5. a combination of any of these.

Regardless of the definition, the aim of all smoke control systems is to reduce the exposure of occupants (and the building) to smoke and heat produced in a fire. To achieve that goal, smoke control systems are dimensioned and designed following a set of rules defined in reference standards. This design process was described thoroughly in [5]. The primary sources of knowledge for the design of traditional smoke control systems are: [2, 3, 6–8]. A common strategy for the design of pressurization systems is given in European standard EN 12101-6 [9], where six different types of PDS are thoroughly described, along with practical design recommendations.

In some approaches, performance-based engineering approach can be used instead of the standardized design methods [10, 11]. In such a case, the requirements for the smoke control system are not predefined, and performance goals take their place. As mentioned in the first pre-design is often verified as a "trial design," to demonstrate its ability to meet the design goals. In the past, a variety of tools was used for this purpose, including hand calculation models, zone models [12] or scale modeling. Today, the common approach is to use computational fluid dynamics (CFD) tools to estimate the performance of the system.

The main difference between traditional solutions for smoke control and innovative solutions presented in this chapter is the use of sensors to determine the state of the fire in the building and provide appropriate system response to the momentary threat. This also allows to reduce dangerous oscillatory feedback loops that can be found in non-automated systems. For ampleness reasons, the solutions for smoke control other than the PDS will not be characterized here in detail. The reader is kindly referred to other sources mentioned in the introduction, and primarily to [2], which provides a thorough introduction to this topic.

11.2 Concepts of Pressurization and Smoke Exhaust

11.2.1 Basic Principles

Pressurization systems [pressure differential systems (PDSs)] are a class of smoke control solutions that limit the spread of smoke by creating an overpressurized region in the building (e.g., stairwell, vestibules, and corridors). As the pressure in this region exceeds the pressure in the smoke-filled compartment, the smoke cannot enter the protected space, as the flow always occurs from higher to lower pressure region. The challenge with the pressurization is that there is a narrow window of pressure difference (sometimes defined as 20–80 Pa), for which the solution is safe. With lower pressures, the protection may fail. With higher pressures, it may be impossible to open doors to the protected space, essentially blocking this escape route, which was described in [13]. Maintaining this pressure is challenging, as the building leakages must be taken into account [14]. In some opinions, the inherent difficulties in achieving this may even prevent the design of PDS [15].

Three distinct modes of operation can be considered for PDS systems:

1. maintaining pressure difference with doors closed (sealed system), usually in the range of 20–80 Pa, Fig. 11.1a;
2. maintaining overpressure difference with some doors open, usually in the range of 1–10 Pa;
3. providing opposed airflow at the doors to the compartment with fire, usually in the range of 0.5–2 m/s, with 1–3 simultaneously open doors, Fig. 11.1b.

The pressurization must be maintained for a variety of states of the stairwell, including scenarios with all of the doors closed and with some of them open. It is impossible to maintain the same pressure with such a variety of leakages in the building. Thus, the system must provide some adaptation to the conditions inside. If one would provide a device that supplies the air to the stairwell at a constant rate at its bottom, and removes it on the top (Fig. 11.2), the resulting pressure distribution

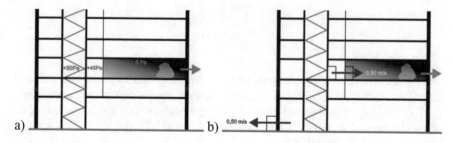

Fig. 11.1 Example of pressurization of a stairwell. (**a**) The pressure difference of 50 Pa maintained when doors are closed, (**b**) 0.50 m/s opposed airflow at the open doors to the fire compartment, with the external doors to the stairwell also open

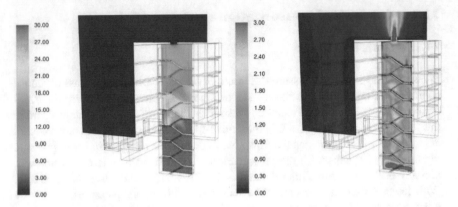

Fig. 11.2 Pressure (left) and flow velocity (right) in a stairwell of a building with a constant air supply at the bottom and an opening on the top. ANSYS Fluent CFD simulations

Fig. 11.3 An example of the most simple pressurization system—a fan with constant velocity and a pressure relief damper installed in the same stairwell

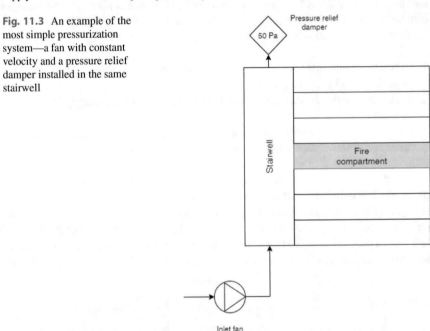

along the stairwell would be an outcome of the pressure losses in the stairwell and the stack effect [16, 17], Fig. 11.3. This solution can be considered useful only for the building of limited heights and does not lead to maintain a specific value of overpressure, but rather "some overpressure" that will depend on momentary leakages and stack effect [18].

For more precise control over the pressure conditions, these systems were equipped with tensometric pressure relief dampers instead of openings that maintain the pressure at the damper at a prescribed level (e.g., 50 Pa), Fig. 11.3. The downside of this solution is that the pressure in the stairwell below the relief opening will change depending on the stack effect and may reach excessively high values if the protected height is too large. To counter this effect the stairwell is separated into multiple fragments of limited height or air is supplied through multiple points, with individual inflow corrected for seasonal stack effect (e.g., more air supplied on top, less on the bottom of a stairwell in winter). The latter solution can be considered as a pinnacle of *traditional* pressurization systems.

11.2.2 Adaptive Pressurization Systems

A new generation of pressurization systems uses adaptive control on the air supply, in place of previously used tensometric pressure relief dampers. The idea of the operation of such systems is to provide the amount of air necessary to maintain a predefined pressure difference between the stairwell and the fire compartment (Fig. 11.4). When all doors are closed, and the stairwell is sealed, the flow required to achieve this effect is minimal—only to cover the flow through the leakages of the building. As a consequence, the system must operate with varying volumetric capacity, which is achieved with inverter powered fans that can quickly change their volumetric flow. The pressure differences between stairwell, vestibule, and a fire compartment are illustrated in Fig. 11.5.

When one opens the doors to the pressurized stairwell, the pressures between stairwell and compartment evens out. The resulting drop in pressure is noticed by the system pressure measurement points, and a signal to increase the volumetric flow on the supply fan is issued. The outcome of this increased flow is the occurrence of opposed airflow in the door between the stairwell and compartment with the fire.

Fig. 11.4 An example of an active pressurization system with a feedback loop between pressure difference sensors and an active, inverter-driven fan

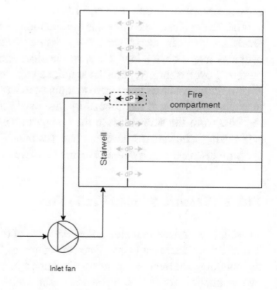

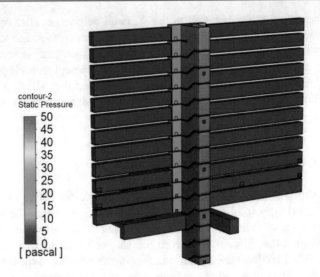

contour-2
Static Pressure

50
45
40
35
30
25
20
15
10
5
0
[pascal]

Fig. 11.5 The pressure difference between stairwell, vestibule, and corridor (fire compartment) for a simple, sensor-driven PDS. ANSYS Fluent CFD simulations

This flow prevents smoke from entering the stairwell through the open doors. For a correct operation of this system, the air supplied to the stairwell must be removed with another system or opening. It is common to install a smoke exhaust system in corridors or other automatic openings through which the smoke may escape (e.g., natural shaft or an open window). It must be noted that this response is almost instantaneous. European standard [9] defines this response time as below 3 s. The ability of a sensor-driven system to meet this is explicitly verified in pre-assessment tests, as described further in Sect. 11.5.1.

The weak point of the system is the situation in which multiple doors on multiple levels are open. In such a case, the pressure difference (and the flow through the opening) may be very low. Thus, mechanical exhaust in the fire compartment is preferred over natural ventilation solutions (windows and dampers). Nevertheless, in some circumstances (e.g., large amount of external glazing is lost in the fire) even mechanical exhaust at the floor of the fire origin may not be sufficient to maintain the flow from the stairwell into the compartment with fire. Therefore, the use of self-closing mechanisms on doors and vestibules is necessary to limit the risk of pressure loss in the stairwell due to an excessive number of open connections.

11.2.3 Pressurization of Vestibules

An effective strategy to increase the resilience of the stairwell pressurization system is to use ventilated vestibules at the connection of the stairwell with the corridors of the building. The vestibules are small compartments that act as a hatch between the two ventilation zones—the pressurized stairwell and the fire compartment. These

Fig. 11.6 An example of an active pressurization system with a feedback loop between pressure difference sensors and an active, inverter-driven fan, and a separate system for additional pressurization of the vestibules

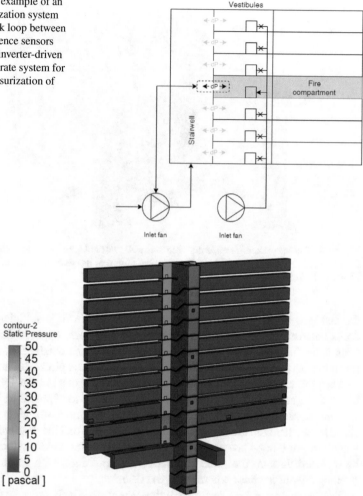

Fig. 11.7 The pressure difference between stairwell, vestibule, and corridor (fire compartment) for a simple, sensor-driven PDS with additional supply point in the vestibule. ANSYS Fluent CFD simulations

vestibules can be additionally equipped with their individual pressurization system, Fig. 11.6, used to maintain a specific pressure difference between the vestibule and the fire compartment (usually ca. 45 Pa). The pressure difference between the vestibule and the stairwell is also maintained (usually ca.—5 Pa). In consequence, the flow is always from the stairwell to vestibule, and then to the fire compartment, Fig. 11.7.

The amount of air supplied to the vestibule is usually much smaller than the stairwell. When the doors to the vestibule are closed, this flow must be sufficient for

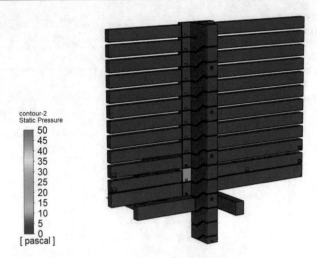

contour-2
Static Pressure

50
45
40
35
30
25
20
15
10
5
0
[pascal]

Fig. 11.8 The pressure difference between stairwell, vestibule, and corridor (fire compartment) for a system with vestibule pressurization, and the failure of the stairwell pressurization. ANSYS Fluent CFD simulations

the leakages of the vestibule only (usually 200–300 m^3/h is sufficient). When the doors between the vestibule and fire compartment open, opposed flow velocity at open doors between vestibule and compartment must be obtained—as an example, for the velocity of 1 m/s and 2.20 × 0.90 m doors this requires 7200 m^3/h flow rate.

The pressure between the vestibule and the compartment is usually measured, as between the stairwell and the compartment. The scenario of operation is the same as with the pressurization of the stairwell, with one exception—the air is supplied only at the level where the fire is detected. As it is common that the vestibules share their location on the floor layout over multiple levels of the building, they can usually be operated through the same ventilation unit and duct, with automatic dampers opening only at the floor where the fire is detected.

The use of pressurized vestibule in the system effectively prevents the possibility of the smoke to penetrate the stairwell, regardless of the status of operation of the stairwell system, Fig. 11.8.

The need for rapid changes in the flow to the vestibule can be problematic, especially that it is expected to operate in evacuation scenarios, where the doors to both vestibule and stairwell will be open for short periods every few second. Furthermore, the compartment where the fire is detected may be equipped with a mechanical exhaust system, which includes the airflow from the vestibule as its source of the make-up air. When the doors to the vestibule are closed, an additional source of air must be provided not to cause excessive underpressure in this region, which could potentially cause significant forces on the doors. Air transfers are used between the vestibule and the compartment to cope with these problems. When the evacuation doors are shut, the majority of air incoming to the vestibule is released through that transfer duct. When the evacuation doors are open, the air is transferred through the door.

There are multiple technologies to maintain the correct operation of this system, among them:

1. An opening or a transfer duct between the vestibule and fire compartment;
2. Pressure relief transfer between the vestibule and fire compartment that opens only when the pressure in the vestibule exceeds a specific value;
3. Automatic, fast damper between the vestibule and fire compartment that opens depending on the state of the doors (or the pressure difference) between the vestibule and the fire compartment;
4. No transfer. The air supply duct is connected to both the vestibule and the fire compartment, both equipped with a fast damper. The damper to the compartment usually is open, and the damper to the vestibule is partially closed (open only to supply enough air to create overpressure). If doors to the vestibule open, the damper in compartment instantly closes, and the damper in the vestibule instantly opens, forcing the air to flow through the vestibule.

The last type of system requires a substantial amount of automation to provide reliable operation. However, it also requires no opening between the vestibule and fire compartment. This is a significant benefit for narrow corridors, in which the air transfer would require installing a duct in the otherwise leasable space.

If air transfer techniques are used, it is important that the openings are located near the floor, and the exit velocity of air is low so that the transfers do not disturb the regular operation of the smoke control systems in the fire compartment (Fig. 11.9).

Fig. 11.9 The idea of an air transfer duct—the air is transferred from the vestibule through a separate duct when the doors to the vestibule are closed so that the vestibule pressurization can act as a source of make-up air to the corridor smoke control

11.3 Countering the Stack Effect

The stack effect (also known as the chimney effect) is a natural phenomenon caused by the difference in the density of the air inside and outside of the building. Depending on the seasonal variations, the air in the buildings may be warmer or colder than the outside, causing buoyancy forces to act and creating an airflow through the stairwell. It is generally assumed that in the stack effect the air will move upwards (air in the building is warmer than outside); while the opposite direction (movement downwards, the air in the building is colder than outside) is often referred as the reverse stack effect. A powerful stack effect is typical for high-rise buildings at higher latitudes, where severe winters are common. A reverse stack effect is typical for warmer climates, with the excessively hot summer period.

The profound impact of the stack effect on the pressurization systems is shown on Fig. 11.10. Consider a system designed for a tall stairwell that is optimized to perfection—in isothermal conditions the system maintains the same pressure difference between the stairwell and any of the floors over the whole height of the building. If a fire occurs on any of the floors, these conditions are maintained. If the same system operates in conditions with a temperature difference between the inside and outside of the building, a natural stack effect will form. If the system is a traditional solution with a tensometric pressure relief damper on the top of the

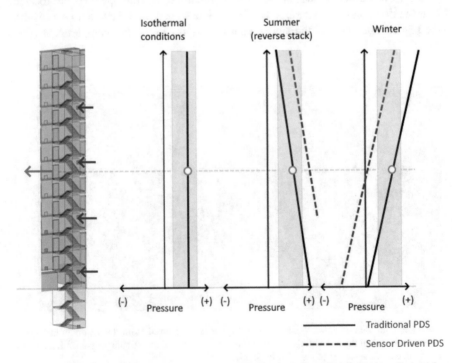

Fig. 11.10 The impact of the stack pressure on pressurization of a stairwell that is optimized for isothermal conditions and no other means to mitigate stack effect are implemented

building the value of pressure difference is set at the location of the damper. The system will maintain the expected overpressure at the damper, and the pressure in the rest of the stairwell will change depending on height and ambient conditions. If the system is sensor-driven, it will maintain the target pressure difference at the floor of the fire compartment, and the pressure in the rest of the stairwell will change with the height. In both cases, the pressure will most likely be outside of the operational range (green area), leading either to excess overpressure and inability to open doors to the stairwell at some of the floors or underpressure (sucking air/smoke to the stairwell). Both can be considered as critical failures of the system. It can be noted that for a certain range of heights, the pressure is in the operational range even with the stack effect. The size of this zone will depend on the chosen point of operation of the system and the height of the stairwell. The stack effect will also significantly depend on the temperature difference; thus, different regions may have different requirements in this regard. In principle, the stack can be calculated [2, 19], and the critical height for which this effect can be disregarded can be determined. Usually, this value will not exceed 30 m, and so the stairwell can be separated into sectors not exceeding this height.

To countermeasure stack effect, some advanced air management strategies can be implemented. These strategies in principle require variable air supply/exhaust at different heights of the building or use of heating/cooling of the supplied air to counteract the stack. With such solutions, the pressure may be maintained over significant heights. As an example, Fig. 11.11 illustrates the simulated pressure difference in winter for ca. 200 m high continuous stairwell equipped with a proprietary adaptive PDS solution.

11.4 PDS and Other Smoke Control Systems in the Building

Due to their inherent complexity, adaptive PDS are usually autonomous from the rest of the building automation systems. The PDS input is limited to receive fire alarm (and in some cases, location of the fire alarm), and usually returns only the signals of activation and fault alarms. It is uncommon for the PDS to feed on the data from sensors of other building systems and to return its own data to the building management system (BMS). This is primarily to the fact that complex PDS are proprietary solutions, tested and certified as "kits." Maintaining autonomy from other system prevents issues with electromagnetic compatibility with other devices or potential issues with data transfer/analysis. It also makes it possible to test this system individually, and thus its supplier to take the full responsibility of the system performance.

This autonomy of the PDS also has significant downsides. PDS is designed to meet their own performance goals (maintain the specific value of pressure or flow), regardless of the smoke control systems in other parts of the buildings. Due to the lack of coordination (and sometimes designer ignorance), the PDSs are often designed for flow velocities exceeding critical values for buoyancy-driven systems commonly used in corridors. In such case, opening doors to a pressurized vestibule

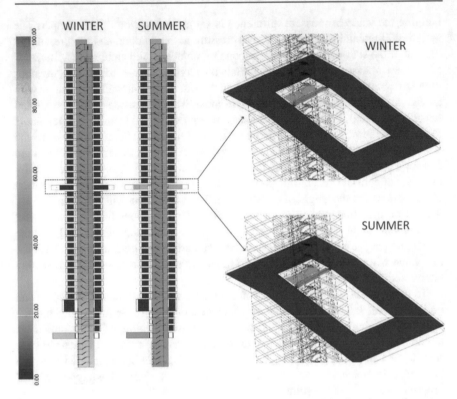

Fig. 11.11 Pressure in a stairwell (left) and in a fire compartment (right) for winter and summer conditions, operation of a PDS with active stack effect countermeasures in a 200 m tall stairwell and fire at 22nd floor, ANSYS Fluent CFD simulations

or stairwell can lead to a release of high-velocity airflow that disturbs otherwise stable smoke layer in the compartment, leading to the sudden decrease of the visibility in the corridor, Fig. 11.12.

This high-velocity flow can also be an unexpected outcome of a "correct" design. If one is designing pressurization of a large stairwell with multiple entrances at one floor (e.g., stairwells in sports arenas), the system capacity is determined to create a flow with a minimum velocity at all of the doors to this stairwell. However, the system is usually unaware of the number of doors open at once—if one opens just one door leaf, instead of all of them, it will trigger the system response and all air supplied will be pushed through this single door, Fig. 11.13. This airstream can reach velocities of approx. 6.5 m/s, which corresponds to a pressure difference of 50 Pa. Furthermore, dynamic flow effects when opening and shutting doors are commonly observed during the commissioning of pressurization systems. To countermeasure this effect, the designers should agree on performance-based system operation goals so that the PDS provides its goal (smoke-free stairwell) without impeding the operation of other systems. Furthermore, PDS system designers may

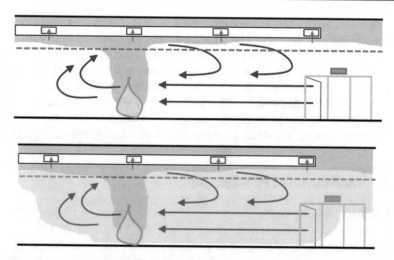

Fig. 11.12 Opening the stairwell doors with high capacity PDS may lead to sudden mixing of the incoming air with otherwise buoyant smoke layer, leading to a quick loss of visibility

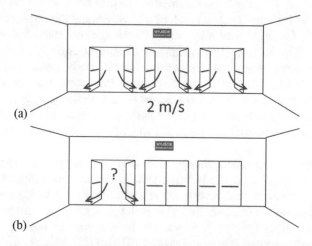

Fig. 11.13 If the PDS system is designed to deliver 2 m/s on six doors simultaneously, opening just one door may trigger system response with full capacity, leading to the excessively high flow velocity

improve the system performance by introducing door contacts, so that the system knows the status of doors, and can adapt its capacity to the one really required. Similar effects can be obtained by setting the pressure difference to a lower set point (e.g., 30 Pa), which can be sufficient to maintain goals of the system operation.

Complicated interactions between active PDS and smoke control systems may occur when more than one stairwell is connected to the same compartment. As mentioned previously, for correct system operation, it is necessary to extract the air from the compartment, in the same amount as it is supplied by the PDS. If this is not explicitly met, it may not be possible to provide a flow on open doors to the stairwell or the vestibule. When just one stairwell is connected, this can be fairly easy maintained. However, when more than one stairwell is connected to the same compartment, some intricate flow patterns may emerge, depending on the status of doors of each of the stairwells. Additional air supply points should be available to compensate for the loss of supply if stairwell doors are shut. It is also possible to use air transfer ducts, as described in Sect. 11.2.3, or adaptive exhaust as in [20].

11.5   Verification of PDS Systems

11.5.1 Laboratory Tests

PDS can be complex solutions that perform measurements on dozens of pressure/temperature sensors throughout the building and continuously adapt the fan unit parameters to match the momentary requirements. Systems like this work in a discrete manner—the adaptation is carried within time steps, which length is usually determined by the temporal resolution of the sensors and the response time of the fan inverters. Common time steps used in PDS systems range between 0.1 and 0.5 s. Shorter time steps rarely give significant improvements and require expensive high fidelity pressure sensors. Longer time steps usually make system response longer, unable to follow the dynamic events during evacuation (opening/shutting doors). The regulation of the system can be done with linear algorithms (only for large leakages and narrow range of regulation), proportional–integral–derivative (PID) controllers, and in case of the most complex solutions—neural network-based controllers.

No matter the regulation system, PDS systems must provide rapid response to events taking place in the building and be resistant to oscillatory behavior. These properties must be demonstrated at a varied range of leakages—from very airtight compartments (e.g., vestibules) to buildings with large leakages (e.g., large stairwells in antique buildings). To provide this range of testing, a special rig has been designed in revised European standard 12101-6 [21], Fig. 11.14.

The test rig consists of two connected compartments (a large "room 1" and a smaller "room 2") connected with a quick response damper. The "room 1" represents the stairwell and "room 2" represents vestibule or fire compartment. The PDS system is connected to the "room 1" in a way that introduces the air with low velocity and allows precise measurement of the flow rate. The "room 1" has a large volume (ca. 140 m³) and acts as a pressure chamber. The leakages of this room can be precisely regulated, in range of ca. 500–40,000 m³/h. The quick damper between rooms simulates doors between stairwell and compartment and is automatically opened and closed. The goal of the PDS is to maintain the designed overpressure in "room 1" (e.g., 50 Pa) and react to dynamic changes in the status of the damper (open/close). When the damper is open, the PDS must supply a designed quantity of air to provide a certain flow on the opening. This response must occur within 3 s

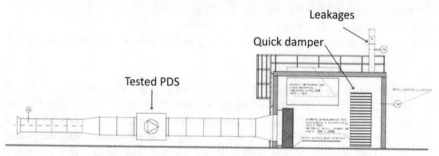

Fig. 11.14 An example of PDS testing facility—room 1 of the PDS test rig at the ITB Fire Testing Laboratory, Warsaw, Poland

from the opening of the damper. Room 2 may be completely open at this point or sealed to mimic a certain level of leakages, depending on the needs.

A single cycle of the test is the so-called dynamic behavior cycle (DBC), Fig. 11.15. This cycle consists of the following:

1. t_1—the opening of the damper (doors), 1 s,
2. t_2—delay to measure the response time of the PDS to rapid change in the pressure, 6 s,
3. t_3—the closing of the damper (doors), 3 s,
4. t_4—time for the pressure to stabilize in the "room 1," 6 s.

The acceptance criteria are as follows:

1. after t_1 the system must obtain 90% of nominal flow rate within 3 s;
2. after t_3 the system must stabilize the pressure in 80–120% of the nominal value.

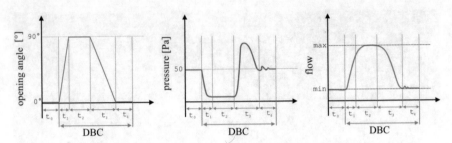

Fig. 11.15 A single dynamic behavior cycle used in the laboratory tests of PDS systems and the expected response of the system

Four types of tests are performed as follows:

1. Functionality (Fu), 20 cycles (DBC), measured.
2. Reliability (Re), 10,000 cycles (DBC), unmeasured.
3. Durability (Du), 20 cycles (DBC), measured.
4. Oscillatory behavior (Osc), rapid opening and closing of the quick damper to various opening levels.

Sample results of Fu and Osc tests are shown in Figs. 11.16 and 11.17. In the Fu test, it can be observed that when the damper fully opens, the pressure in the "room 1" falls almost to zero, and afterwards increases to approx. 5 Pa as the system increases its volumetric flow. After 6 s, the damper closes and the pressure peaks. The PDS shuts down the flow and then finds a new value of the flow that will allow maintaining 50 Pa. In the Osc test, the system behavior is slightly different, as the damper opens and closes in rapid succession. The system is forced to change the flow rate in fractions of seconds, multiple times in a row. After the last oscillation, the damper remains closed, and it is expected that the PDS is still able to maintain 50 Pa overpressure within 3 s of that event. In this case, this requirement was met.

11.5.2 Electromagnetic Compatibility (EMC)

Electromagnetic compatibility (EMC) is, in general, the ability of the device (in this case, the whole PDS kit) to withstand disturbances in the electrical network, and not change the state of operation of the system in such a disturbance. The disturbances can be the signal from another system in the same environment (or part of the system), electromagnetic noise, the unwanted signal caused by natural phenomena (e.g., lighting or electromagnetic discharge, a short nonstationary state in the circuit, electrostatic discharge) or induced by a presence of changing electromagnetic field [22]. It is the principal that the PDS system does not change its state of operation or operational parameters when exposed to electromagnetic disturbance. An effect of EMC failure of the system can be a change in the output fan control signal, leading

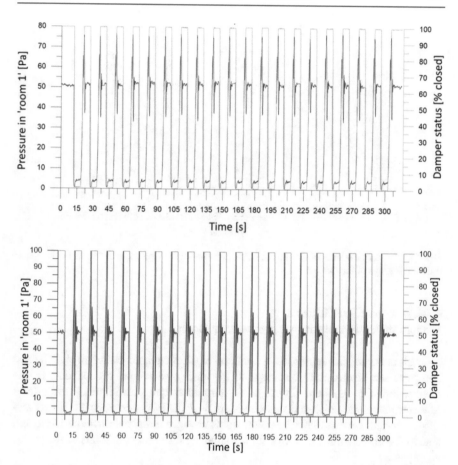

Fig. 11.16 Sample results of Fu test of a PDS (upper—passed, bottom—failed), illustrating how system regulates to reach target pressure after damper closes

to change (increase/decrease) of the fan capacity. Such a change can inevitably lead to changes in the flow, and pressure in the stairwell, potentially leading to life-threatening conditions within the protected space. Therefore, EMC testing is critical for the correct system operations, and must be performed in two states—idle and during operations.

As a part of EMC evaluation, the whole kit (control panel, sensors, fans, and auxiliary components) must go through a series of exposures, that include the following:

- electrostatic discharge (ESD);
- electromagnetic fields at radio frequencies (GTEM chamber, Fig. 11.18);
- induced disturbances at radio frequencies (SINUS);
- fast transient bursts (BURST);

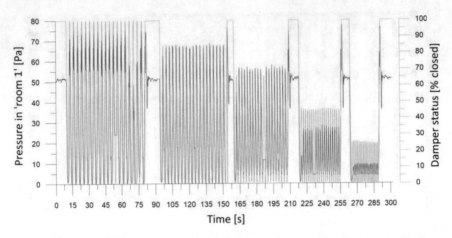

Fig. 11.17 Sample results of Osc test of a PDS, illustrating how pressure in "room 1" changes at different settings of the damper

- slow surges (SURGE); and
- rapid shutdown and changes in the power supply.

Each of these tests represents a common electromagnetic disturbance. ESD test simulates the electrostatic discharge to the casing and metal components of the device. Induced and field disturbances at radio frequencies simulate the effects of electromagnetic noise and other emitters near the PDS components. Bursts and surges are phenomena that accompany quick changes in the electrical network, and shutdowns and changes in power supply simulate different conditions of the power supply unit.

A group of particularly interesting tests are the tests for resistance to electromagnetic fields at radio frequencies. The PDS kits are often installed at the roofs of buildings, in proximity to common sources of electromagnetic fields—4G and 5G transmitters, inverters of ventilation devices, and complex arrays of power supply to other devices. Such fields may affect the electronic components of PDS systems, especially the auxiliary components such as pressure sensors, smoke sensors, etc., and the control units within control panels. These components are tested in gigahertz transverse electromagnetic cell (GTEM cell, Fig. 11.19), which can produce uniform electromagnetic field (in case of ITB GTEM—80 MHz/3000 MHz, up to 30 V/m). The sample is tested for the whole range of frequencies, in idle and working status, and for three orientations. The test is considered successful if no change to operation is observed. It is common that at particular frequencies, the systems start to disturb the pressure measurements, turn on/off or issue a failure signal. Obviously, such events during operation in a fire could lead to significant problems with maintaining the correct operation of the PDS system.

Fig. 11.18 Interior of GTEM chamber of the ITB Fire Testing Laboratory, Warsaw, Poland

11.5.2.1 Commissioning and In Situ Testing

The certification and approval tests performed on a PDS family are necessary to determine the feasibility of the device to perform a particular role in the fire safety of the building. However, even the most perfect laboratory testing method does not immediately guarantee that the performance of a system built into a building will be correct. Thus, there is a need for in-situ testing and commissioning of PDS systems to confirm that the system is configured correctly, it does meet the design goals, and no unexpected (e.g., oscillatory) behavior is observed during typical operation. For this purpose, we have adapted the testing procedure described in Sect. 3.1 to the in situ testing. The tests are performed over an extended time (e.g., 1–2 h), which is in contrast to the standard way such systems are commissioned with the use of single-point measurements of the pressure difference. The advantage of the

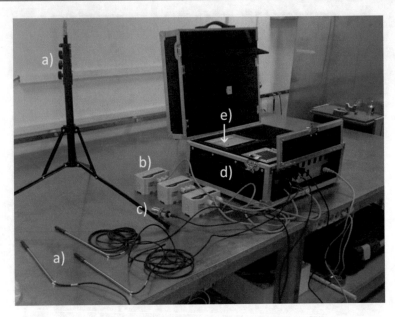

Fig. 11.19 Test kit for in situ testing of PDS systems: (a) Thermo anemometer, (b) pressure meters, (c) ultrasound distance meter, (d) power supply/control unit, and (e) remote visualization/control panel

continuous approach is the lesser impact of external disturbances (human traffic on the stairwell, accidental opening of doors, etc.) on the final result of the test.

The test setup, Fig. 11.19, consists primarily of (a) velocity, (b) pressure, and (c) distance meters. Velocity meters (thermo-anemometers) are placed in the doors to the stairwell or vestibule at three heights to measure the average velocity in the door opening. The pressure measurements are performed between the stairwell and fire compartment, stairwell and vestibule, and vestibule and fire compartment. The distance meter is placed near the door hinges, in a way that allows recording of the door positions. The measurements share the same timeline, and are saved by the control unit, and may be viewed on a Wi-Fi-connected remote panel. The device has its own power supply and is entirely independent of the building systems. Additional measurement devices are used by the person carrying the test—force meters for measuring the opening forces on doors, hand anemometers to measure airflow velocity on air supply and exhaust points and remote weather station usually fit at the roof of the tested building.

The main goal of this approach is to replace common "point measurements" performed to capture a singular value of the pressure difference between stairwell and fire compartment with a dynamic test that allows investigation of the behavior of all of the PDS components. The change of the state of the building elements (e.g., doors, the start of smoke exhaust) can also be flagged in the measurement system in a timeline shared with the measurements. After the initial setup, the personnel

Fig. 11.20 A hot smoke test used in parallel with PDS tests, to determine the system integration and impact of the PDS system on the performance of the smoke control in the fire compartment

needs to carry only the remote panel and is free to take notes and observations. This means that the person may perform a walk through the building and attempt to open doors at different levels to identify the system response. Furthermore, if combined with other methods such as hot smoke testing [23, 24], it is possible to determine the optimal system parameters—pressure difference, maximum flow, that do not interfere with the performance of other safety systems in the building, Fig. 11.20.

An example commissioning scenario for a PDS includes the following:

1. Measurement system deployment and the start of the measurements;
2. Qualitative assessment of the integration of PDS and fire alarm system (FAS);
3. Determination of the ways how air is supplied and removed from the building, and choice of the most onerous pathway;
4. Verification of the reading of the PDS measurement system with the use of precise pressure meters;
5. DBC tests for chosen connections between stairwell and compartments, with the smoke exhaust system operational;
6. Assessment of the system reaction time (in buildings the expected reaction time is 5 s)
7. Oscillation test following the Osc laboratory test procedure;
8. Assessment of the system behavior in unexpected scenarios (e.g., opening doors at different levels at the same time, opening multiple windows in the fire compartment, blocking doors with a fire hose). These are meant to yield important information on system behavior and show areas where some improvements may be done.

An example result of a DBC test of a PDS installed in a building (the same system as shown in Sect. 3.1) is shown in Fig. 11.21. The proposed commissioning

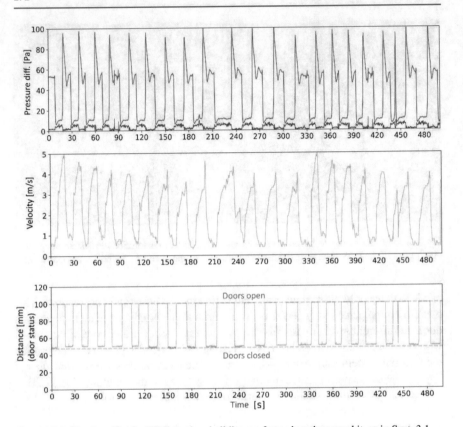

Fig. 11.21 Results of in situ DBC test in a building performed on the same kit, as in Sect. 3.1

method allows not only to regulate the PDS, but also allows for in-depth analysis of the system performance and interaction with other safety systems in the building. Such effort significantly improves system reliability.

11.6   Conclusions

Pressurization systems (PDS) are an essential element of the safety of modern buildings. They are paramount to provide safe escape routes. Many challenges are associated with their design, installation, and commissioning. A new generation of PDS based on active measurement of the pressure difference between the protected and fire compartments provides new possibilities for designing the building safety infrastructure. These systems are less fragile to changes in leakages or the stack effect, providing better performance than traditional systems based on the tenso-metric pressure relief dampers. As complex, autonomous systems, their use should be limited only to tested and certified kits, that passed rigorous performance and

EMC tests. Furthermore, to ensure that systems meet their performance objectives, the in situ testing method is proposed.

The use of PDS still faces some challenges, for which future research is required. These gaps include the following:

- consequences of the change of leakages during the life of a building;
- PDS response to opening the stairwell at multiple levels at the same time, and provision of a "minimal" level of safety in such an event;
- interaction between the PDS and smoke control systems in the building;
- performance of the PDS during a failure of other safety systems of the building, and
- performance of the PDS system after substantial fire damage to the building envelope.

Modern PDS systems can undoubtedly meet their performance objectives; however, they rarely cooperate with other safety systems. One additional research gap is the holistic design of the PDS as a part of the building safety system.

Acknowledgments This research was funded by the Building Research Institute statutory grant financed by the Ministry of Science and Higher Education, grant number NZP-093/2020

References

1. Węgrzyński W, Sulik P (2016) The philosophy of fire safety engineering in the shaping of civil engineering development. Bull Polish Acad Sci Tech Sci 64:719–730: https://doi.org/10.1515/bpasts-2016-0081
2. Klote JH (2016) Smoke Control. In: SFPE Handbook of Fire Protection Engineering. Springer New York, New York, NY, NY, pp 1785–1823
3. NFPA (2015) NFPA 92 Standard for Smoke Control Systems 2015 Edition
4. BSI (2003) BS 7974 Part 4: Components for smoke and heat control systems. Functional recommendations and calculation methods for smoke and heat exhaust ventilation systems, employing steady-state design fires. Code of practice
5. Węgrzyński W (2018) Can smoke control become smart? Arch Civ Eng 64:201–208: https://doi.org/10.2478/ace-2018-0040
6. Klote JH, Milke JA (2002) Principles of Smoke Management, American Society of Heating. Refrigerating and Air-conditioning Engineers Inc., Atlanta
7. Milke JA (2016) Smoke Control by Mechanical Exhaust or Natural Venting. In: SFPE Handbook of Fire Protection Engineering. Springer New York, New York, NY, pp 1824–1862
8. NFPA (2015) NFPA 204 Standard for smoke and heat venting
9. (2007) EN 12101-6:2007 Smoke and heat control systems - Part 6:Specification for pressure differential systems – Kits.
10. Alvarez A, Meacham BJ, Dembsey N, Thomas J (2013) Twenty years of performance-based fire protection design: challenges faced and a look ahead. J Fire Prot Eng 23:249–276: https://doi.org/10.1177/1042391513484911
11. Hurley MJ, Rosenbaum ER (2016) Performance-Based Design. In: SFPE Handbook of Fire Protection Engineering. Springer New York, New York, NY, pp 1233–1261

12. Tofiło P, Węgrzyński W, Porowski R (2016) Hand Calculations, Zone Models and CFD – Areas of Disagreement and Limits of Application in Practical Fire Protection Engineering. In: 11th Conference on Performance-Based Codes and Fire Safety Design Methods. SFPE

13. Bellido C, Quiroz A, Panizo A, Torero JL (2009) Performance assessment of pressurized stairs in high rise buildings. Fire Technol 45:189–200: https://doi.org/10.1007/s10694-008-0078-0

14. Fryda M, Brzezińska D, Dziubiński M (2020) High rise buildings stairwells pressure differential systems tests and improvement solutions. Build Serv Eng Res Technol 014362442096431: https://doi.org/10.1177/0143624420964313

15. Lay S (2014) Pressurization systems do not work & present a risk to life safety. Case Stud Fire Saf 1:13–17: https://doi.org/10.1016/j.csfs.2013.12.001

16. Shi WX, Ji J, Sun JH, Lo SM, Li LJ, Yuan XY (2014) EXPERIMENTAL STUDY ON INFLUENCE OF STACK EFFECT ON FIRE IN THE COMPARTMENT ADJACENT TO STAIRWELL OF HIGH RISE BUILDING. J Civ Eng Manag 20:121–131: https://doi.org/10.3846/13923730.2013.802729

17. Zhao G, Beji T, Merci B (2017) Study of FDS simulations of buoyant fire-induced smoke movement in a high-rise building stairwell. Fire Saf J 1–8: https://doi.org/10.1016/j.firesaf.2017.04.005

18. Kubicki G, Cisek M (2019) How to Protect Staircases in Case of Fire in Mid-Rise Buildings.Real Scale Fire Tests. Saf Fire Technol 54:6–20: https://doi.org/10.12845/sft.54.2.2019.1

19. Klote JH, Milke JA (2002) Principles of smoke management. In: Heating AS of (ed) Principles of Smoke Management. Refrigerating and Air-conditioning Engineers Inc., Atlanta

20. Węgrzyński W (2017) Transient characteristic of the flow of heat and mass in a fire as the basis for optimized solution for smoke exhaust. Int J Heat Mass Transf 114:483–500: https://doi.org/10.1016/j.ijheatmasstransfer.2017.06.088

21. (2020) prEN 12101-6 Smoke and heat control systems - Part 6: Specification for pressure differential systems - Kits

22. Flasza J, Barasiński A (2012) The influence of electromagnetic interference of electrical work in fire. Prz Elektrotechniczny 88:243–244

23. AS (1999) Smoke Management Systems - Hot Smoke Test

24. Węgrzyński W, Krajewski G (2015) Systemy wentylacji pożarowej garaży. Projektowanie, ocena, odbiór, 493/2015. Instytut Techniki Budowlanej

Hybrid Fire Testing: Past, Present and Future 12

Ana Sauca

12.1 Introduction

The impact of climate change, including extreme weather events, more and more limited access to resources, and the increase and aging of population are the world's present and future challenges [16]. The technology is rapidly developing; the population is migrating to larger and prosperous cities, leading to a rapid growth in size and density of the cities. In general, the impact of fire on health, safety, climate, and the economy is an area that must be re-evaluated due to the present and future changes. In particular, the engineering world will adapt to these challenges by developing new materials, new type of structural solutions, or new fire-based evaluation tools.

The new developed materials or structural solutions will require an extensive experimental campaign to comprehend the behavior at elevated temperatures. Preferably, the experimental campaign should be performed under more realistic testing conditions than the current practice of standards testing, and here are just few of the reasons supporting this necessity. Globally, the thermal expansion of the structural elements exposed to fire induces important changes in the structural behavior during the fire test. The changes are dependent on the fire characteristics as the mechanical boundary conditions adopted in the test. Locally, specific phenomenon occurs for some materials when exposed to fire such as the spalling of the concrete. This phenomenon is yet challenging to be numerically modeled; therefore, the fire tests in the appropriate conditions are required to comprehend the action along with the structural behavior. Generally, the fire tests are performed on single members, under standard fire exposures, e.g., ISO834 [10], and constant mechanical boundary conditions (e.g., [44]). However, full-scale testing (e.g., [24])

A. Sauca (✉)
The Danish Institute of Fire and Security Technology (DBI), Hvidovre, Denmark
e-mail: as@dbigroup.dk

© Springer Nature Switzerland AG 2022
M. Z. Naser, G. Corbett (eds.), *Handbook of Cognitive and Autonomous Systems for Fire Resilient Infrastructures*, https://doi.org/10.1007/978-3-030-98685-8_12

of full structural systems revealed different behavior of the structural members, but the high cost makes the practice uncommon. Ideally, a new method is needed to capture the realistic behavior of tested members but at the same time to be affordable and reproducible. Keeping the advantage of testing only parts of the structure, but at the same time to consider the effect of the remainder structure, thus reproducing the results of the full-scale testing, Hybrid Fire Testing (HFT) methodology is a promising technique. HFT uses a hybrid model to update the mechanical boundary conditions of the tested component *online*, that is, during the execution of the fire test. In detail, such a hybrid model combines a physical substructure and a numerical substructure (PS and NS, respectively). The PS is tested in the laboratory by means of servo-controlled actuators equipped with force transducers, whereas the NS is instantiated in a finite-element analysis software or simply represented by a stiffness matrix. An online coordination algorithm ensures compatibility of interface displacements and balance of interface forces between PS and NS. HFT builds upon hybrid simulation, extensively developed for testing the seismic response of structures [38]. The structural response to fire is static and time-dependent, which prevents from crude reuse of methods and algorithms developed for seismic testing.

New fire-based evaluation tools are starting to be developed based on machine learning (ML), which shows high potential to facilitate performance-based fire design of structures [22]. Machine learning (ML) is a subset of artificial intelligence (AI), which after receiving a set of data, is comprehending these data, thus learning to identify the key features for representing certain phenomena embodied within the data set. ML can be a powerful tool in scenarios where mathematical or conventional methods, e.g., finite-element modeling, are limited.

The new emerging testing method and fire-based evaluation tool, i.e., HFT and ML, can provide means in evaluating all of the present and future changes, and a conceptual description of overlapping the two is discussed here.

This chapter is organized as follows. Section 12.2 describes in detail the HFT fundamentals covering the components and procedure, advantages, and challenges, followed by the state of the art. Section 12.3 presents the HFT framework developed at the Danish Institute of Fire and Security Technology (DBI) covering both algorithmic and experimental implementation details, for exemplification purposes. Then, Sect. 12.4 describes the ML fundamentals, followed by Sect. 12.5 where the overlap between the HFT an ML is conceptually discussed. Finally, Sect. 12.6 summarizes the conclusions of the study and provides an overlook to future research directions.

12.2   Hybrid Fire Testing Fundamentals

12.2.1 The Influence of the Mechanical Boundary Conditions on the Fire Test Results

The design of more unique buildings is now enabled by performance-based design methodologies. The modern architectural design is more challenging, and the fire safety community is moving away from the prescriptive methods toward alternative methods that prove compliance with fire safety performance objectives. Computational methods provide invaluable support for this purpose, but there is always a need of experimental test data to validate the models or to provide inputs about new materials or structural system with corresponding failure modes. The testing procedures should therefore provide more realistic conditions and be less costly, which can help engineers make informed decisions about the conditions that can induce severe damage or collapse under fire.

The structural fire behavior of members is dependent on the mechanical boundary conditions provided in the tests or numerical models, and various researchers used various means to underline the statement.

A very comprehensive such example is presented in [28] where the fire performance of steel columns is analyzed under various mechanical boundary conditions. It is well known that the axial restrain prevents the column from axial elongation while rotational restrain opposes any rotations induced due to thermal exposure. As an effect, the rotational restraint to a steel column increases its critical temperature, whereas the axial restrain lowers its critical temperature. Therefore, the paper investigated the effect of various boundary conditions (axial and rotational end restraints), by first performing a detailed literature review of fire tests conducted on steel columns in the past decades, followed by performing new comparative numerical analysis of structural steel columns with various mechanical boundary conditions. The review of the literature indicates that the columns are sensitive to the complex combination of both axial and rotational restraints. Moreover, the restraint degree in real fire events varies because the stiffness of the surrounding structure degrades when exposed to elevated temperatures. Next, the test results of a planar moment-resisting frame when only a singular column was exposed to fire were compared with the simulated results obtained from isolated column models with constant mechanical boundary conditions and static springs at the ends, which mimics the stiffness of the surrounding structure. The results showed that the models with the equivalent static springs experienced buckling sooner compared to the full frame. The static (linear) springs were able to mimic the effect of the surrounding frame on a column in the initial stage of the fire event, but not at the later stage close to failure. The results presented by the authors showed that the simplified mechanical boundary conditions at the end of a steel column (e.g., ideal free or linear spring conditions) do not accurately reproduce the complex restraints of the surrounding frame during fire events. Generally, the mechanical boundary conditions provided in the worldwide performed fire tests of structural members consist either of fixed (constant) degree-of-freedom (DoF) or a constant applied load. Such constant or static boundary conditions are not proper for representing the structural system behavior.

Another such example was presented in [34]. A reinforced concrete beam, part of a concrete structure, was numerically modeled under various idealized constant mechanical boundary conditions, and the results were compared with the numerical model of the concrete frame. The constant mechanical boundary conditions did not capture the full model response.

Previous contributions suggested the use of linear (springs) mechanical boundary conditions so that the limitations posed by the constant mechanical boundary conditions are overcome. Nevertheless, in [35] is shown that a linear mechanical condition also has shortcomings. Using the HFT methodology, the stiffness of the surrounding structure was represented by a constant stiffness matrix. The results showed that the obtained solution began to diverge from the expected solution at the time when the material behavior of the surrounding became nonlinear. A correct solution was reproduced only when the surrounding remained linear for the full duration of the analysis.

Similar type of analysis was also presented in [37]. The global analysis of a 10-story 2D steel frame was investigated when exposed to design fire, which was spreading between the compartments. An isolated beam member was next analyzed as an individual member under various constant mechanical boundary conditions (e.g., roller supports, hinged ends, fixed ends), exposed to the same elevated temperature as in the 10-story 2D steel frame analysis. The results showed that no failure was achieved under the various constant mechanical boundary conditions, as opposed to the global analysis of the 10-story 2D steel frame, where the collapse was initiated. This study underlined the importance of the proper mechanical boundary conditions in general, especially when the fire spreads between the compartments in particular.

One answer of the aforementioned discussion is to perform fire tests on individual members with "active" boundary conditions (also known as HFT), where actuators can be provided at the boundary ends to replicate the interaction of the member-structure for the duration of fire loading. During each time step within the time history, a computer simulation of the surrounding structure can return the forces or displacements to be applied to the tested member by the actuators.

A more detailed description of the HFT methodology is next presented.

12.2.2 Introduction to HFT

There are several available testing methods to assess the fire behavior of the structural systems.

The most common one is the standard fire testing method (or component testing), which assumes testing individual members under constant mechanical boundary conditions, most of the times exposed to standard fires curves (e.g., ISO 834 [10]). The main purpose of these tests was to produce results that are comparable between laboratories around the world. Such tests are relatively easy and economical to execute compared with the other fire testing methods, but the mechanical boundary condition and the loading rate do not reproduce the reality. During standard fire tests, the thermal expansion is free; therefore, no thermal reaction force develop in the member. The main question when performing such tests is whether the fire resistance of the tested member is under- or overestimated for a specific project situation.

Full-scale testing is a second form of performing fire tests. These tests consider the real mechanical boundary conditions and loading rates, and provide important data on the fire response of the structural system. So far this practice is uncommon because of the high cost and only few such tests were performed (e.g., [24]). Building the structural system, installing the measurement instrumentation, and the result analysis require an important amount of work.

Another emerging method is HFT, inspired from substructuring methodology. The hybrid simulation method in general has been widely used in seismic field [32] but not much in other fields. The idea of hybrid simulation is to combine the advantages of component testing method and full-scale testing, meaning that the individual component(s) will be tested while taking into consideration the effect of the reminder global structure (referred to as the surrounding structure). The surrounding structure can be modeled via nonlinear numerical software or a predetermined matrix [34], and its effect on the tested member is taken into account in real time, continuously, with the help of actuators provided at the interface of the tested specimen (the physical point where the substructures were separated). The part of the structure tested in the furnace will be referred to as the physical substructure (PS), while the surrounding as the numerical substructure (NS). The method then consists in ensuring equilibrium and compatibility between these two substructures over the duration of the test. At each time step, data (displacements or forces) are measured at the substructure interfaces and manipulated in such a way that the global response of the structure is reproduced. Due to the fire exposure of the PS and NS, equilibrium and compatibility at the interface are generally not satisfied at the end of the time step. To restore the equilibrium and compatibility at the interface, new computed data, based on the previous time step measured data (forces or displacements) are imposed as the mechanical boundary conditions at the interface of the substructures. At frequent intervals (referred to as time step Δt), the displacements and the forces at the interface of the PS are measured while for the NS are computed, and used to compute the boundary conditions to be updated for the next time step. The procedure to control the PS and NS can be either force control procedure (FCP) or displacement control procedure (DCP). The FCP refers to the situation when the reaction forces are imposed on the PS and NS, and displacements are the measured reactions. The DCP assumes imposing displacements on the PS and NS while measuring the reaction forces.

The key components in hybrid simulation will be next presented in detail along with the HFT procedure.

12.2.3 Components and Procedure

Several components interact with each other during a HFT.

- **Physical substructure (PS)** is the experimentally tested member of the structure during the HFT. The PS is tested because of its unknown behavior, which cannot be analyzed via other available tools such as finite-element method (FEM).

- **Numerical substructure (NS)** is the surrounding structure of the PS, which in most cases is numerically modeled during the HFT or represented by a predetermined matrix if possible [34]. Because the HFT must be performed in real time, the calculation time of the NS must be as reduced as possible.
- **Transfer system** refers to the actuators used to control the mechanical boundary conditions of the PS during the test, and they represent the stiffness of the NS. DCP or FCP can be used depending on the case study.
- **Data-acquisition system** is responsible with the real-time acquisition of the data during the test. The most common sensors used during the HFT are:

    (i) Thermocouples for temperature measurements.
    (ii) Displacement transducers and/or inclinometers for displacement and rotation measurements.
    (iii) Load cells for force measurements.

DCP is the most commonly used procedure, and this is because the FCP might be instable when there is a change in the behavior of the tested element such as for example the softening of the material. Nevertheless, the FCP seams to be the best approach for the rigid structures, when the interface displacements are smaller than the resolution/precision of the transducers; therefore, it would be very difficult to control [38].

DCP is next described in detail, since it is later used in Sect. 12.3. A similar approach is also applicable to FCP. We define the interface as the physical location where the PS and NS are separated.

Step 1 Before the start of the HFT, the entire structure (with the PS included in the analysis) is analyzed at ambient conditions under the fire load combination (called here "external loads"). The interface forces (e.g., bending moment, axial force, shear force, torsion moment) and displacements (including rotations) are extracted at the interface level.

Step 2 Run the analysis for the NS subjected to the external loads and the interface displacements resulted in Step 1.

Step 3 The PS is placed in a furnace (or other setup, depending on the heating conditions of the laboratory) and loaded with the external loads and the interface displacements computed in Step 1. The PS is now ready to be exposed to fire. Note that when the PS is loaded before the start of the fire, it may be possible to get a different value of reaction forces than the ones computed in Step 1. In this situation some additional calculations are needed to restore the interface equilibrium and compatibility at ambient temperature, before the start of the fire.

Step 4 Start the fire exposure.

Step 5 During one time increment Δt, the interface displacements of the PS are kept constant, and this induces the variation of the interface reaction forces. At the end of the time increment, the interface reaction forces of the PS are measured.

Step 6 During one time increment Δt, the interface displacements of the NS are constant, and the interface reaction forces of the NS are computed.

Step 7 Use the measured and computed interface forces and displacements of the PS and NS to compute the new boundary conditions to be imposed on the PS and NS.

Step 8 Impose the new computed displacements on the PS and NS. At the end of this step, the equilibrium and compatibility at the interface must be ensured.

Step 9 Repeat Steps 5–8 for each time increment Δt for the entire period of the test including the cooling phase if relevant.

The following problems must be solved when modeling the behavior of structures subjected to fire: (i) fire (temperature) development in the compartment; (ii) the thermal response of the structure; and (iii) the mechanical response of the structure. A comprehensive review of coupling strategies between the mentioned parts is provided in [43]. Most of the times the fire development in Computational Fluid Dynamics (CFD) software is performed first, followed by the thermal response of the structure and then the mechanical response of the structures. This is mainly because the spatial scale and the time scale are different to solve problems within CFD and finite-element (FE) method. For the same reasons, the HFT presented in the literature focused mainly on performing one-way hybrid fire tests. One-way coupling HFT refers to the cases when the fire development in the compartment(s) and the thermal response of the NS are performed before the start of the HFT. Two-way coupling HFT implies that the fire development, the thermal and the mechanical response of the NS are performed in real time during the HFT. Two-way coupling would be possible with the use of performant computers for the CFD software, thus, allowing performing the hybrid fire test in real time. Another solution to simulate the fire development would be to develop and use other tools that are less computationally costly, such as for example ML.

In the current study, one-way coupling is considered. Only limited applications are available in the literature when considering the two-way coupling (e.g., [48]) and most of the time is applied to simplified systems without involving CFD.

12.2.4 Advantages and Challenges

HFT technique combines the analytical and the experimental approaches to investigate the behavior of the structural system.

The most important advantages are summarized here below:

- HFT consists in subdividing a structure into two substructures: (i) the NS characterized by a well understood behavior reliably modeled aside and (ii) the PS characterized by an unknown behavior to be modeled aside, thus obtained from a physical test in a laboratory. The uncertainties will be reduced if the numerical model of the unreliable parts is replaced with actual physical components.

- Experimental and numerical models can take advantage of the different capabilities available in different laboratories using geographically distributed hybrid
simulation.
- The global behavior of structures can be reproduced only by means of HFT. This
method presents the advantage of providing the real failure time and mode.

The HFT method presents advantages compared with the other existing methods,
but several challenges still need to be addressed by conducting further research in
this field.

The challenges of using HFT are next presented:

- During the HFT, data are measured and imposed on the PS every time step.
The considered data-acquisition system and transfer system must be suitable for
various case studies in various configurations.
- Each implementation of HFT done so far is dependent on the testing site.
The software framework is difficult to be adapted to general various structural
problems and even to various laboratories.
- There is a need for a common software framework for developing and performing
HFT. The object-oriented software framework should be robust, transparent,
easily extensible, and environment independent.
- Using HFT, there are different sources of errors that can affect the accuracy of the
results: (a) modeling errors (e.g., discretization process); (b) experimental errors
generated by the control and transfer system; and (c) unappropriated method of
performing hybrid simulations. All these errors can lead to instability during the
test.
- Depending on the stiffness of the substructures, tests with stiff PS can be difficult
to execute under DCP. One solution to stiff structures can be the use of FCP.
This means that the method should be suitable for this specific problem and the
measurement instruments must be able to measure data characteristics for these
specific cases, e.g., stiff PS yield small displacements, difficult to be measured
with the traditional instruments.
- If the future geographically distributed hybrid simulation can be used, then the
networks delays and breakdowns should be avoided, and the communication
speed needs to be improved as much as possible. A delay in communication
can have a crucial effect on the results, since the HFT can be only performed in
real time.

12.2.5 State of the Art

Schwarz in [40] applied the strategy "divide and conquer" to a mathematical
problem where a complex domain was divided into two simple parts (a circle and
a rectangle) in order to find a solution for the associated differential equations of
the combined domains. The analytical solutions of both subdomains were known.
To find the solutions on the interfaces, an iterative way was used to converge to

the solution on the complex domain. Therefore, the idea of domain decomposition can be seen as the ancestor of substructuring, where the subdomains are in fact the components of the total structural system.

Searching for new methods to evaluate the dynamic behavior of structures, especially the seismic performance of large-scale structures, the hybrid simulation (HS) testing technique started to be developed and widely used in seismic field [38].

Some differences between the fire versus seismic approach are next discussed.

One of the main challenges in the fire field is related to the necessity to conduct the HFT in real time; except for metallic elements in which a uniform temperature distribution can develop, the temperature distribution in most elements is highly non-uniform and time-dependent and cannot be scaled down in time (this is particularly significant for concrete or timber elements) as opposed to seismic field, where slow tests, rapid tests, real-time tests, and smart shaking table tests are possible. The real-time HFT requires a real-time testing facility with suitable transfer and data-acquisition system and interface communication.

In fire field, the PS is exposed to elevated temperatures, therefore, the transfer system and data-acquisition system must be protected from fire, which can induce some limitations when developing the test setup.

Another difference between the two fields is the equation that is solved during the hybrid simulation. When the PS is exposed to fire the expansion develops slowly in time; therefore, usually a static equation is solved. The static approach is nevertheless characterized by the fact that the stiffness of the PS is required in the calculations. This is a challenge because the PS's stiffness changes due to elevated temperatures and is challenging to be measured. Moreover, for floating subdomains (which experience rigid body moves) additional calculations need to be done to find the invert of the singular stiffness matrix. For the seismic field a dynamic equation is solved during the hybrid simulation. Depending on the type of the solver, the evolution of the stiffness matrix during the test is not always requested.

All the aforementioned challenges in the fire field made the progress slower; nevertheless, an increase interest on HFT is observed.

The first HFT was performed in 1989 as reported in [12], but the results are not publicly available.

The first reported HFT [13] was performed on a column specimen, which was experimentally tested as part of a simulated building environment. The mode of action between both parts was exemplified on a single-DoF basis, i.e., the axial column force was measured and adjusted continuously to the model force, which was represented through a—not necessarily constant—stiffness, in displacement control. A realistic case study based on a restrained column was tested a few years later using the same methodology in collaboration with the University of Coimbra, Portugal [14].

The next HFT was performed in CERIB, using actuators in force-control mode. This experimental campaign was documented in [31] and the PS consisted of a concrete slab whereas the NS was a surrounding one floor concrete building. The behavior of the NS was modeled by a constant predetermined matrix, which was calculated before the test. One axial DoF and two rotational DoFs were controlled.

The first attempt to extend the HFT to nonlinear NS was done by Mostafaei and presented in [20, 21]. The HFT was performed on the first-floor central column part of a 3D concrete frame. The column was tested in a furnace (PS) while the surrounding (NS) was exposed to fire and was numerically modeled in the nonlinear finite-element software SAFIR [7]. For each time step, the interaction between the PS and NS was manually enforced, and axial force in the column was controlled.

However Sauca et al. in [34, 35] showed that the HFT methodology considered in the aforementioned cases performed on full size members (referred to as the first generation method) can be unstable. The methodology was only stable for certain ratios between the NS and PS stiffness. A new modified Newton–Raphson algorithm to solve the static equilibrium equation of the hybrid model relying on an estimate of the initial tangent stiffness matrix of the PS was proposed by Sauca et al. in [33], which is stable independently of the stiffness ratio between the NS and PS. A continuous iteration is necessary to achieve equilibrium due to the continuous fire effect on the PS and NS. The stability of the method was numerically proven and validated. The same method was later used in [37] when the communication between the nonlinear finite-element software SAFIR was enabled with OpenFresco [2] and OpenSees [1]. A prototype case study was first experimentally tested, where the PS was cold and the NS was exposed to fire. The simplified experiment helped validating the communication framework. Next, a 10-story steel frame exposed to spreading design fire was analyzed in the numerical (virtual) environment using the HFT.

Pinoteau et al.[25] performed a series of HFT using the first generation method. The PS was a steel column and the numerical substructure was a steel frame, modeled in SAFIR. Only one degree-of-freedom (i.e., the horizontal displacement of the column) was controlled during the test, and the communication between the PS and NS was done via text files. The exposure of the PS and the values of the time step were varied in the experimental campaign. The NS was also exposed to fire. The tests confirmed that the first generation method is unstable, depending on the stiffness ratio between the NS and PS.

As an improvement, Wang et al. in [47] proposed a single-iteration variant to the modified Newton–Raphson method whose static balance equation accounts for residual forces from the previous time step. The same paper reported an experimental verification where the NS is linear.

Memari et al. [17] performed a hybrid test of a small-scale braced frame using the same communication as in [47]. The test showed the applicability of hybrid fire simulation for fire following earthquake.

A HFT based on the Newton–Raphson algorithm and performed on a small-scale specimen is presented in [39]. The PS, i.e., steel coupon specimen, was tested in a universal testing machine, inside an electric furnace while the NS was analyzed in ABAQUS [4] FE software. The work provided a thorough discussion of all numerical and experimental issues related to the coupling of substructures and reported a comprehensive verification of the testing procedure based on linear and nonlinear NS implementations. The study showed that updating the tangent stiffness matrix of the PS strongly reduces the number of iterations required by the Newton–

Raphson solver to achieve convergence. However, the author pointed out that the tangent stiffness matrix estimate of the PS is already noisy for a single-DoF system, suggesting that multiple-DoF implementations should rely on a constant stiffness matrix.

In [48], another HFT performed on a small-scale steel coupon is presented. The objective was to extend the mechanical hybrid simulation of OpenFresco and OpenSees (software commonly used for seismic hybrid simulation) by introducing the temperature DoF. The NS was modeled in OpenSees while OpenFresco was the framework used for the experimental setup and control.

A static solver for HFT was presented in [42], a method based on the finite-element tearing and interconnecting (FETI) algorithm. The validation of the method was done in a numerical environment considering a moment-resisting steel frame.

The proportional (P) and proportional integral (PI) controller for HFT was discussed in [19]. It was shown that the P controller was stable, and the stability was formally expressed. The PI controller was developed to eliminate the major drawback of the P controller, i.e., the inability to adapt to changes of stiffness in the system. The two controllers were compared in a virtual hybrid fire test performed on a 2D steel frame. The effects on stability of the main sources of errors, i.e., delays, experimental errors, and estimation of the stiffness of the PS, when using the PI controller mentioned above were investigated in [18]. The case study consisted of a steel beam (PS) that interacted with a steel frame (NS), exposed to spreading fire in the horizontal adjacent compartments. It was shown that the underestimation of the stiffness can lead to instability. The experimental errors must also be limited, nevertheless, the impact on the stability can be neglected compared to the effect of the stiffness estimation. The study also showed that the PI controller is robust to high values of delays.

A method to account for the continuously changing stiffness of the heated physical substructure by applying a Broyden update of the stiffness matrix to various formulations for actuator control was described in [27]. The presented method was validated in a virtual environment for displacement and force controlled procedure.

The analysis of a composite beam exposed to fire in several approaches was presented in [11], and one of them was hybrid simulation. The composite beam was modeled in 3D while the surrounding was modeled in 2D. For the hybrid simulation approach, OpenFresco and OpenSees were considered. The Cardington restrained beam was also modeled, and it was shown that the hybrid simulation approach is an effective method to study the global behavior of the structure.

A novel testing procedure using an adaptive controller was presented in [30]. The testing framework assumed that the controller is not aware of the displacement of the physical substructure, instead it only reacts to the measured non-equilibrium forces between the two substructures. The authors stated that the advantage of the adopted methodology is the achievement of a better precision in some cases, compared with the other existing methods and there is no need to estimate the tangent stiffness of the physical substructure. A case study is presented where the potential sources of the experimental errors were taken into account and a small-scale preliminary experiment was done, where the PS was cold and the NS was exposed to elevated temperatures.

A real-time hybrid fire simulation method based on the dynamic relaxation and partitioned time integration was proposed in [5]. As an equivalent dynamic solution method, dynamic relaxation allows for coupling substructure equations of motion by using a partitioned time integration approach. Minimal data exchange between substructures and negligible computational overhead plus ease of reusability of verified finite-element software makes the proposed algorithm suitable for coordinating real-time hybrid fire simulations. The hybrid fire simulation of a virtual steel frame case study was reported as a validation example.

12.3   Hybrid Fire Testing Case Study

For exemplification purposes, this section presents the HFT capability developed at DBI and the results of a benchmark case study performed on the gypsum based wall assemblies. First, the considered algorithm is shortly described, followed by the description of the test setup, including the quantification of the experimental uncertainties. Finally, experimental results of a single-DoF benchmark problem are presented at the end of this section. More details about this case study can be found in [36].

12.3.1 Time Integration Algorithm

A real-time HFT method based on the dynamic relaxation and partitioned time integration is considered in this study. A simplified description is given here and the reader is addressed to [5] for more details.

As an equivalent dynamic solution method, dynamic relaxation allows for coupling substructure equations of motion by using a partitioned time integration approach. Minimal data exchange between substructures and negligible computational overhead plus ease of reusability of verified finite-element software makes the proposed algorithm suitable for coordinating real-time HFT.

The dynamic relaxation (DR) method as a static solution algorithm approximates the solution of a static problem by means of an equivalent dynamic system with both mass and damping matrices derived from the initial tangent stiffness matrix [45]. An explicit Newmark scheme integrates the time-history response of the equivalent dynamic system avoiding the factorization of the tangent stiffness matrix. The first advantage of DR is that, as an equivalent dynamic solution method, it allows for coupling PS and NS equations of motion by using a two-stage solution approach based on partitioned time integration [9]. In detail, the first stage is referred to as a free solution and entails the calculation of PS and NS responses by neglecting the coupling conditions. This means that PS and NS equations of motions are solved independently in parallel. The second stage of the solution is referred to as a link solution and entails the calculation of a set of Lagrange multipliers that restore interface velocity compatibility between substructures. The link problem involves the solution of a system of linear equations with interface velocities as unknowns, which adds a small computational overhead to a simulation time step. Furthermore, the limited data exchange between substructures makes the proposed algorithm suitable for implementation of the method in existing hybrid simulation middleware, e.g., OpenFresco. The second advantage of DR is the possibility to overcome issues related to singular stiffness matrices caused by material or geometric nonlinearities. Namely, simulation using a conventional static solution scheme may stop when buckling of a member in statically indeterminate structures occurs. This, however, does not mean that a global instability of the entire model occurred. For example, the SAFIR finite element software implements a dynamic analysis based on the Newmark time integration scheme to solve the aforementioned issue.

The coupled equations of motion of the hybrid system are solved with a partitioned time integration approach. An interface force represented by a Lagrange multiplier imposes velocity compatibility between PS and NS. The coupled solution is computed with a two-stage approach where interface forces are computed by solving a linearized dual problem.

For a detailed analysis of the HFT coordination algorithm, the reader is addressed to [5].

12.3.2 Description of the Test Setup and Quantification of Experimental Errors

The DBI-HFT setup was conceived to test a gypsum plasterboard (GPB)-based wall assembly (PS) of the approximate size of 200×950 mm, of which 500 mm only are exposed to elevated temperatures ensuring that fixtures stay outside the heated zone. Figure 12.1 provides an overview of the setup with the main components highlighted. As it can be observed, the PS is installed on a loading frame that mates a 20 kW electric furnace. A screw electric actuator controls the PS elongation, which is measured using displacement transducers. The temperature controller of the furnace relies on three shielded 3 mm Type K thermocouples. No temperature correction for bead radiation is performed. One additional Type K thermocouple measures the temperature on the unexposed side of PS. As a result, the PS is assimilated to a single-DoF bar element subjected to a uniform temperature. Both PID displacement and temperature controllers are implemented in the same INDEL SAM4 real-time system that solves the equation of motion of the hybrid model and simulates the NS. Also, data acquisition is performed by the same real-time system using INDEL COP-IT modules for thermocouples and displacement transducers. The coordination algorithm presented in detail in [5] and briefly described in Sect. 12.3.1 is implemented in Python, where a dedicated API module provides communication with the INDEL real-time system. For additional information, the reader is addressed to the INDEL AG website (www.indel.ch/).

Figure 12.2 presents an overview of the experimental setup used to apply or measure temperatures, displacements, and force on the PS (1). The designed loading

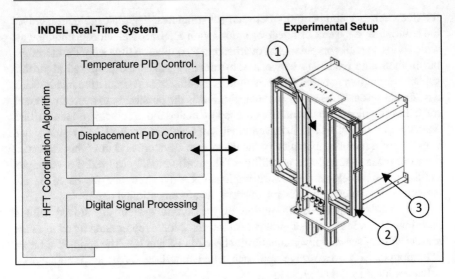

Fig. 12.1 DBI-HFT setup: (1) PS; (2) loading frame; (3) electric furnace

frame is attached to the furnace, allowing exposure on one side of the PS. The furnace test apparatus was developed by DBI and designed to perform fire tests on small-scale specimens, having the exposure dimensions of 500×500 mm. The DBI furnace can deliver a maximum of 20 kW by electrical heating and is capable of running multiple fire curves. The furnace temperature is controlled using three shielded 3 mm type K thermocouples positioned close to the specimen at various positions inside the furnace. The skeleton of the loading frame is symmetric and is made out of aluminum members (3). The PS is screwed to the fixations (4) at both ends. The bottom fixation is connected to the rotary motor (7), which controls the axial displacement of the PS. In between the rotary motor and the bottom fixation, the S-load cell (6) measures the axial restoring force of the PS. The axial displacement of the PS is measured using linear displacement transducers (5). Three displacement transducers are used to measure the displacement of the board at predefined locations while one displacement transducer is used to measure the thermal elongation of the aluminum profile. In order to keep a low temperature of the aluminum members, calcium silicate boards (2) are fixed on the test rig and along with the PS will be the only elements directly exposed to heat. The measurement instruments are placed at the bottom of the test rig, away from the hot zone to minimize the effect of temperature on the measurements. The temperature of the unexposed side of PS is measured using type K thermocouple model FT-50511102-0200.12.GG.D. The length of the PS is cut to 950, where only 500 mm is exposed to elevated temperatures. The GPB is cut longer than the heated zone dimension so that the fixations are outside of the heated zone.

The sensors used for the HFT were characterized prior the start of the testing program, and the summary is presented in Table 12.1.

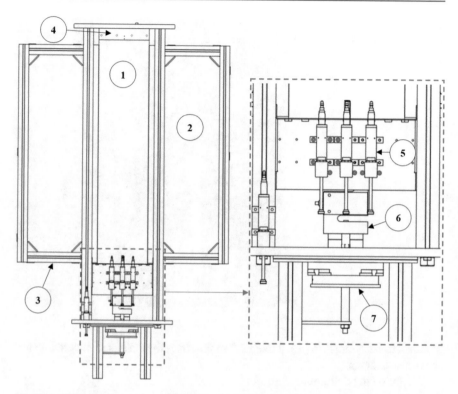

Fig. 12.2 HFT experimental setup at DBI: (1) PS; (2) Calcium silicate board; (3) Aluminum frame; (4) PS fixation; (5) Linear displacement transducer; (6) S-load cell; (7) Rotary motor

Table 12.1 Sensors used in the experiment

Sensor description	Manufacturer	Model	Full-scale value
Specimen thermocouples	GUENTHER	FT-50511102	0–1250 °C
S-beam load cell	PT-global	PT4000	0–500 kg
Linear transducer	Megatron	SPR 18	0–50 mm

Prior the HFT tests all sources of experimental errors were defined conform with Taylor and Kuyatt [41]. Type A and/or Type B uncertainties, combined standard uncertainties, and total expanded uncertainties were estimated for each sensor and actuation device. Type A uncertainty is evaluated using statistical methods, and Type B uncertainty is estimated by other means such as the information available in the manufacturer's specifications, from past-experience, or engineering judgment. The combined standard uncertainty is estimated by combining the individual uncertainties using "root-sum-of-squares." The expanded uncertainty is then computed by multiplying the combined uncertainty by a coverage factor of 2 corresponding to an approximately 95% confidence interval. All uncertainties are

Table 12.2 Uncertainty in the experimental data

Measurement component	Estimation method of uncertainty	Component standard uncertainty	Combined standard uncertainty	Total expanded uncertainty ($k = 2$)
Temperature (specimen)				
Calibration	Type B	±0.55%		
Installation	Type B	±5.60%	±5.73%	±11.47%
Random	Type A	±1.10%		
Load				
Zero	Type B	±0.33%		
Linearity	Type B	±0.02%		
Installation	Type B	±0.38%	±1.17%	±2.34%
Random	Type A	±1.00%		
Repeatability	Type A	±0.34%		
Displacement				
Linearity	Type B	±0.10%		
Installation	Type B	±0.38%	±0.40%	±0.80%
Random	Type A	±0.06%		
Repeatability	Type A	±0.07%		

assumed to be symmetric (±). Table 12.2 summarizes the estimated uncertainties of the measurements.

The following definitions are used:

- Calibration, zero, linearity: Uncertainties from known sources of error and derived from instrument specifications (Type B).
- Installation: Uncertainty due to installation and estimated based on engineering judgment (Type B).
- Random: Uncertainty due to random, unpredictable variations in the measurement process during a typical steady-state period and derived using the standard deviation of the residuals from the mean value of the measurements (Type A).
- Repeatability: Uncertainty when measuring the same point multiple times during a typical steady-state period (Type A).

The thermocouples are glued on the unexposed side of the PS with the help of a pad. The installation effect on temperature measurements was evaluated considering the following parameters: welding technique of the thermocouples; weight, thickness, and density of the pad; and type of the glue, respectively. The installation effect on the load and displacement measurements was evaluated considering an installation error of 5 degrees.

12.3.3 Verification of the HFT Setup

Displacement control and restoring force measurement accuracy are crucial to the stability of the coupled simulation. Noteworthy thermal loading can be assimilated to an equivalent mechanical load from the perspective of the HFT. Accordingly, the verification of the DBI-HFT setup was performed at ambient temperature using an M8-threaded steel rod of 950 mm length clamped at both ends as PS. In this regard, Fig. 12.3 shows the steel rod installed on the DBI-HFT setup. The NS consisted of a linear single-DoF spring of $K^N = 300$ N/mm stiffness (same as the PS) connected in parallel to the PS.

The hybrid model was subjected to a compressive load of $f_{ext} = 2000$ N applied as a linear ramp in 200 steps. The integration time step was set to 2 s. Such a load was sufficiently large to cause buckling of the PS and, therefore, a highly repeatable nonlinear response of the hybrid model. The HFT coordination algorithm reported in Sect. 12.3.1 was used to solve the coupled equation of motion.

Figure 12.4 presents the total internal force developed by PS and NS altogether to the applied external load. As can be appreciated from the plots, internal and external forces are almost perfectly balanced demonstrating the effectiveness of the proposed coordination algorithm in presence of experimental errors. Note that $f_{int} = r^P + K^N u^N$ where r^P is the restoring force of the PS while $K^N u^N$ is the restoring force of the NS, with K^N being the stiffness of the NS and u^N is the displacement of the NS.

Fig. 12.3 Threaded steel rod (PS) installed on the DBI-HFT setup

Steel bar

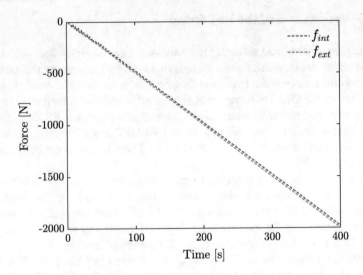

Fig. 12.4 Time histories of the hybrid model restoring force and the applied external loading

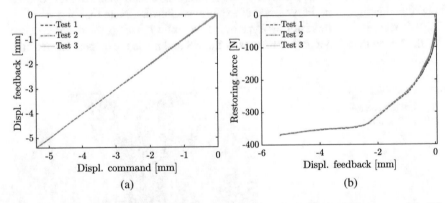

Fig. 12.5 Verification of the DBI-HFT setup: (**a**) command vs. feedback displacement of the PS (control accuracy); (**b**) feedback displacement vs. restoring force of the PS (test repeatability)

A series of three nominally identical HFTs tests were performed as presented in Fig. 12.5.

Figure 12.5a compares the command displacement versus feedback displacement of the PS for all the three tests. The DBI-HFT setups provide a fairly accurate displacement control accuracy.

Figure 12.5b shows the feedback displacement versus restoring force of the PS for all three tests, testifying the high repeatability of the tests. The restoring force plateau indicates the development of buckling in the PS.

12.3.4 HFT of a GPB-Based Wall Assembly

12.3.4.1 Description of the Case Study

The case study illustrated in Fig. 12.6 is next analyzed by means of HFT using the DBI-HFT setup.

The wall assembly is made of GPB, insulation, and steel studs. The GPB is characterized by 12.5 mm thickness and 8.3 kg/mm^2 surface density. The initial tangent stiffness of the board is $K_{PS} = 4140$ N/mm, and it was evaluated by performing a series of monotonic tests at ambient conditions. The cross-section area of the C.100.50.12 steel stud is 185 mm^2 [3].

The hybrid model of the GPB-based wall assembly corresponds to a singe-DoF system, free to displace in the axial direction. During the HFT, the GPB (PS) is tested because of its unknown thermomechanical behavior while the steel studs (NS) are represented by a constant elastic stiffness because of their fairly well known

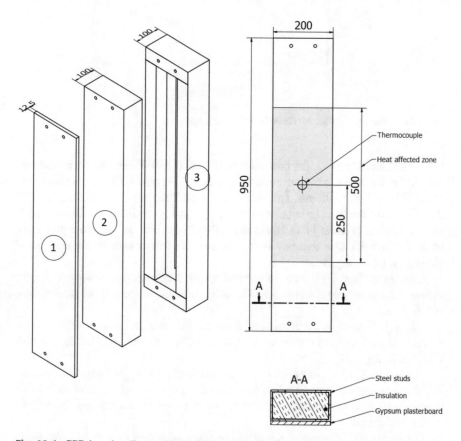

Fig. 12.6 GPB-based wall assembly (hybrid model): (1) Gypsum plasterboard (PS); (2) Insulation; (3) Steel studs (NS). The gypsum plasterboard (PS) is tested using the DBI-HFT setup, while the two steel studs (NS) are numerically simulated by means of a single-DoF linear elastic spring

Table 12.3 Parameters of the HFT tests

	GPB	K^N	Fire
ID	type	(N/mm)	exposure
Test 1	Standard	117600	ISO 834
Test 2	Standard	117600	Ramp-&-Hold-600
Test 3	Standard	117600	Ramp-400
Test 4	Standard	10000	Ramp-400

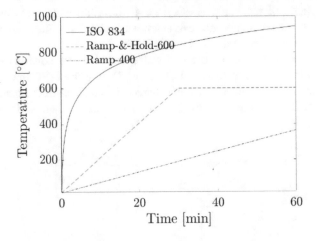

Fig. 12.7 Time–temperature curves in the HFT campaign

behavior. The stiffness of the two studs is $K^N = 117600$ N/mm, and it is assumed that the NS remains at ambient temperature during the entire HFT. The fire exposure of the PS will induce thermal strains/stresses in the PS, which will be restrained by the stiffness of the NS. During every time step, which is equal to 2 s in this example, the mechanical boundary conditions are updated based on the real-time properties of the PS and NS. One thermocouple measures the temperature on the unexposed side of the PS during the HFT.

A series of four HFT were performed under various fire exposures and with different stiffness values of the NS. The characteristics of the test are summarized in Table 12.3.

Figure 12.7 presents the considered time–temperature curves for the experimental campaign. Three fire curves were used, namely, ISO-834 [10], Ramp-&-Hold-600, and Ramp-400.

The Ramp-&-Hold-600 fire curve assumes heating the specimen with around 20 °C/min up until 600 °C and then the temperature will be kept constant for another 30 min. This type of heating condition allows reaching the steady-state conditions across the section.

The Ramp-400 fire curve exposes the specimen at heating rates of approximately 6 °C/min for one hour long.

12.3.4.2 Discussion of Experimental Results

This section presents the experimental results of the HFT described in the Sect. 12.3.4.1.

Figure 12.8 shows the evolution of temperature on the unexposed side of the PS. During Test 1, the thermocouple detached from the specimen at around 20 min, where a drop in temperature can be observed, and this is because the paper on the unexposed side of the PS burned. During the Test 2, the temperature on the unexposed side of PS was constant for the last 20 min of the test, indicating that a steady-state condition was reached across the section. For Test 3 and 4 the maximum temperature on the unexposed side was around 100 °C.

Restoring force and displacement histories of both PS and NS during the Test 1 are reported in Fig. 12.9. An initial expansion of the PS was observed during the Test 1 (for approximately 10 min since the start of the test), which led to the build of compression force in the PS. The compression force was induced due to the

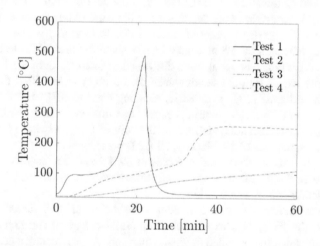

Fig. 12.8 Measured temperature on the unexposed side of the PS

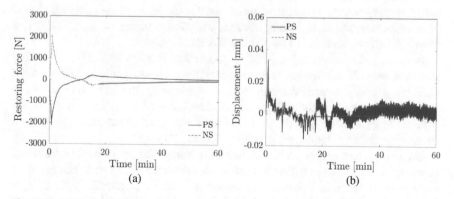

Fig. 12.9 Test 1 time histories of PS and NS (**a**) restoring forces; (**b**) displacements

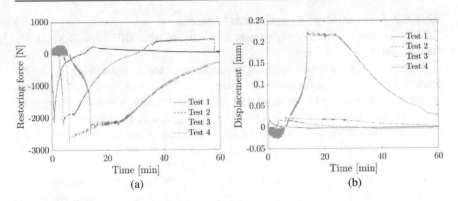

Fig. 12.10 Time histories: (**a**) restoring forces of the PS; (**b**) displacements

restraint given by the stiffness of the NS. With the continuation of the temperature rise, the board started shrinking and this led to the build of tensile force. The tensile force along with the degradation of material due to high temperatures led to the formation of cracks in the PS at around 16 min since the start of the test. Once the crack started forming, the axial force decreased in intensity until the board could not carry any load. From the standpoint of the accuracy of the HFT, the restoring forces of the PS and NS are symmetric, meaning that the HFT model satisfies the static equilibrium. However, control accuracy was limited for such a small range of displacement, causing a noisy residual on the compatibility between PS and NS, as shown in the same Fig. 12.10. Notably, all HFTs were stable.

Figure 12.10 shows the restoring forces of the PS and displacements for all the four HFT tests. For simplicity reasons, only the restoring forces of the PS are plotted, but more details can be found in [36].

From the standpoint of the structural behavior of GPB-based wall sub-assemblies, the PS always experienced compressive load in the initial phase of the experiment due to restrained thermal expansion. After reaching a peak of 2100, 2300, and 2750 N in Test 1, 2, and 3, respectively, a restoring force drop was systematically observed because of shrinking. Tensile force, along with degradation of material strength at high temperatures, triggered cracking of the GPB, which further developed until failure, for Test 1 and Test 2 only.

In detail, failure occurred at about 16 min during Test 1, whereas crack opening started at about 36 min during Test 2. Paper on the unexposed side of the PS ignited during Test 1 only, causing complete loss of load-bearing capacity. No tensile force and PS failure occurred during Test 3. It is interesting to note that Test 3 is associated with the largest restoring force peak and the lower temperature rate suggesting a possible strain/temperature rate dependency in the PS response.

Both Test 3 and 4 were performed using the Ramp-400 time–temperature curve. However, a much smaller stiffness characterized the NS during Test 4. The restoring force peak of Test 4 is delayed with respect to Test 3 counterpart. Also, a larger displacement response during Test 4 is measured, which is consistent with the

(a)   (b)

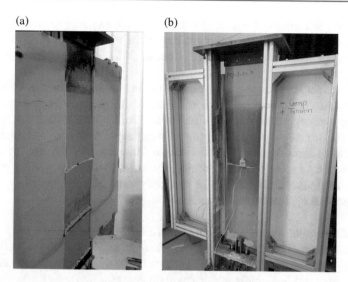

Fig. 12.11 PS at the end of the Test 2: (**a**) exposed and (**b**) unexposed sides

reduced NS stiffness. At the displacement range of Test 4, the control accuracy of the HFT-DBI setup was fairly sufficient to ensure negligible residuals on the displacement compatibility between PS and NS.

Figure 12.11 shows pictures of the exposed and unexposed sides of the PS at the end of Test 2.

The fire behavior of the GPB-based wall assemblies is highly dependent on the GPBs thermomechanical properties. Nevertheless, the available numerical tools to model the thermomechanical behavior of GPB-based walls have a limited applicability, i.e., these tools typically cover only heat and mass transfer processes, with limited capability of performing structural analysis [29]. Therefore, the fire resistance of GPB-based walls is mostly assessed by performing expensive large-scale fire tests.

The DBI-HFT setup enables the investigation of the GPB-based walls behavior at a reduced cost by testing only parts of the GPB-based wall but at the same time considering the effect of the surrounding structure in real time, thus simulating the results of the full-scale test.

12.4   Cognitive and Autonomous Systems

In recent years, machine learning (ML) algorithm was used in solving domain specific problems in various fields of engineering, such as monitoring natural disasters [26] or evaluating fire resistance and identifying damage/spalling magnitude in reinforced concrete (RC) structures [22].

ML field is a subset of artificial intelligence (AI), and deep learning is a subset of machine learning. A good understanding of the topic can be found in [8]. Data science is the field that uses techniques from AI, ML, deep learning, and computer science.

ML handles large amounts of data to predict an outcome or extract insight from the data. The specialized literature describes three general categories of ML: supervised, unsupervised, and reinforcement learning. Supervised learning uses labeled data to develop a relationship between input data and a desired output. Unsupervised learning uses unlabeled data to determine how the input data relates to the output. In reinforcement learning, the learning is developed by providing rewards or punishments related to the previous results from the process.

The general approach of ML is as follows:

- Definition of the scope, goals, and constraints of the problem
  Such as the desired inputs and outputs from the ML model, the expected limits of each input and output, and the topology of the data (such as individual point in space or full-field, individual point in time, or resolved).
- Data gathering, combining, evaluating
  The data may come from the literature, historic data, new experiments, point measurements in a system, or generated using simulations. The evaluation of the available data is done with statistical techniques to develop a high-level understanding of its form.
- Model selection
  The optimal model for a specific problem will be dependent on its scope, goals, and constraints, the form of the input and outputs, and the availability of data.
- Pre-processed data
  The data is pre-processed to enable input into the ML. In addition, the data is separated into three data sets with similar statistical properties: training, validation, and testing sets.
- Conduct Learning (training and validation data sets)
  The training data is used to optimize the model parameters while the validation data set is used periodically during the training process to determine how well the model predicts the unseen scenarios and to avoid the overfitting with the training data set. Modeling parameters are tuned until the model predictions on both the training and validation data sets meet the desired performance criteria.
- Model Validation (test data set)
  After the optimization of the model parameters through the learning process, the model is then evaluated using the testing data set that was not included in the learning. Therefore, applying the testing data set to the model provides an unbiased estimate on the model generalization.

ML is starting to be a relatively new area of study within fire safety engineering community. The following applications can be mentioned: the use of ML in physics-based simulation of fire [15], modeling and interpreting pre-evacuation decision-making using ML [49], or fire resistance evaluation through artificial intelligence of timber structures [23].

ML requires a comprehensive data set, preferably collected from fire tests. Nowadays, there are still not many available data base models around the world, and few of the reasons are arising from the industry confidential testing, scarce number of tested specimens, reliability of the testing methods, or the destruction of the instrumentation during the test because of the severe nature of fire testing.

The creation of the data base is thus crucial; therefore, the tests must represent the real life situations so that the models created using the ML algorithm can capture the realistic physical phenomenon. This is where a new testing methodology such as HFT can be helpful in creating such a data base.

12.5   Future of HFT in the Context of ML

HFT and ML have a similar goal: prediction of a system behavior using as tools, testing, numerical modeling or mathematical modeling, and data analysis. On the one side, testing and numerical modeling simulation is commonly used in engineering field, where the derivation of causal relationships plays an important role, or for the evaluation of reactions, stresses, and displacements regarding a structure. On the other side, ML is successfully used in fields where causal relationships are often sparsely given, but huge amounts of data are available, e.g., image classification or language processing.

However, some applications can benefit from combining ML and testing/simulation in testing/simulation-assisted ML, ML assisted testing/simulation, and a combined approach with a strong and mutual interplay as presented in [46]. In terms of HFT and ML, a more detailed combination between them is next discussed.

12.5.1 HFT-Assisted ML

HFT offers additional source of information for ML that goes beyond typically available data and that is rich of knowledge. Fire tests are particularly useful for creating additional training data in a controlled environment or can be integrated into the hypothesis set or learning algorithm or the validation of the final hypothesis.

A significant progress was reached within the HFT field and the popularity among researchers is increasing. Nevertheless, there is still a long way to use HFT at a comparable scale with standard fire testing. This is mainly because there are still many developments to be done in order to enable the possibility of performing large numbers of HFT in an acceptable time frame. This requires developing universal methodologies, frameworks for experimental setup and control, independently on the testing facilities. These developments are complex and take time, but once it is enabled, a large variety of configurations and structural members can be tested, leading to better reality representation results, which can be further used in the ML data bases.

12.5.2 ML Assisted HFT

ML can be used in HFT with the intention to support the solution process or to detect patterns in the tested or numerically modeled data. ML can be used for the initial numerical model, the input parameters, the numerical method, and the final simulation results.

A type of ML techniques into HFT is the identification of simpler models for representing the NS for example, such as surrogate models. This provides a cheap model, especially when the solution of the original NS, is very time- or resource-consuming. The surrogate models can be developed with ML techniques either with data from experiments or with data from high-fidelity simulations. This approach is very valuable in the context that the results from the NS must be provided to the PS in real time, every time step (often every few seconds), which can become problematic if the computation time of the NS is time consuming.

Another area where ML can possibly enhance HFT is the data assimilation, which includes the calibration of constitutive models and the estimation of system states.

ML can also be used to perform a study of the parameters that influence the results. After a sequence of HFT or simulations is executed, the ML can be used to detect different behavioral modes in the results and thus reduce the analysis effort during the engineering process. Therefore, this supports the selection of the parameter setting for the next HFT, for which active learning techniques can also be employed. In the context of response to extreme loads, the structure, its boundary conditions, and the fire exposure itself are all subjected to large parameter variability. However, in current practice, HFT campaigns rely on a few prototype structures with fixed parameters subjected to fire loads of different intensity. While this approach effectively reveals structural weaknesses, it does not reveal the sensitivity of structure's response. This thus far missing information could support the planning of further experiments as well as drive modeling choices in subsequent analysis and evaluation phases of the structural design process. As an example, Abiatti et al. in [6] describes a Global Sensitivity Analysis framework for Hybrid Simulation. The framework, based on Sobol' sensitivity indices, is used to quantify the sensitivity of the response of a structure tested using the Hybrid Simulation approach due to the variability of the prototype structure and the excitation parameters. Polynomial Chaos Expansion is used to surrogate the hybrid model response. Thereafter, Sobol' sensitivity indices are obtained as a by-product of polynomial coefficients, entailing a reduced number of Hybrid Simulations compared to a crude Monte Carlo approach. An experimental verification example highlights the excellent performance of Polynomial Chaos Expansion surrogates in terms of stable estimates of Sobol' sensitivity indices in the presence of noise caused by random experimental errors. This type of approach paves the road in using the ML to perform global sensitivity analysis for HFT.

Another application of ML is on the results in order to detect patterns.

12.5.3 Combined Approach

A combined approach is when ML and HFT are combined in an optimal way. Such an approach can automatically decide when and where to apply surrogate models or high-fidelity simulations. Surrogate models are efficiently organized and reused through the use of transfer learning. Parameter and design optimization is an integral component of the learning simulation engine, and active learning methods allow the efficient reuse of costly high-fidelity computations.

ML and HFT are thus able to enable future fire safety methods, where a better representation of reality can be captured.

12.6   Conclusions and Future Work

Hybrid fire testing (HFT) is a new emerging structural fire testing technique that allows using realistic mechanical boundary conditions during testing, leading to a better understanding of thermomechanical behavior of the members. The HFT method is described in detail, and a case study is illustrated for a better understanding of the concept. The results provided by the HFT reflect a more realistic behavior of members exposed to fire, compared with the other testing methods. Therefore, these results can be further explored with another emerging concept, which is machine learning (ML).

ML enables computers to gather knowledge from experience without the need of a human computer operator formally to specify all of the knowledge needed by the computer. The hierarchy of concepts allows the computer to learn complicated concepts by building them out of simpler ones, and a graph of these hierarchies would be many layers deep. ML is thus a technique that is becoming more explored within the fire engineering field and can enable new fire safety tools.

The possible interactions between HFT and ML are conceptually described, and the advantages of doing so are also mentioned. Both methods are currently emerging in fire safety engineering, but promising possibilities are identified.

References

1. Open System for Earthquake Engineering Simulation.
2. OpenFresco. Library Catalog: openfresco.berkeley.edu.
3. SFS C Studs I Drywall Steel Sections.
4. ABAQUS. ABAQUS Documentation, 2011. ZSCC: 0000003.
5. G. Abbiati, P. Covi, N. Tondini, O. Bursi, and B. Stojadinovic. A Real-Time Hybrid Fire Simulation Method Based on Dynamic Relaxation and Partitioned Time Integration. *ASCE Journal of Engineering Mechanics*, 2020.
6. G. Abbiati, S. Marelli, N. Tsokanas, B. Sudret, and B. Stojadinović. A global sensitivity analysis framework for hybrid simulation. *Mechanical Systems and Signal Processing*, 146:106997, January 2021.
7. Jean-Marc Franssen and Thomas Gernay. Modeling structures in fire with SAFIR®: theoretical background and capabilities. *JSFE*, 8(3):300–323, September 2017.

8. Ian Goodfellow, Yoshua Bengio, and Aaron Courville. *Deep learning*. Adaptive computation and machine learning. The MIT Press, Cambridge, Massachusetts, 2016.

9. A. Gravouil and A Combescure. Multi-time-step explicit–implicit method for non-linear structural dynamics. *Int. J. Numer. Methods*, 50 (1):199–225, 2001.

10. ISO. Iso 834: Fire-resistance tests - Elements of building construction, 1975.

11. Mustesin Ali Khan, Liming Jiang, Katherine A. Cashell, and Asif Usmani. Analysis of restrained composite beams exposed to fire using a hybrid simulation approach. *Engineering Structures*, 172:956–966, October 2018.

12. M. Kiel. Entwicklung einer intelligenten Prüfmaschine für brandbeanspruchte Gesamttragwerke, 1989. ZSCC: 0000004.

13. M. Korzen, G. Magonette, and P Buchet. Mechanical loading of columns in fire tests by means of the substructuring method. *Zeitschrift für Angewandte Mathematik und Mechanik*, 79:S617–S618, 1999.

14. M. Korzen, J. Rodrigues, and A. Correia. Thermal restraint effects on the fire resistance of steel and composite steel and concrete columns. *Appl. Struct. Fire Eng.*, pages 512–517, 2009.

15. B. Lattimer, J. Hodges, A. Lattimer, and Russia) International Seminar On Fire And Explosion Hazards (9; 2019; Saint Petersburg. Using Machine Learning in Physics-based Simulation of Fire. 2019.

16. Margaret McNamee, Brian Meacham, Patrick van Hees, Luke Bisby, W.K. Chow, Alexis Coppalle, Ritsu Dobashi, Bogdan Dlugogorski, Rita Fahy, Charles Fleischmann, Jason Floyd, Edwin R Galea, Michael Gollner, Tuula Hakkarainen, Anthony Hamins, Longhua Hu, Peter Johnson, Björn Karlsson, Bart Merci, Yoshifuni Ohmiya, Guillermo Rein, Arnaud Trouvé, Yi Wang, and Beth Weckman. IAFSS agenda 2030 for a fire safe world. *Fire Safety Journal*, 110:102889, December 2019.

17. Mehrdad Memari, Xuguang Wang, Hussam Mahmoud, and Oh-Sung Kwon. Hybrid Simulation of Small-Scale Steel Braced Frame Subjected to Fire and Fire Following Earthquake. *J. Struct. Eng.*, 146(1):04019182, January 2020.

18. E. Mergny, G. Drion, and J.-M. Franssen. Stability in Hybrid Fire Testing Using PI Control. *Exp Tech*, May 2020.

19. E. Mergny, G. Drion, T. Gernay, and J.M. Franssen. A PI-controller for hybrid fire testing in a non-linear environment. In *10th International Conference on Structures in Fire*, Belfast, 2018.

20. Hossein Mostafaei. Hybrid fire testing for assessing performance of structures in fire—Application. *Fire Safety Journal*, 56:30–38, February 2013.

21. Hossein Mostafaei. Hybrid fire testing for assessing performance of structures in fire—Methodology. *Fire Safety Journal*, 58:170–179, May 2013. ZSCC: 0000038.

22. M. Z. Naser. Autonomous Fire Resistance Evaluation. *J. Struct. Eng.*, 146(6):04020103, June 2020.

23. M.Z. Naser. Fire resistance evaluation through artificial intelligence - A case for timber structures. *Fire Safety Journal*, 105:1–18, April 2019.

24. Gerald M. Newman, Jef T. Robinson, and Colin G. Bailey. *Fire safe design: a new approach to multi-storey steel-framed buildings*. Number 288 in SCI publication Fire and steel construction. Steel Construction Inst, Ascot, 2. ed edition, 2006.

25. Nicolas Pinoteau, Duc Toan Pham, Hong Hai Nguyen, and Romain Mège. Development of hybrid fire testing by real-time subdivision of physical and numerical substructures. *JSFE*, ahead-of-print(ahead-of-print), July 2020.

26. Alexander L. Pyayt, Ilya I. Mokhov, Bernhard Lang, Valeria V. Krzhizhanovskaya, and Robert J. Meijer. Machine Learning Methods For Environmental Monitoring And Flood Protection. June 2011. Publisher: Zenodo.

27. R. Qureshi and N. Elhami-Khorasani. Instantaneous stiffness correction for hybrid fire testing. In *10th International Conference on Structures in Fire*, Belfast, 2018.

28. Ramla Karim Qureshi, Negar Elhami-Khorasani, and Thomas Gernay. Adaption of active boundary conditions in structural fire testing. *JSFE*, 10(4):504–528, December 2019.

29. I Rahmanian. *Thermal and Mechanical Properties of Gypsum Boards and their Influences on Fire Resistance of Gypsum Board Based Systems*. PhD thesis, University of Manchester, 2011.

30. S. Renard, J.-C. Mindeguia, F. Robert, S. Morel, and J.-M. Franssen. An Adaptive Controller for Hybrid Fire Testing. *Exp Tech*, June 2020.
31. F Robert, S Rimlinger, and C Collignon. PROMETHEE, FIRE RESISTANCE FACILITY TAKING INTO ACCOUNT THE SURROUNDING STRUCTURE. *1st International Workshop on Concrete Spalling Due to Fire Exposure*, page 7, 2009.
32. Victor E. Saouma and Mettupalayam V. Sivaselvan, editors. *Hybrid simulation: theory, implementation and applications.* Balkema-proceedings and monographs in engineering, water, and earth sciences. Taylor & Francis, London; New York, 2008.
33. A. Sauca, E. Mergny, T. Gernay, and J.M. Franssen. A method for hybrid fire testing: Development, implementation and numerical application. In Martin Gillie and Yong Wang, editors, *Applications of Fire Engineering*, pages 225–234. CRC Press, September 2017.
34. Ana Sauca. *Development and implementation of a methodology for hybrid fire testing applied to concrete structures with elastic boundary conditions.* Doctoral Thesis, University of Liege, Liege, Belgium, 2017.
35. Sauca, A., Gernay, T., Robert, F., Tondini, N. and Franssen, J.-M. (2018), Hybrid fire testing: Discussion on stability and implementation of a new method in a virtual environment, *Journal of Structural Fire Engineering*, 9(4):319–341. https://doi.org/10.1108/JSFE-01-2017-0017
36. Ana Sauca, Nils Mortensen, Anders Drustrup, and Giuseppe Abbiati. Experimental validation of a hybrid fire testing framework based on dynamic relaxation. *Fire Safety Journal*, 121:103315, May 2021.
37. Ana Sauca, Chao Zhang, Artur Chernovsky, and Mina Seif. Communication framework for hybrid fire testing: Developments and applications in virtual and real environments. *Fire Safety Journal*, 111:102937, January 2020.
38. Andreas H Schellenberg, Stephen A Mahin, and Gregory L Fenves. Advanced Implementation of Hybrid Simulation. Pacific Earthquake Engineering Research Center PEER 2009/104, College of Engineering University of California, Berkeley, 2009.
39. P. Schulthess. *Consolidated fire analysis—Coupled numerical simulation and physical testing for global structural fire analysis.* Doctoral Thesis, ETH Zurich, 2019.
40. H. A. Schwarz. *Gesammelte Mathematische Abhandlungen: Erster Band.* Springer Berlin Heidelberg, Berlin, Heidelberg, 1890.
41. B. Taylor and C. Kuyatt. Guidelines for Evaluating and Expressing the Uncertainty of NIST Measurement Results. Technical Report NIST, Technical Note 1297, 1994.
42. N. Tondini, G. Abbiati, L. Posidente, and B. Stojadinovic. A static partitioned solver for hybrid fire testing. In *9th International Conference on Structures in Fire*, Princeton, NJ, USA, 2016.
43. N. Tondini and J.-M. Franssen. IMPLEMENTATION OF A WEAK COUPLING APPROACH BETWEEN A CFD AND AN FE SOFTWARE FOR FIRES IN COMPARTMENT. In *V International Conference on Computational Methods for Coupled Problems in Science and Engineering*, pages 185–192, Ibiza, Spain, June 2013.
44. N. Tondini, V.L. Hoang, J.-F. Demonceau, and J.-M. Franssen. Experimental and numerical investigation of high-strength steel circular columns subjected to fire. *Journal of Constructional Steel Research*, 80:57–81, January 2013.
45. P. Underwood. Dynamic relaxation techniques: a review. *Computational Methods for Transient Analysis*, 1983.
46. Laura von Rueden, Sebastian Mayer, Rafet Sifa, Christian Bauckhage, and Jochen Garcke. Combining Machine Learning and Simulation to a Hybrid Modelling Approach: Current and Future Directions. In Michael R. Berthold, Ad Feelders, and Georg Krempl, editors, *Advances in Intelligent Data Analysis XVIII*, volume 12080, pages 548–560. Springer International Publishing, Cham, 2020.
47. Xuguang Wang, Robin E. Kim, Oh-Sung Kwon, and Inhwan Yeo. Hybrid Simulation Method for a Structure Subjected to Fire and Its Application to a Steel Frame. *J. Struct. Eng.*, 144(8):04018118, August 2018.

48. Catherine A. Whyte, Kevin R. Mackie, and Bozidar Stojadinovic. Hybrid Simulation of Thermomechanical Structural Response. *J. Struct. Eng.*, 142(2):04015107, February 2016.

49. Xilei Zhao, Ruggiero Lovreglio, and Daniel Nilsson. Modelling and interpreting pre-evacuation decision-making using machine learning. *Automation in Construction*, 113:103140, May 2020.

Realistic Fire Resistance Evaluation in the Context of Autonomous Infrastructure

13

Liming Jiang, Xiqiang Wu, and Yaqiang Jiang

13.1 Introduction

For decades, the community of fire safety engineers have been dedicated to applying appropriate measures to improve the fire resistance of modern structures. However, estimation of structural resilience in fires is almost narrowed down to the practice of using standard fire curve and prescriptive design (i.e., applying fire protection simply according to the rate of fire resistance). Even with the recent spread of performance-based design (PBD) approaches within the fire safety community, the tools for and understanding of structural fire engineering still lag severely behind. In the 1990s, the ground-breaking Cardington fire tests have demonstrated the significance of investigating system-level responses in evaluating the fire resistance, from which the floor system was found to survive in a post-flashover fire through a tensile membrane action. A similar breakthrough was made after rigorous investigations on the tragic collapse of World Trade Center (WTC) buildings on September 11, 2001, which, by contrast, has shown the system-level vulnerability of modern designed structures. Had the towers not been hit by the aircraft and only set on fire, comprehensive investigations showed that they would have still collapsed. With the

L. Jiang (✉) · X. Wu
Department of Building Environment and Energy Engineering, The Hong Kong Polytechnic University, Hong Kong, China

Research Centre for Fire Safety Engineering, The Hong Kong Polytechnic University, Hong Kong, China

Department of Building Environment and Energy Engineering, The Hong Kong Polytechnic University, Hong Kong, China
e-mail: liming.jiang@polyu.edu.hk; xiqiang.wu@polyu.edu.hk

Y. Jiang
Sichuan Fire Research Institute, Chengdu, China
e-mail: jiang.yaqiang@scfri.cn

© Springer Nature Switzerland AG 2022
M. Z. Naser, G. Corbett (eds.), *Handbook of Cognitive and Autonomous Systems for Fire Resilient Infrastructures*, https://doi.org/10.1007/978-3-030-98685-8_13

decades of research, many fundamental mechanisms of structural behaviour in fires have been identified. The advance of modern technologies and techniques has been utilised in understanding the structural behaviour in fires and ultimately preventing the fire induced failure and collapse. This chapter begins with a brief introduction of these fire induced collapses to underline the complexity of fire–structure interaction analyses, which is followed by a summary of the latest established design fire scenarios and structural failure mechanisms. While highlighting visionary application of autonomous infrastructure in evaluating structural fire resistance, smart technologies such as artificial intelligence (AI) have been summarised in the context of predicting fire behaviour and structural responses. As a pioneering attempt to investigate the collapse criteria for early warning, the tests in Sichuan Fire Research Institute (SCFRI) are presented to demonstrate the use of advanced technologies to monitor the critical events leading to fire induced structural collapse. At the end of this chapter, the prospect of future structural fire resistance evaluation is discussed, while such a vision will always focus on improving the safety performance of built environment considering fire threats.

13.2   Complexity Embedded in Fire Induced Collapses

The collapse of WTC twin towers on September 11, 2001 in New York shocked the word for the magnitude of life loss, which had also induced grave impact to global society and the structural engineering community. Two modern skyscrapers fell down mainly owing to the spread of fires after the crash of hijacked airplanes, whereas a more unexpected collapse occurred at the neighbouring WTC 7 building as a result of a fire triggered by the flaming debris from the WTC tower. There have been other several major fire accidents (e.g., the recent Plasco Building collapse in a fire in January 2017 and the Grenfell Tower fire in June 2017) causing many casualties, and the risks of collapse alarmingly exist. When intending to understand the collapse mechanisms of these tragic events, the complexity has been worryingly recognised and the existing knowledge of structural fire evaluation has been substantially challenged.

13.2.1 Collapse of WTC Towers

The collapse of WTC Tower 1&2 caused 2749 fatalities including occupants and fire fighters on that day. The 110-storey WTC 1 tower was hit by a Boeing 767 between Floors 93 and 97 at 8:46 AM and the collapse began at 10:28 AM. The south face of WTC 2 was hit by another Boeing 767 aircraft between floor 77 and floor 85, and the WTC2 began to collapse at 9:58:59, approximately 1 h after hitting [1]. Forensic investigation was performed by the Federal Emergency Management Agency [2] and the National Institute for Standards and Technology [3], followed by a number of research efforts aiming to reveal the true collapse mechanisms [4–7]. It has been recognised that the structural damage induced by

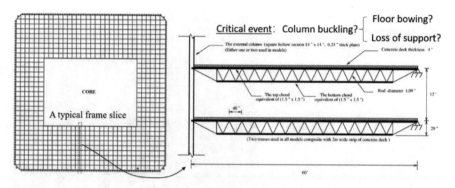

Fig. 13.1 Collapse mechanism of WTC tower illustrated. (Adapted from [7])

the aircraft impact was not sufficient leading to the structural collapse. The fire started by the burning of jet fuel carried by the aircraft was the major reason to the system-level failure, while the critical events leading to the eventual collapse were believed to be the failure of external columns. However, the failure mechanism was differently explained in various investigation papers and reports mentioned before. As summarised by McAllister et al. [1], a variety of hypothetical mechanisms were proposed in explaining the collapse of WTC towers: (1) failure of fire heated core columns initiating the overall building collapse; (2) a progressive collapse due to sequential failure of composite floors in fire after the initial loss of connection after the impact loading; (3) thermally induced buckling of outrigger trusses leading to the collapse; (4) a global collapse led by the buckling of exterior columns due to the large deflection sagging of the composite floor system. The typical floor of WTC tower is illustrated in Fig. 13.1. The collapse of this building comprising 94 similar floors certainly involved numerous unexpected failures, and to find those critical ones becomes essentially important to prevent the catastrophic collapse.

13.2.2 Collapse of WTC7 Building

On the same day of 9/11, WTC 7, a 47-story steel framed building of the World Trade Center Complex, collapsed solely due to the fire ignited by the burning debris from the WTC towers. Fortunately, all occupants of WTC 7 evacuated the building after the terrorist attack to the WTC towers. However, the falling debris from WTC1 struck the southwest façade of WTC 7, which did not pose severe damage to structural components but started a fire lasting for 7 h inside the building. The fire was totally uncontrolled as the water supply to the sprinklers from the city mains was disrupted. Moreover, fire fighters were not deployed to WTC 7 because of the completed evacuation and the overwhelmed priority in rescuing injuries from WTC 1&2. Fires started at 10:28 AM were observed near the southwest corner and the horizontal and vertical spreading of fires were observed and later reproduced in

Fig. 13.2 Horizontal fire
spreading in WTC 7 [8]

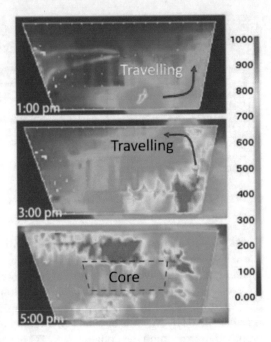

the NIST investigation (Fig. 13.2). At 5:20:45 PM, the east penthouse on the roof
began to descend as the earliest observed signal of the collapse.

Massive efforts have been devoted to investigate the collapse of WTC 7 because
this building without severe impact induced damage was demolished only by fire.
It is unprecedented as a modern building complying with the codes and standards
failed in an extensively spread fire. The earlier reports released by NIST [8], Arup
[9], and Weidlinger [4] suggested that the collapse was a sequence of failures but
the mechanism of such progressive collapse was differently stated in each report.
The initial failure was alleged to be either a connection failure or the buckling of a
column, whereas localised collapse of composite floors was reckoned as a key factor
causing the following sequential damage. A research team led by Toreo and Usmani
believed that the global collapse was induced by the failure of transfer trusses and
composite floor failure was of minor effect [10]. Hulsey et al. [11] published a
report concluding that the collapse of WTC7 "was a global failure involving the
nearly simultaneous failure of all columns" and "not a progressive collapse". These
hypotheses explaining the collapse were pointed to different collapse mechanisms
and the occurrence of these identified critical events (Fig. 13.3) become the keys in
preventing such failure.

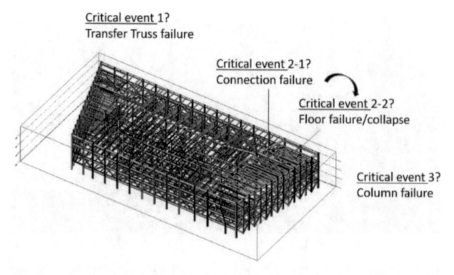

Fig. 13.3   A partial model of WTC 7

13.2.3 Plasco Building Collapse

More recently, the Plasco building in Tehran collapsed in a fire on January 19, 2017. The building was constructed in 1962, consisting of a 16-storey tower and a 5-storey podium (Fig. 13.4). Initially, the low-rise podium part of the building and the lower floors of the tower were used as a shopping centre, while the upper floors of the tower were occupied as offices. At the time that the fire occurred, however, most of the units in the tower were used as garment manufacturing units. Fire safety measures were not adopted in the building regardless of the extreme vulnerability of the structural steel components to heat. Neither passive nor active fire safety measures were implemented in the building. Hajiloo et al. [13] studied various aspects of the Plasco fire including the condition of the tower before the fire, initiation of the fire, the evacuation process, and the progressive collapse of the building. It was concluded that the fire spread horizontally and vertically and then led to the structural collapse. The catastrophic loss of firefighters could have been prevented if appropriate measures could be used with sufficient knowledge of the structural fire behaviour.

The fire started at the tenth floor and was found travelling along the floor horizontally and through the staircase and windows vertically. The fire load was estimated as 452 MJ/m^2 and localised distribution of combustible materials existed in the building. The plan configuration of Plasco tower is shown in Fig. 13.5. Four internal columns carried the loads transferred by the primary beams and regularly distributed box columns were installed around the exterior of the building. It was observed that the first stage of collapse (critical event) initiated on the 11th floor due to the failure of the truss (beam). The second stage of collapse caused the falling of

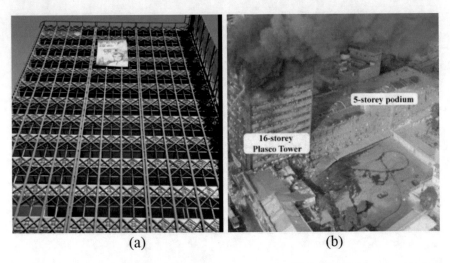

(a)   (b)

Fig. 13.4 Plasco building: (**a**) Steel frame and bracing system; (**b**) the attached 5-storey shopping centre [12]

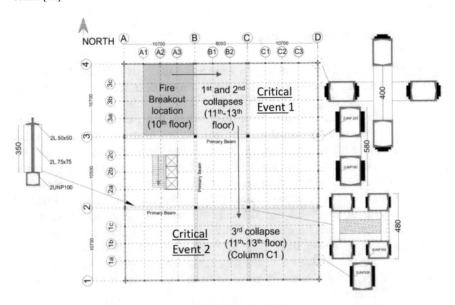

Fig. 13.5 A schematic plot of Plasco fire and structural layout. (Adapted from [12])

the concrete slabs and left a void space between the tenth and the 13th floors (critical events). A preliminary analysis using OpenSees [12] suggested that the failure of exterior columns induced by the deflected composite floor slabs triggered the chain of failure leading to the eventual collapse.

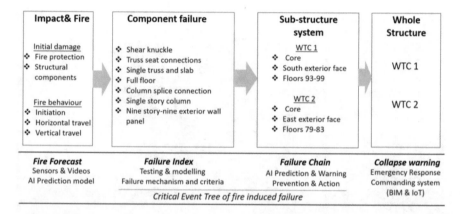

Impact& Fire	Component failure	Sub-structure system	Whole Structure
Initial damage ❖ Fire protection ❖ Structural components **Fire behaviour** ❖ Initiation ❖ Horizontal travel ❖ Vertical travel	❖ Shear knuckle ❖ Truss seat connections ❖ Single truss and slab ❖ Full floor ❖ Column splice connection ❖ Single story column ❖ Nine story-nine exterior wall panel	**WTC 1** ❖ Core ❖ South exterior face ❖ Floors 93-99 **WTC 2** ❖ Core ❖ East exterior face ❖ Floors 79-83	WTC 1 WTC 2
Fire Forecast Sensors & Videos AI Prediction model	*Failure Index* Testing & modelling Failure mechanism and criteria	*Failure Chain* AI Prediction & Warning Prevention & Action	*Collapse warning* Emergency Response Commanding system (BIM & IoT)
		Critical Event Tree of fire induced failure	

Fig. 13.6 The complexity of WTC 1&2 collapse investigation (Adapted from [1]) and smart measures in future

13.2.4 Resolve the Complexity of Fire Induced Structural Damage, Failure, and Collapse

Taking the investigation of WTC tower collapse as an example, various effects were analysed using numerical models of different scales based on the collected design data and the recorded failure patterns. These analyses as shown in Fig. 13.6 include: estimates of aircraft impact, immediate damage due to the impact, damage developed in the passive fire protection (thermal insulation), fire spreading behaviour, thermal action towards the structure members, structural response, and collapse sequence. To characterise the collapse mechanisms involving nearly a hundred floors, analyses were performed extensively from individual components to major subsystems and to global systems, because it was not feasible to complete the investigation embedding enormous complexities within one computational model. From these collapse investigations, lessons have been learnt and it turns out to be essentially important to develop and implement advanced technologies in modern buildings to intelligently predict the development of fires and to prevent the possible critical events potentially leading to larger extent of structural damage and failure.

For smart prediction of fire development, an autonomous platform should be ideally able to forecast the fire growth from the ignition to the travelling horizontally along the floor and vertically through the floors. On-site data such as temperature, smoke density, CO density, and captured images can be obtained in real time to enhance the accuracy of prediction (Fig. 13.7) [14]. The FireGrid project as a pioneering attempt, it transferred the sensor-collected data to an expert system to extract the useful information including the key parameters of fire size, fire location, travelling speed, and local buckling. Based on these analyses, a forecast of the fire scenarios, structural collapse, and any pre-defined critical events can be potentially realised. The predicted information can be transferred and displayed on terminals to provide valuable references for decision-making staff. In total, the platform contains

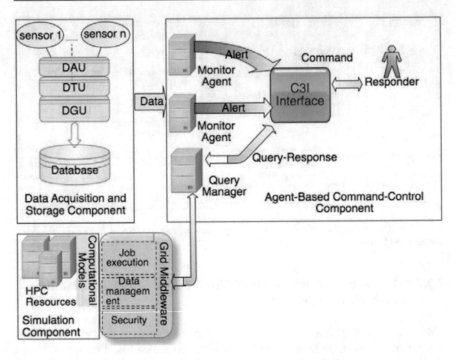

Fig. 13.7 The FireGrid architecture for emergency response support [14]

three main modules responsible to data collection, development forecasting, and terminal displaying.

Once the development of fire and structural response are predicted, the prediction and warning should be displayed to facilitate the on-site guidance of evacuation and firefighting, which can be achieved via a Building Information Model (BIM) base. Moreover, it has been advocated to build up digital twins of the important infrastructures in megacities, such as the WTC towers, which can be equipped with fire emergency modules as well. The contour of the smoke or high temperature distribution can be visualised directly, and the key information alongside suggestions could also be provided for respondents. Internet-based devices or phone applications (APP) are often developed for terminal displaying in these digital infrastructures. As example, the current platforms developed by several research groups are available and are discussed in details later.

13.3   Realistic Fire Impacts in Modern Built Environment

Normally a fire initiated in a compartment behaves as a localised fire in the pre-flashover stage. As the flames spread over the surface, the localised fire grows and hot smoke accumulates, which may be followed by a flashover phenomenon as the room combustibles are spontaneously ignited due to intense heating. If the

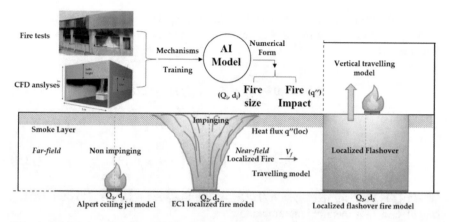

Fig. 13.8 Potential AI fire model based on fire size and fire impact models

compartment is large enough, a travelling fire may occur which is characterised by the moving of intense burning zone (localised burning). Moreover, fires can vertically travel through the internal openings and façade. These phenomena have been observed in the above-mentioned cases, which underline the importance of broader fire behaviour prediction in evaluating building fire safety rather than traditional single compartment fires. Researchers usually conduct fire tests and CFD analyses to explore the dependencies between fire impact and fire characteristics (i.e., Heat release rate of fire Q, fire dimension D). Due to the complexity of fire behaviour, predicting fire development is a challenging task and AI models may be employed on the basis of tests and simulation cases as training samples (Fig. 13.8). As the CFD field model is extremely resource-demanding, engineering fire models are feasible to represent the fire impact. Thus, predicting complex fire behaviour yields to estimate limited number of variables, which enables the use of AI models deriving from the sufficient samples and an AI predicted fire model can be disseminated as localised fire models, horizontally travelling fire models, and vertically travelling fire models, since these represent the natural phases of fire development in a modern building.

13.3.1 Localised Fires

When the fire remains at the surface of the initial fire source after ignition, it is generally described as a localised fire. In built environment, structural fire safety engineering usually does not consider a localised fire as a major design fire scenario because the heating impact is much lower than a post-flashover fire. Such pre-flashover fires representing an earlier stage of fire development are usually considered in the design of fire detection systems. When the localised fire flames have not impinged a ceiling, the Alpert Ceiling Jet fire model [15] can be used to estimate the weak-plume fire action along the ceiling. The correlations between gas-phase temperature T_g at the ceiling level and the radius distance to the fire centre r are given as

$$T_g = 16.9 \frac{Q^{2/3}}{H^{5/3}} + T_a, \quad (r/H \leq 0.18) \tag{13.1}$$

$$T_g = 5.38 \frac{(Q/r)^{2/3}}{H} + T_a, \quad (r/H > 0.18) \tag{13.2}$$

where Q is the heat release rate (HRR) of the fire source in kW. H is the ceiling height, and T_a is the ambient temperature.

When the localised fire grows to impinge the ceiling, the heating impact was studied extensively with experimental tests and numerical models. A series of small-scale pool fire tests were conducted and analysed by Wakamatsu and Hasemi [16] and Hasemi et al. [17], which measured the heat fluxes underneath the ceiling and proposed a parameter y to describe the localised heating impact:

$$y = \frac{r + H + z'}{L_h + H + z'} \tag{13.3}$$

Various correlations between the heat fluxes and parameter y were proposed on the basis of localised fires test data. In structural fire engineering, the most commonly used localised fire correlation is the Eurocode Localised Fire model mathematically described as follows:

$$\begin{cases} q'' = 100 & \text{if } y \leq 0.3 \\ q'' = 136.3 \text{ to } 121y & \text{if } 0.3 < y < 1.0 \\ q'' = 15y^{-3.7} & \text{if } y \geq 1.0 \end{cases} \tag{13.4}$$

Noticed these localised fire tests or models assumed unconfined condition to the smoke layer, which may cause underestimated heating impact in building compartments. Based on the localised fire tests with smoke layer presented by Wakamatsu et al. [18], a smoke induced additional heat flux was suggested by Jiang et al. [19] to modify the Eurocode localised fire model in the context of localised fires in compartments (Fig. 13.9). To consider the heating impact of localised fires to vertical structural members, Lange and Sjöström [20] measured the vertical distribution of heat fluxes on the column surfaces engulfed by a localised fire. Tondini and Franssen [21] then proposed mathematical models similar to the ceiling model to estimate vertical heat fluxes at various heights based on the measured results from dozens of localised fire tests. Although localised fires are of relatively lower gas temperature, it should be noticed that the localised heating impact on critical components may cause damage or deformation before the fire spreads.

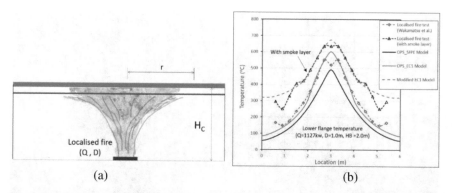

Fig. 13.9 Fire impact of localised fire impinging on ceiling: (**a**) a schematic plot; (**b**) heat transfer using modified heat flux distribution considering smoke layer [19]

13.3.2 Horizontally Travelling Fires

The travelling fire concept was raised after reviewing the past experimental data in modern compartments. Stern-Gottfried et al. [22] found that the intense burning region was moving in large compartments. This echoes a statement that Drysdale [23] mentioned in his classic textbook: most of our knowledge regarding compartment fires comes from experiments in near-cubical compartments of characteristic dimensions ranging from 0.5 to 3 m. In a review over the large-scale structural fire tests, Bisby et al. [24] pointed out that natural fire exposure should be introduced to the contemporary fire safety design rather than simply applying a standard fire curve. In recent years, experimental investigations towards the real fire behaviour in large compartments were conducted, such as the full-scale test in BRE, UK, and another test in Malveira, Spain [25, 26]. The tests successfully identified and demonstrated the typical modes of a fire in large compartments (Fig. 13.10): namely a growing mode (localised fire), a travelling mode, a flashover mode, and a decay mode. The first edition of travelling fire model to describe the fire impact was proposed by Stern-Gottfried and Rein [27, 28], which was later extended by Dai et al. [29, 30] and Heidari et al. [31] to introduce localised fire models for near field. These models focused on describing the fire impact, yet with assumptions adopted for the travelling behaviour including the travelling trajectory and velocity. Nan et al. [32] have proposed various travelling velocity models as shown in Fig. 13.11, which were extracted from the existing travelling fire test data. Alternatively, travelling fires can be modelled using CFD simulation package (e.g., Fire Dynamics Simulator). A recent attempt was reported by Charlier et al. [33], which modelled the travelling fire test carried out by Ulster University in the frame of TRAFIR project. The model disseminated the fire load of wood cribs in test as cubes in the FDS representation and the calibrated model has presented similar travelling behaviour and fire impact compared to the test data.

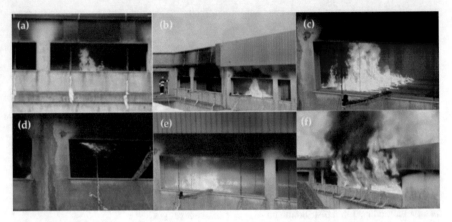

Fig. 13.10 (**a–f**) Various stages of a large compartment fire test in Malveira [26]

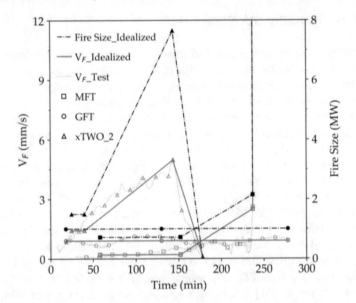

Fig. 13.11 Different travelling models in comparison with large compartment fire tests [32]

13.3.3 Vertically Travelling Fires

When a fire breaches the compartmentation and travel through multiple floors, a larger extent of structural damage and failures could occur simultaneously in multiple floors. This fire scenario is not conventionally considered but it has been observed in various fire accidents including previously mentioned cases. The vertical travelling of fire can occur either along the facade after breaking the glass or through the internal initial or fire induced openings inside the buildings. The NIST report [8] suggested that the local failure of composite floors occurred after the

Fig. 13.12 Vertical travelling of fire in Plasco building [12]

breakage of the connections. The failed floor slabs or other partition components can lead to the deactivation of compartmentation, allowing fire spread through the floor openings to upper floors. The preliminary investigation of Plasco building collapse [12] found a similar vertical travelling of fire after the local failure of the floor system (Fig. 13.12). When investigating the collapse of WTC towers, Usmani et al. [7] proposed a global failure due to the deflection of composite floor slabs can take place in multiple floor fires. The vertical spreading behaviour was assumed as a sequence of time delays (Fig. 13.13) associated with parametric fire models at each floor. The similar methodology was adopted by Kotsovinos et al. [34] in discussing the effect of vertically travelling fires on the collapse of tall buildings. In reality, vertical travelling is closely dependent on the integrity of passive compartmentation, which once again underlines the importance of preventing local failure as critical events to prevent disproportionate collapse.

13.4   Failure Patterns and Mechanisms of Structures in Fires

As a fire develops and spreads in a building, damage and failure of structural members may occur due to material degradation and restraints to thermally induced expansion [35]. The structure with local failure may experience further damage and multiple component failures, which may become a chain of failure as the fire horizontally and vertically travels. Among these failures, some are critical events dominating the development to partial collapse or global collapse (Fig. 13.14).

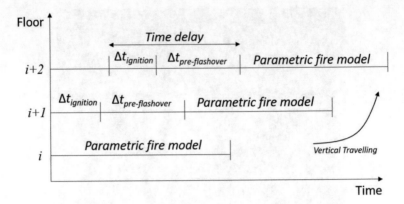

Fig. 13.13 Vertically travelling fire modelled using time delays. (Adapted from [34])

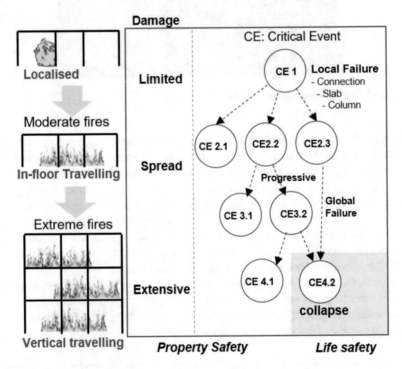

Fig. 13.14 Damage and failure tree describing structural responses to fires

Along with the advances of structural fire engineering research, various failure mechanisms have been identified, which could provide important input to the global behaviour prediction models and potentially enable prevention of progressive collapses or global failures.

13.4.1 Local Failure

Local failure, in contrast to a global failure or collapse, might be caused by localised fire heating or localised load redistribution in a compartment fire. For bare steel structural members, it may fail at the region heated by a localised fire because of the locally weakened load bearing capacity [36, 37]. A similarly local failure may occur when the fire protection of the steel structural member is partially damaged [38]. In a full-scale eight-story steel framed building, the Cardington fire tests have provided fruitful data on the local failure of structural members in the context of a real structural system subjected to compartment fires [39]. Six fire tests were performed, which include the restrained beam test (Test 1), the plane frame test (Test 2), two corner compartment tests (Test 3&4), and two large compartment fire tests (Test 5&6). When a beam in a structural system is restrained at both ends, local buckling in the beam section was found in the heating stage due to the restraint of thermal expansion [40, 41]. In the cooling stage, the contraction of the beam may cause the fracture in the vicinity of connection, which was observed in the Cardington restrained beam test (Fig. 13.15a). The beam-column connection failure (Fig. 13.15b) in different fire scenarios were also investigated by various researchers, such as Al-Jabri et al. [42], Dai et al. [43], Wang et al. [44]. In the Cardington Plane Frame test, local buckling of columns occurred when the three-bay (6m-9m-6m) plane frame was heated uniformly. The column sections were unprotected near the ceiling causing high temperature rises in steel and then local failures occurred near the column top (Fig. 13.15c). In Cardington fire tests, one major finding was that floor slabs could survive in intense fires by developing tensile membrane action, which explains the necessity of including floor slabs in structural fire analyses. On the other hand, the floor systems in WTC [3, 8] buildings and Plasco [12] were found to fail in spread fires possibly following local failures of connections. Currently, such mechanisms from local failure to the floor system failure remains unclear and require more studies.

Fig. 13.15 Local failure: (**a**) local buckling of steel beam [41]; (**b**) connection failure in Cardington Test; (**c**) column failure in Cardington Test [39]

13.4.2 Progressive Collapse

The earlier version of "progressive collapse" concept was systematically discussed in the US design guidelines [45] and described in the UK code [46] as "disproportionate collapse". Many research efforts have been devoted to understanding the mechanism by which localised damage develops into a chain of failures eventually leading to the collapse of the building [47–49]. Sudden column loss is the most examined scenario and often modelled as the artificial removal of column [50, 51]. Vlassis et al. [52] studied the progressive collapse resulting from additional load due to debris from collapsed top floors using a simplified framework. When considering the collapse in 3D, Pham and Tan [53] suggested that the membrane action of RC slabs may help mitigate the progressive collapse of building structures. The risk of progressive collapse can be obviously seen in fire incidents. The robustness of steel-composite building structures was examined in the case of a central column heated by a localised fire [54]. It was pointed out that the failure of beam-column joints over-heated by fire is a key factor with composite floor systems, and a ductility supply/failure criteria is employed to assess the robustness of joints at elevated temperature based on the work by Al-Jabri et al. [42]. When exposed to fire, steel columns may experience column failure in the form of buckling [55]. To consider the effect of column buckling, progressive collapse analyses were performed using a 2D frame model in Vulcan [56]. The work was highlighted for its combined static–dynamic procedure, aimed at overcoming the numerical singularity or structural instability of static analysis due to failure of localised members. Agarwal and Varma [47] modelled two ten-story buildings with composite floor systems and pointed out that the gravity columns play a significant role in fire induced progressive collapse. More recent studies on the progressive collapse mechanisms induced by column losses in localised fires were discussed by Jiang and Li [57] and Gernay and Gamba [58], which showed the fire induced load redistribution was a key factor to the structural collapses.

13.4.3 Global Failure Mechanism

Unlike the progressive collapse theories, the fire induced collapse is explained as a system-level or global failure of structures, in which the localised damage or failure is of minor effect. One hypothetical cause to the WTC collapse was explained as the failure of core columns as a result of heating [59]. The loss of core column led to the redistributed vertical load to other columns exceeding its load bearing capacity, which quickly induced the collapse. In 2019, a similar failure description based on sequential core column failures was proposed by Hulsey et al. [11] in describing the collapse of WTC 7. Another hypothesis of global failure also emerging from the WTC collapse investigations is the outer column failure due to the deflection of multiple floor composite slabs, which was first proposed by Usmani et al. [7] using plane frame finite element analyses on the WTC model subjected to multiple floor

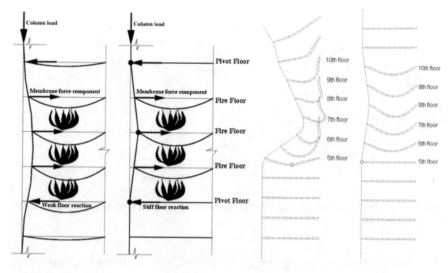

Fig. 13.16 Weak/strong floor collapse mechanisms [5, 60]

fires. The combined action of lateral support and pulling action of the fire impacted floors was reckoned as the cause of instability of exterior columns [60]. This global failure mechanism was later further elaborated by Kotsovinos and Usmani [5] with the weak floor and strong floor mechanisms (Fig. 13.16). The weak floor collapse is initiated by the failure at the lowest floor (the "weak floor") due to the "pull-in" forces in fire heated floors, as the thermal bowing of these floors is resisted by the outer column. On the contrary, the strong floor mechanism describes the collapse initiated at the outer column due to the "push-out" action of the "strong floor". These global collapse mechanisms are usually associated with the failure of columns, and the critical events leading to column failure should be prevented in fire incidents. Similar to the other mechanisms, more fundamental investigations are required to disclose how the fire action cause the thermo-mechanical responses leading to global structural collapse.

13.5   AI Application in Structural Fire Engineering

13.5.1 Automatic Identification of Fire Scenarios

During a fire, timely and accurate identification of the fire characteristics is of critical importance in guiding firefighting and rescue operations as well as mitigating infrastructure damage and other losses. Various types of fire detection techniques have been adopted in the practice of fire protection systems. For instance, smoke particles produced during fires can be detected by light extinction sensors, although their reliability is questionable in the cases of poor air quality and visibility

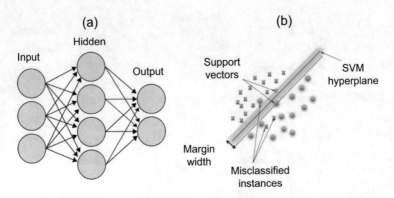

Fig. 13.17 The architecture of: (**a**) typical ANN model. (Adapted from [65]); (**b**) SVM model (adapted from [66])

[61]. Video-image-processing systems have been increasingly used in recent years and have shown the outstanding capability for early fire detection of flame and smoke [62–64].

Benefitting from the recent evolutions of artificial intelligence and big data, data-driven methods especially artificial neural network (ANN) (Fig. 13.17a) have been applied in fire safety engineering. Most of the fire detection methods are developed based on consecutive images captured by on-site cameras [67]. Characteristics of fire flames or smoke can be extracted from visible or infrared images for detection [68]. To improve the performance of fire protection systems in identifying fire aiming to reduce false alarm rate, researchers often make a comprehensive utilization of the colour, shape, and motion of fire [69, 70]. Töreyin et al. [71] extracted the feature of flame colour and motion by setting thresholds of RGB visual images and background motion [72], respectively. The flame flicker was also detected by a three-state hidden Markov model. Experimental tests showed that this method is generally robust and computationally efficient while false alarm may be induced by a moving object mimicking flame flicker process. Ko et al. [73] also obtained the features of fire colour and motion, and a luminance map was used to remove the noise of non-fire pixels. A temporal fire model with wavelet coefficients was then created and applied to a support vector machine (SVM) classifier as illustrated in Fig. 13.17b, which adopts a radial basis function kernel. Testing results proved the robustness of this method against noise, while more efforts are needed for practical application to further reduce false alarm rate. Comparing with flame, smoke is a better media for fire detection but the reliability of only smoke based detection is lower because smoke is generally mixed with its background [74, 75]. Currently, there have been many vision-based methods to detect fires under various situations. Calibrations is needed as suitable colour threshold should be set beforehand to extract the fire features from its background. In view of the image background changing drastically in compartment fires, it may be favourable to design specific detection systems to overcome the visibility and heat issues.

13.5.2 AI Prediction of Infrastructure Fires

The sensing technologies are used to provide input data for AI prediction of fire behaviour, which were primarily based on heat, gas, flame, smoke, and some other important fire characteristics. Xue [76] built up a three-layer ANN to identify fire in tunnels from the temperature, smoke density, and density of CO measured by sensors. The neural network was trained using numerical simulation results. The accuracy, generalization ability, and correct recognition rate were demonstrated. Wu et al. [77] proposed a fire detection method using an LSTM RNN model to identify the source information of the fire including fire location, heat release rate, and ventilation condition. Numerical modelling results were obtained to form a large dataset to train the model and high accuracy of fire prediction was attained with the proposed approach.

For predicting fires in buildings, sensor data from monitoring smoke, CO, and heat was employed in AI models using multiple types of NN [76, 78, 79], because of the low cost and maturity of these sensor technologies. Camera images are more difficult to achieve fast processing, and they are not sufficiently installed in existing infrastructures. The difficulty even arises as the visibility inside the buildings decreases significantly after the development of fire. Hodges et al. [65, 80] adopted a transpose convolutional neural network (TCNN) (Fig. 13.18) to predict the steady-state temperature distribution inside a compartment and the model was trained by a numerical database. A framework was proposed to collect real-time sensor data and to predict the evolution of compartment fires based on a zone model [14]. Carvel et al. [81] employed Bayesian methods to forecast the probability of downstream flame extension length. Dexters and Coile [82] proposed an LR model to predict the occurrence of flashover in a compartment utilizing the information of fire intensity, fire duration, and the thickness of combustible materials from past experiments.

The "FireGrid" project was a pioneering attempt for technologies-aided decision-making during fire emergencies [14]. Real-time data, including temperature, smoke, and CO, were measured continuously by multiple sensors. Then, the fire scenario and fire risk rank can be given to the firefighters. Though the fire within a compartment can be detected, this framework cannot predict the fire development or structural response. This project later fostered the Hong Kong theme-based research scheme "SureFire: Smart Urban Resilience and Firefighting", which is led by Prof. Asif Usmani aiming to improve the resilience of megacities to fire hazards by forecasting the critical events of a fire. Although the structural responses and failure modes are currently not included in the objectives of this project, the feasibility of using cutting-edge technologies, including IoT sensor network, big data, and AI have been preliminarily demonstrated [77, 83], providing valuable information for the practical firefighting (Fig. 13.19).

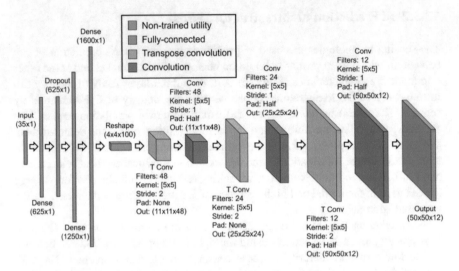

Fig. 13.18 The general architecture of TCNN model [65]

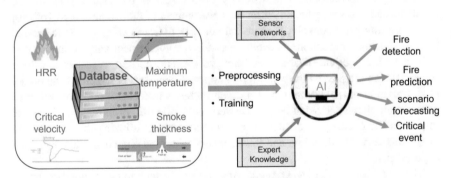

Fig. 13.19 The technical route of smart firefighting. (Adapted from [83])

13.5.3 AI Prediction of Structural Response

Currently, limited studies have been carried out on the prediction of the structural behaviour of buildings under fire. Naser [84–86] conducted several studies to apply AI algorithms on the identification of the material properties and the structural response of concrete and steel components at elevated temperatures. Although these AI methods were still used for solving relatively simple problems, their feasibility in predicting structural behaviour was demonstrated. For the first time, Fu [87] conducted the real-time structural failure prediction of whole steel framed building structures under fire leveraging machine learning. The failure of structural components was defined according to critical temperatures and the work demonstrated the capability of predicting the failure pattern of selected structures exposed to fires. However, the work adopted standard fire rather than some real fire scenarios as its

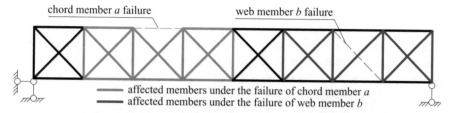

Fig. 13.20 A truss model for safety monitoring system [88]

fire scenario, which was not ideal for predicting the real behaviour of structures in fires. Jiang et al. [88] developed a monitoring system to predict the collapse of steel truss structures under fire (Fig. 13.20). With the real-time temperature data collected from the preinstalled sensor network, and the safety of structure was assessed by calculating the failure index on each component of the truss system iteratively. The remarkable coincidence between experimental and predicting results revealed the feasibility of evaluating the state of the structure in a fire using sensor captured data. Nevertheless, it should be noticed that the safety assessment was based on a manually tuned index, which may not be suitable for a real building comprising many structural components. Key information indicating the state of the structural responses, such as component deflection, local buckling, cracks, and collapses are of many interests for structural engineers. Certainly, AI methods can be deployed to predict the fire induced structural failures, yet the key is a database of sufficient and reliable samples to train the AI models. Meanwhile, the trained AI model regarded as a surrogate model should subject to extensive verification before being widely applied in real buildings.

13.6   A Full-Scale Fire Test Demonstrating Failure-to-Collapse Mechanism

The full-scale tests in Sichuan Fire Research Institute (SCFRI) were recent attempts in demonstrating the collapse prediction and warning of fire induced structural collapse. The work investigated the failure criteria of structural members for predicting the structural collapse in a fire. A novel approach for early warning of building collapse was realised by remotely monitoring primary structural parameters (column axial deformation, deformation rate) of critical structural members. The full-scale fire tests were conducted on two steel portal frame structures, which triggered 3.5–7 min earlier warning ahead of the actual collapses. For any collapse warning techniques, the criteria to trigger a warning signal should be critical and reliable. The approach employed in this experimental study is based on a hypothesis that the failure criteria for its structural members (i.e., steel columns) do not differ from those in isolated situations, i.e., furnace tests. In the tests, simple and robust failure criterion such as those present in ISO standard [89] for columns were adopted, which allows fire fighters only to focus on the deformation of columns subjected to fire.

13.6.1 Experimental Design and Instrumentation

The first structure was a one-story, two-bay steel portal frame with a 4 m × 6 m fire compartment. Stone wool sandwich panels were used for the walls and roofs of the compartment. An inner separation wall of 3-h fire resistance rate was built to prevent fire propagation. The second structure was also a one-story portal frame but with longer spans. The plan and elevation views of the two test frames are shown in Fig. 13.21. Sand bags were uniformly distributed on each of the steel beams to simulate the vertical load of 1100 kg, and the lateral wind load was also added on the columns located at the left end of the two structures. The sizes of the fire compartments were identical in the two structures. Wood cribs were used to simulate post-flashover fires with peak heat release rates around 20 MW and 32 MW, respectively, in the two tests, while the fire load densities were prescribed as 3412.5 MJ/m^2 and 5118.75 MJ/m^2, respectively. These values were taken in accordance with the warehouse storing wooden furniture [15].

The tested structures were heavily instrumented with conventional thermocouples and linear variable differential transducers (LVDT) to measure temperatures and deformations of critical regions and members. The LVDT (Fig. 13.22) values were used as benchmark data against the measurements from other remote monitoring devices. Three kinds of remote monitoring systems were developed, including the Radar Interferometry Monitoring System (RIMS), the Laser Array Monitoring System (LAMS), and the Electronic Distance Measuring System (EDMS) (EDMS was based on commercially available total stations). These systems were used to

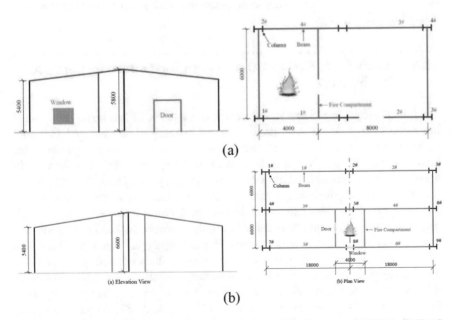

(a)

(b)

Fig. 13.21 Plan and elevation views of the real-scale structural collapse tests: (**a**) Test 1; (**b**) Test 2

Fig. 13.22 Mounting of LVDT on the structure

measure the structural deformation at multiple critical positions, and their accuracy and performance in hash environment with smoke and flame would be compared against the LVDTS values. RIMS was used to monitor the deformation of column and the displacement of beams at locations adjacent fire compartment, while other remote devices were used at locations that were less affected by fire smoke.

13.6.2 Experimental Results

In both tests, the steel structures were not fire protected, which gradually deformed within the first 10 min as fire was developing. Local collapses of the two structures were observed at 19 min in Test 1 and 14 min in Test 2, as shown in Fig. 13.23. It should be noted that the longer fire exposure to collapse in Test 1 was due to the presence of a 3 h fire partition wall, which could resist partial loading from the beams and roof. During the tests, it was observed that the structural deformation began to accelerate just before their collapses. Therefore, only visual observation of

Fig. 13.23 Structural collapse in Test 1 and Test 2

a structure is not feasible for predicting the fire induced collapse, which provides very limited time window to fire fighters in deciding a retreat from the building.

13.6.3 Collapse Warning Based on LAMS

For a long span portal frame structure, the top of perimeter columns would firstly move outward due to thermal expansion of the beams, and then move inward resulting from large displacements at the beams. This was observed in Test 2 with the aid of LAMS which was not deployed in Test 1. As shown in Fig. 13.24, LAMS was used to produce ten parallel laser beams of equal distance of 10 cm, and four of them were just spotted on the top of the column. As the laser array was kept constant, it is similar to measure the relative displacement of a slowly moving object using a ruler. It can be seen in Fig. 13.24 that the top of the column was moving outward during the first 11 min 51 s, and then gradually moved inward. If the moment at which the outward movement just reverse to inward is defined as the critical point of collapse warning, a valuable time window of 2 min 9 s would be obtained to arrange emergency evacuation of on-site firefighters. An interesting phenomenon was observed from the video footage that the structure demonstrated intense shaking in the last 2 min before collapse, which may be additional index of collapse. Although the LAMS technique is visually easy to follow, it should be pointed out that this approach is still crude owing to its accuracy and range, which could not allow for capturing the whole deformation process.

Results from Test 1 showed that the axial deformation of the steel column (column 1#) in the fire compartment started at around 7 min after ignition, as given in Fig. 13.25. Then the column seemed to deform faster and it reached the critical deformation rate at 14 min and the critical deformation value at 16 min. This suggests that the deformation rate may be a more safer collapse criterion for the 2 more minutes of early warning. After analysing the column deformation data in Test 2, similar results were found. The critical deformation rate was reached at 10 min, while critical deformation was found at 10.5 min, as given in Table 13.1. The prediction enabled 3.5–4 min earlier warning before the final collapse at 14 min. Thus, by monitoring the column deformation in real time, it is possible

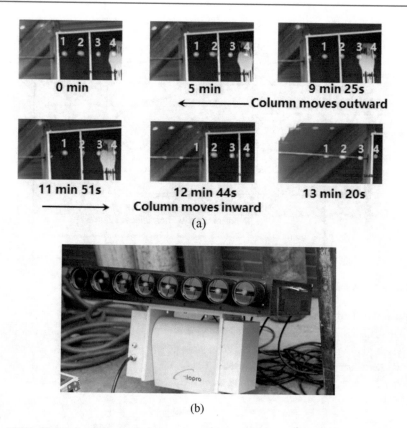

Fig. 13.24 Monitoring column movement with LAMS: (**a**) Column movement; (**b**) LAMS system

to release collapse signal for early warning based on axial deformation and rate of axial deformation.

13.6.4 Performance of Different Remote Monitoring Systems

By comparing the measured values from those remote monitoring systems and LVDT system, it is found that data from RIMS agreed well with the LVDT measurements, while the EDMS measurements of fluctuations seemed very sensitive to hot smoke. Figure 13.26 shows the temporal vertical displacement at the mid-span of the 6# beam, from which we can see that measured displacement curves from LVDT and RIMS were very close in the first 10 min (600 s). From 10 to 12 min, the differences between two curves arise. Nonetheless, the RIMS in general seems to be a reliable remote monitoring approach in view of its accuracy and the ability against the interference of flame and hot smoke.

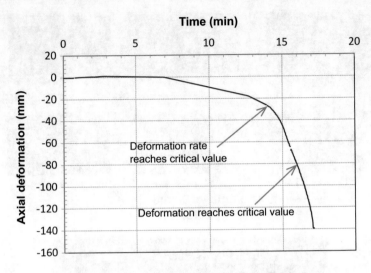

Fig. 13.25 Axial deformation of steel column subjected to fire in Test 1

Table 13.1 Early warning time from different collapse criteria for Test 2

Collapse criteria (column)	Warning point (min)	Available time window (min)
Axial deformation: h/100 mm	10.5 (Column 8#, LVDTS)	3.5
Axial deformation rate: 3 h/1000 mm/min	10 (Column 8#, LVDTS)	4

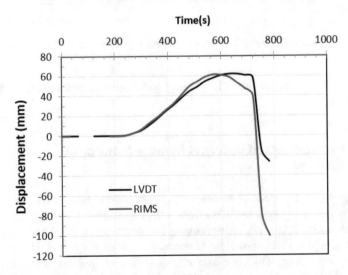

Fig. 13.26 Displacement at the mid-span of the 6# beam

Displacement was also recorded at the mid-span of the 5# beam using both LVDT and EDMS. Figure 13.27 shows the comparison of results from the two approaches, which suggests that the EDMS measured displacements were much lower than the

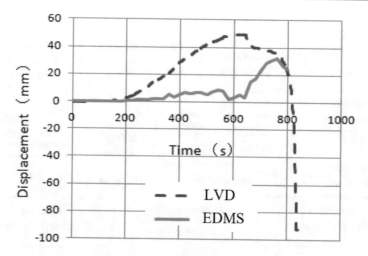

Fig. 13.27 Displacement at the mid-span of the 5# beam

LVDT results prior to 12 min. The EDMS results may have been significantly affected by the hot smoke escaped from the fire compartment. The commercially available video-based displacement measurement was also used in the two tests. However, it was found that this kind of device were greatly affected by the smoke and flames filled in the compartment. For future applications of sensor devices, the effect of smoke and flames along with the development of fire might be a major challenge to the prediction and warning of failure and collapse.

13.7 Concluding Remarks on a Vision of Structural Fire Safety Evaluation

13.7.1 Potential Estimation Framework for Fire Impact

The complexities of fire behaviour and impact to the structures have been briefly reviewed in the previous sections. Although computational fluid dynamic (CFD) models can be used to predict the fire behaviour in modern buildings, the high computational cost and requirement of expertise certainly limit the appropriate use of CFD tools in structural fire safety assessment. Facilitated by the advances of AI techniques, sensor technologies, and computational modelling approaches, the future version of fire safety design can be conducted with smart prediction models featuring realistic fire behaviour and structural responses. A natural fire model should be able to describe the fire behaviour including localised fire impact and travelling behaviour, which are associated with the characteristics of building geometry and fuel distribution. In the work by Jiang et al. [19], the calibration of such models will be achieved by using the thermal analysis infrastructure and optimisation packages available in the Python and AI communities. Full-scale fire

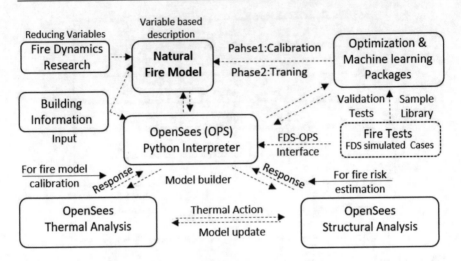

Fig. 13.28 A framework for smart fire design [19]

test measurement, computer vision (CV) data, and CFD simulation cases become samples to obtain various calibrated fire models (See Fig. 13.28). Furthermore, the research outcome on the fundamental mechanism of large compartment fires will help reduce the required parameters and validate the optimisation results. While such calibration is repeatedly conducted for many times, the machine learning packages can thereafter "learn" from the calibration procedure and the AI prediction model can be trained. The long-term goal is that the engineering natural fire model can be automatically given by the software after reading the input of the building geometry information and fuel distribution. The model for fire safety design should be able to suggest the worst cases of fire development and structural failures with the aid of thermal and structural analyses in simulation packages such as OpenSees for fire [90, 91]. Taking advantage of the smart fire design framework, it can be eventually possible to provide design guidance in the thermal response stage such as controlling the ventilation condition or fuel distribution, aiming to prevent high intensity of fire exposure, e.g. flashover.

Possibilities may be also extended to employ such a natural fire model in the fast response application such as on-site fire fighting. Once a critical event is triggered and detected, the further development of fire and structural failure can be immediately displayed to commanders and fire fighters to assist them making timely decisions. The real-time data obtained from various sensors can simultaneously correct the AI prediction model to ensure the reliability of making relevant decisions. Multiple types of sensors including temperature sensor, optical sensor, smoke detector, and camera can be mounted at certain location of an infrastructure or as wearable gears. A wireless sensor network (IoT) is preferably be formed to provide a robust data collection router, while 5G or Bluetooth technologies can be adopted for communication between sensors.

13.7.2 Future Application for Predicting Structural Behaviour in Fire

Fire accidents such as WTC buildings and Plasco building collapse have drawn great attention to fire brigades and structural fire engineers regarding the identification of critical failure events and true mechanisms induced by fire. The following aspects may require further research and development: (1) Sophisticated construction materials should be developed for future generation of buildings to achieve better fire safety performance and occupant comfortability; (2) A more integrated simulation tool is desirable, which should address the realistic fire impact and realistic structural responses to fires in buildings of innovative forms and materials; (3) Critical events such as large structural deflection or local buckling/yielding leading to further damage and failure should be identified, which can be used to establish the early warning of collapse in fire and to guide fire respondents to prevent the occurrence of such critical events; (4) AI prediction models for structural responses to fires can be employed to forecast the potential risks during fire fighting and to suggest the engineers to strengthen the components during the design stage; (5) The use of advanced monitoring devices should be exploited to provide critical data and to co-work with AI models.

In terms of investigation tools, computational simulation/prediction and various types of sensors will be predominately used. While the simulation becomes more integrated across multiple disciplines and scales, the numerical models would become inclusively complex and smart. The hybrid use of numerical model and physical measurement/monitoring may be enhanced as the technical improvement being achieved. Video systems alongside computer vision techniques can provide input information to simulation models, and can be used as virtual reality (VR) or augmented reality (AR). On-site and remote measuring systems such as electrical/optical sensors and laser scanning can be employed to monitor the building status and critical failures, which are expected to be resilient to challenging environment (e.g., hot flames, smoke, and heat). Meanwhile, IoT and BIM technologies have been developed and implemented to enable seamless data transfer and visualization. Alongside the development and implementation of multi-disciplinary technologies, it remains essential to understand the behaviour and mechanisms of structures responding to fires, which relies on the joint efforts of the researchers, engineers, and regulators in the field of fire safety engineering.

References

1. McAllister, T. P., Sadek, F., Gross, J. L., Kirkpatrick, S., MacNeill, R. A., Bocchieri, R. T., ... Sarawit, A. T. (2013). Structural Analysis of Impact Damage to World Trade Center Buildings 1, 2, and 7. Fire Technology, 49(3), 615–642.
2. FEMA. (2002) World Trade Center Building performance study: data collection, preliminary observations and recommendations. New York, 1(May), 1–92.
3. NIST. (2005). NIST NCSTAR 1: Final Report on the Collapse of the World Trade Center Towers. NIST NCSTAR 1, WTC Investigation.

4. Abboud, N. N. (2010) WTC7 Collapse Analysis and Assessment Report. Weidlinger Associates Inc.
5. Kotsovinos, P., & Usmani, A. (2013). The World Trade Center 9/11 Disaster and Progressive Collapse of Tall Buildings. Fire Technology, 49(3), 741–765.
6. Quintiere, J. G., di Marzo, M., & Becker, R. (2002). A suggested cause of the fire-induced collapse of the World Trade Towers. Fire Safety Journal. https://doi.org/10.1016/S0379-7112(03)00058-4
7. Usmani, A. S., Chung, Y. C., & Torero, J. L. (2003). How did the WTC towers collapse: A new theory. Fire Safety Journal, 38(6), 501–533. https://doi.org/10.1016/S0379-7112(03)00069-9
8. NIST. (2008). NIST NCSTAR 1A: Final Report on the Collapse of the World Trade Center Building 7. NIST NCSTAR 1A, WTC Investigation.
9. Arup. (2009) WTC -7 Structural Fire Analysis. London.
10. Empis, C. A., Cowlard, A. J., Jowsey, A. I., Usmani, A. S., & Torero, J. L. (2015) A Novel Methodology for the Analysis of Tall Buildings in Fire: WTC-7.
11. Hulsey, J. L., Quan, Z., & Xiao, F. (2019) A Structural Reevaluation of the Collapse of World Trade Center 7. Fairbanks.
12. Yarlagadda, T., Hajiloo, H., Jiang, L., Green, M., & Usmani, A. (2018) Preliminary modelling of Plasco Tower collapse. International Journal of High-Rise Buildings, 7(4), 397–408. https://doi.org/10.21022/IJHRB.2018.7.4.397
13. Hajiloo, H., Adelzadeh, M., and Green, M. F. (2017) "Col- lapse of the Plasco Tower in Fire." Proc., The Second International Conference on Structural Safety under Fire & Blast, London, UK.
14. Han, L., Potter, S., Beckett, G. et al (2010) FireGrid: An e-infrastructure for next-generation emergency response support. J Parallel Distrib Comput 70:1128–1141.
15. SFPE. (2016). SFPE Handbook of Fire Protection Eng. In M. Hurley, J. Hall, K. Harada, E. Kuligowski, M. Puchovsky, J. Torero, … C. Wiezorek (Eds.), SFPE Handbook of Fire Protection Engineering.
16. Wakamatsu, T., & Hasemi, Y. (1988). Heating Mechanism of Building Components Exposed to a Localized Fire -FDM Thermal Analysis of a Steel Beam under Ceiling-. In Fire Safety Science 3 (pp. 335–346).
17. Hasemi, Y., Yoshida, M., Yokobayashi, Y., & Wakamatsu, T. (1997) Flame heat transfer and cocurrent flame spread in a ceiling fire. In Fire Safety Science–Proceedings of the Fifth International Symposium (pp. 379–390). https://doi.org/10.3801/IAFSS.FSS.5-379
18. Wakamatsu, T., Hasemi, Y., Kagiya, K., & Kamikawa, D. (2003) Heating Mechanism of Unprotected Steel Beam Installed Beneath Ceiling and Exposed to a Localized Fire: Verification using the real-scale experiment and effects of the smoke layer. In Fire Safety Science–Proceedings of the Seventh International Symposium (pp. 1099–1110). International Association for Fire Safety Science.
19. Jiang, L., Jiang, Y., Zhang, Z., Usmani, A. (2021a) Thermal Analysis Infrastructure in OpenSees for Fire and its Smart Application Interface Towards Natural Fire Modelling, Fire Technology.
20. Lange, D., & Sjöström, J. (2014). Mechanical response of a partially restrained column exposed to localised fires. Fire Safety Journal, 67, 82–95.
21. Tondini, N., & Franssen, J. M. (2017). Analysis of experimental hydrocarbon localised fires with and without engulfed steel members. Fire Safety Journal, 92(May), 9–22. https://doi.org/10.1016/j.firesaf.2017.05.011
22. Stern-Gottfried, J., Rein, G., Bisby, L.A. & Torero, J.L. (2010) Experimental review of the homogeneous temperature assumption in post-flashover compartment fires. Fire Safety Journal.45 (4), 249–261.
23. Drysdale, D. (2011) An Introduction to Fire Dynamics. Third Edition. Chichester, Wiley.
24. Bisby, L., Gales, J. & Maluk, C. (2013) A contemporary review of large-scale non-standard structural fire testing. Fire Science Reviews.2 (1), 1–27.
25. Hidalgo, J.P., Cowlard, A., Abecassis-Empis, C., Maluk, C., et al. (2017) An experimental study of full-scale open floor plan enclosure fires. Fire Safety Journal.89, 22–40.

26. Hidalgo, J. P., Goode, T., Gupta, V., Cowlard, A., Abecassis-Empis, C., Maclean, J., ... Torero, J. L. (2019) The Malveira fire test: Full-scale demonstration of fire modes in open-plan compartments. Fire Safety Journal, 108, 102827.
27. Stern-Gottfried, J., & Rein, G. (2012a). Travelling fires for structural design–Part I: Literature review. Fire Safety Journal, 54, 74–85. https://doi.org/10.1016/j.firesaf.2012.06.003
28. Stern-Gottfried, J. and Rein, G. (2012b). Travelling fires for structural design-Part II: Design methodology, Fire Safety Journal. 54, pp. 96–112. doi: https://doi.org/10.1016/j.firesaf.2012.06.011.
29. Dai, X., Jiang, L., Maclean, J., Welch, S., et al. (2016) A conceptual framework for a design travelling fire for large compartments with fire resistant islands. In: Proceedings of the 14th International Interflam Conference. 2016 pp. 1039–1050.
30. Dai, X., Welch, S., Vassart, O., Cabova, K., Jiang, L., Maclean, J., ... Usmani, A. (2020) An Extended Travelling Fire Method (ETFM) Framework for Performance-Based Structural Design. Fire and Materials, 44(33), 1–21.
31. Heidari, M., Kotsovinos, P., & Rein, G. (2020) Flame extension and the near field under the ceiling for travelling fires inside large compartments. Fire and Materials, 44, 423–436. https://doi.org/10.1002/fam.2773
32. Nan, Z., Khan, A., Jiang, L., Chen, S., Usmani, A. (2021). Effect of travelling behaviour to structural members subjected to fire. Building Simulation (under review)
33. Charlier, M., Vassart, O., Dai, X., Welch, S., Anderson, J., Sjostrom, J., Nadjai, A. (2020) A simplified representation of travelling fire development in large compartment using CFD analyses. In Proceedings of the 11th International Conference on Structures in Fire, 2020:526-536, Brisbane.
34. Kotsovinos, P., Jiang, Y., & Usmani, A. (2013). Effect of vertically travelling fires on the collapse of tall buildings. International Journal of High-Rise Buidings, 2(1), 49–62.
35. Usmani, A. (2005). Understanding the Response of Composite Structures to Fire. Engineering Journal, Second Qua, 83–98.
36. Jiang, L. & Usmani, A. (2018a) Computational performance of beam-column elements in modelling structural members subjected to localised fire. Engineering Structures. 156, 490–502.
37. Ramesh, S., Choe, L. & Zhang, C. (2020) Experimental investigation of structural steel beams subjected to localized fire. Engineering Structures. 218 (May), 110844.
38. Wang, W.Y. & Li, G.Q. (2009) Behavior of steel columns in a fire with partial damage to fire protection. Journal of Constructional Steel Research, 65 (6), 1392–1400.
39. British Steel (1999) The Behavior Of Multi Storey Steel Framed Buildings In Fire, A European Joint Research Program.
40. Khan, M.A., Jiang, L., Cashell, K.A. & Usmani, A. (2018) Analysis of restrained composite beams exposed to fire using a hybrid simulation approach. Engineering Structures.172, 956–966.
41. Li, G., & Guo, S. (2008). Experiment on restrained steel beams subjected to heating and cooling, 64, 268–274. https://doi.org/10.1016/j.jcsr.2007.07.007
42. Al-Jabri, K. S., Burgess, I. W., & Plank, R. J. (2004) Prediction of the degradation of connection characteristics at elevated temperature. Journal of Constructional Steel Research, 60(3–5), 771–781.
43. Dai, X. H., Wang, Y. C., & Bailey, C. G. (2010) Numerical modelling of structural fire behaviour of restrained steel beam-column assemblies using typical joint types. Engineering Structures, 32(8), 2337–2351.
44. Wang, Y. C., Davison, J. B., Burgess, I. W., Plank, R. J., Yu, H. X., Dai, X. H., & Bailey, C. G. (2010) The safety of common steel beam/column connections in fire. Structural Engineer, 88(21), 26–35.
45. GSA (2003) Progressive Collapse Analysis and Design Guidelines. Buildings. doi:https://doi.org/10.1061/40700(2004)156.
46. ODPM (2004) The Building Regulations 2000, Part A, Schedule 1: A3, Disproportionate Collapse.

47. Agarwal, A. & Varma, A.H. (2014) Fire Induced Progressive Collapse of Steel Building Structures: The Role of Interior Gravity Columns. Engineering Structures. 58, 129–140.

48. Fu, F. (2009) Progressive collapse analysis of high-rise building with 3-D finite element modeling method. Journal of Constructional Steel Research. 65 (6), 1269–1278.

49. Vlassis, A.G., Izzuddin, B. a., Elghazouli, A.Y. & Nethercot, D.A. (2008) Progressive collapse of multi-storey buildings due to sudden column loss—Part II: Application. Engineering Structures. 30 (5), 1424–1438.

50. Izzuddin, B.A., Vlassis, A.G., Elghazouli, A.Y. & Nethercot, D.A. (2008) Progressive collapse of multi-storey buildings due to sudden column loss — Part I: Simplified assessment framework. Engineering Structures. 30 (5), 1308–1318.

51. Izzuddin, B.A. (2012) Mitigation of Progressive Collapse in Multi-Storey Buildings. Advances in Structural Engineering. 15 (9), 1505–1520. doi:https://doi.org/10.1260/1369-4332.15.9.1505.

52. Vlassis, A.G., Izzuddin, B. a., Elghazouli, A.Y. & Nethercot, D.A. (2009) Progressive collapse of multi-storey buildings due to failed floor impact. Engineering Structures. 31 (7), 1522–1534.

53. Pham, X.D. & Tan, K.H. (2013) Membrane actions of RC slabs in mitigating progressive collapse of building structures. Engineering Structures. 55, 107–115.

54. Fang, C., Izzuddin, B. a., Elghazouli, a. Y., & Nethercot, D. a. (2011) Robustness of steel-composite building structures subject to localised fire. Fire Safety Journal, 46(6), 348–363.

55. Shepherd, P.G. & Burgess, I.W. (2011) On the buckling of axially restrained steel columns in fire. Engineering Structures. 33 (10), 2832–2838.

56. Sun, R., Huang, Z. & Burgess, I.W. (2012) Progressive collapse analysis of steel structures under fire conditions. Engineering Structures. 34, 400–413.

57. Jiang, J., & Li, G. Q. (2017) Progressive collapse analysis of 3D steel frames with concrete slabs exposed to localized fire. Engineering Structures, 149, 21–34.

58. Gernay, T., & Gamba, A. (2018) Progressive collapse triggered by fire induced column loss: Detrimental effect of thermal forces. Engineering Structures, 172(June), 483–496.

59. Bažant, Z. P., & Zhou, Y. (2001) Why did world trade center collapse? - Simple analysis. Archive of Applied Mechanics, 71(12), 802–806.

60. Lange, D., Röben, C. and Usmani, A. (2012) 'Tall building collapse mechanisms initiated by fire: Mechanisms and design methodology', Engineering Structures, 36, pp. 90–103.

61. Aralt TT, Nilsen AR (2009). Automatic fire detection in road traffic tunnels. Tunn Undergr Sp Technol 24:75–83.

62. Cho, B.H., Bae, J.W., Jung, S.H. (2008) Image processing-based fire detection system using statistic color model. Proc - ALPIT 2008, 7th Int Conf Adv Lang Process Web Inf Technol 245–250.

63. Diehl, D.A. (1976) Fire Detection Systems. In: Plumbing Eng. pp 23–26

64. Noda, S., Ueda, K. (2002) Fire detection in tunnels using an image processing method. pp 57–62

65. Hodges, J.L., Lattimer, B.Y., Luxbacher, K.D. (2019) Compartment fire predictions using transpose convolutional neural networks. Fire Safety Journal 108:102854.

66. Mountrakis, G., Im, J., Ogole, C. (2011). Support vector machines in remote sensing: A review. ISPRS Journal of Photogrammetry and Remote Sensing. 2011 May 1;66(3):247-59.

67. Muhammad K, Ahmad J, Mehmood I, et al (2018) Convolutional Neural Networks Based Fire Detection in Surveillance Videos. IEEE Access 6:18174–18183. https://doi.org/10.1109/ACCESS.2018.2812835

68. Vandecasteele, F., Merci, B., Verstockt, S. (2016) Reasoning on multi-sensor geographic smoke spread data for fire development and risk analysis. Fire Safety Journal 86:65–74. https://doi.org/10.1016/j.firesaf.2016.10.003

69. Torabnezhad, M., Aghagolzadeh, A., Seyedarabi, H. (2013) Visible and IR image fusion algorithm for short range smoke detection. Int Conf Robot Mechatronics, ICRoM 2013 38–42. https://doi.org/10.1109/ICRoM.2013.6510078

70. Yuan, F., Fang, Z., Wu, S., et al (2015) Real-time image smoke detection using staircase searching-based dual threshold AdaBoost and dynamic analysis. IET Image Process 9:849–856. https://doi.org/10.1049/iet-ipr.2014.1032

71. Töreyin, B.U., Dedeoğlu, Y., Çetin, A.E. (2005) Flame detection in video using hidden Markov models. In: Proceedings - International Conference on Image Processing, ICIP. pp 1230–1233.

72. Chen, T.H., Wu, P.H., Chiou, Y.C. (2004) An early fire-detection method based on image processing. Proc - Int Conf Image Process ICIP 3:1707–1710.

73. Ko, B.C., Cheong, K.H., Nam, J.Y. (2009) Fire detection based on vision sensor and support vector machines. Fire Safety Journal 44:322–329.

74. Alamgir, N., Nguyen, K., Chandran, V., Boles, W. (2018) Combining multi-channel color space with local binary co-occurrence feature descriptors for accurate smoke detection from surveillance videos. Fire Safety Journal 102:1–10.

75. Zhou, Z., Shi, Y., Gao, Z., Li, S. (2016) Wildfire smoke detection based on local extremal region segmentation and surveillance. Fire Safety Journal 85:50–58.

76. Xue, C.J. (2010) The road tunnel fire detection of multi-parameters based on BP neural network. CAR 2010 - 2010 2nd Int Asia Conf Informatics Control Autom Robot 3:246–249. https://doi.org/10.1109/CAR.2010.5456677

77. Wu, X., Park, Y., Li, A., et al (2020) Smart Detection of Fire Source in Tunnel Based on the Numerical Database and Artificial Intelligence. Fire Technology,57, 657-682.

78. Pei, Y., Gan, F. (2009) Research on data fusion system of fire detection based on neural-network. Proc 2009 Pacific-Asia Conf Circuits, Commun Syst PACCS 2009 665–668.

79. Yao, Y., Yang, J., Huang, C., Zhu, W. (2010) Fire monitoring system based on multi-sensor information fusion. 2010 2nd Int Symp Inf Eng Electron Commer IEEC 2010 448–450. https://doi.org/10.1109/IEEC.2010.5533209

80. Hodges, J.L. (2018) Predicting Large Domain Multi-Physics Fire Behavior Using Artificial Neural Networks. Virginia Polytechnic Institute and State University

81. Carvel, R.O., Beard, A.N., Jowitt, P.W. (2005) Fire spread between vehicles in tunnels: Effects of tunnel size, longitudinal ventilation and vehicle spacing. Fire Technology 41:271–304.

82. Dexters, A., Coile, R. (2020) Testing for knowledge: Application of machine learning techniques for prediction of flashover in a 1/5 scale ISO 13784-1 enclosure. Fire Mater 1–12.

83. Zhang, X., Wu, X., Park, Y., et al (2020) Perspectives of big experimental database and artificial intelligence in tunnel fire research. Tunnelling and Underground Space Technology, 108, 103691. https://doi.org/10.1016/j.tust.2020.103691

84. Naser, M.Z. (2019a) Properties and material models for modern construction materials at elevated temperatures. Computational Materials Science 160:16–29.

85. Naser, M.Z (2019b) Fire resistance evaluation through artificial intelligence - A case for timber structures. Fire Safety Journal 105:1–18. https://doi.org/10.1016/j.firesaf.2019.02.002

86. Naser, M.Z (2019c) AI-based cognitive framework for evaluating response of concrete structures in extreme conditions. Engineering Applications of Artificial Intelligence 81:437–449. https://doi.org/10.1016/j.engappai.2019.03.004

87. Fu, F. (2020) Fire induced progressive collapse potential assessment of steel framed buildings using machine learning. J Constr Steel Res 166:105918.

88. Jiang, S., Zhu, S., Guo, X., Chen, C., Li, Z. (2020) Safety monitoring system of steel truss structures in fire. Journal of Construction Steel Research, 172:106216. https://doi.org/10.1016/j.jcsr.2020.106216

89. ISO (1999) Fire-resistance tests-Elements of building construction-Part 1: General requirements. ISO 834-1:1999.

90. Jiang, L., & Usmani, A. (2018b) Towards scenario fires – modelling structural response to fire using an integrated computational tool. Advances in Structural Engineering, 21(13), 2056–2067.

91. Jiang, L., Anwar, O., Jiang, J. & Usmani, A. (2021b) Modelling concrete slabs subjected to fires using nonlinear layered shell elements and concrete damage plasticity material. Engineering Structures.234, 111977.

Printed in the United States
by Baker & Taylor Publisher Services